Italian Cadastre of Artificial Cavities

Part 1

HYPOGEAN ARCHAEOLOGY
Research and Documentation of Underground Structures
Edited under the aegis of the Federazione Nazionale Cavità Artificiali
(FNCA)
No 1

Italian Cadastre of Artificial Cavities

Part 1

(Including introductory comments and a classification)

Roberto Basilico - Luigi Bavagnoli
Stefano Del Lungo - Gianluca Padovan
Klaus Peter Wilke

Translation by Ivana Micheli

BAR International Series 1599
2007

Published in 2016 by
BAR Publishing, Oxford

BAR International Series 1599

Hypogean Archaeology No. 1
Italian Cadastre of Artificial Cavities Part 1

ISBN 978 1 4073 0052 8

BAR Publishing is the trading name of British Archaeological Reports (Oxford) Ltd. British Archaeological Reports was first incorporated in 1974 to publish the BAR Series, International and British. In 1992 Hadrian Books Ltd became part of the BAR group. This volume was originally published by John and Erica Hedges Ltd. in conjunction with British Archaeological Reports (Oxford) Ltd / Hadrian Books Ltd, the Series principal publisher, in 2007. This present volume is published by BAR Publishing, 2016.

Printed in England

BAR
PUBLISHING

BAR titles are available from:

BAR Publishing
122 Banbury Rd, Oxford, OX2 7BP, UK
EMAIL info@barpublishing.com
PHONE +44 (0)1865 310431
FAX +44 (0)1865 316916
www.barpublishing.com

HYPOGEAN ARCHAEOLOGY

Research and Documentation of Underground Structures

The study and registration of artificial cavities means the documentation of underground structures. Just as Man started creating buildings on the surface of the Earth, over the course of time, he also perforated the surface thus creating new spaces and handing down structures which are essentially intact, which can be studied, restored and even utilised. In fact there exists an underground heritage, consisting of structures both built and buried underground over the passing of time. Our interpretation and understanding of such structures is a source of interesting information on our past, in favour of the present.

This series was created under the aegis of the Federazione Nazionale Cavità Artificiali (F.N.C.A.). Both the trademark and the title were especially created by the editors of this series (info@hypogeanarchaeology.com; c/o BAR Publishing, e-mail: info@barpublishing.com) and their use is reserved for the sole purpose of this product.

The aim is to create a base for the disclosure of relevant, scientific research studies, whether monographs, the works of various authors or documentation from conferences and conventions and a series of easily consultable tools for the development of artificial cavity research.

Text: Roberto Basilico, Luigi Bavagnoli, Stefano Del Lungo, Gianluca Padovan, Klaus Peter Wilke

Translation: Ivana Micheli

Editorial co-ordination: Klaus Peter Wilke

Photographs: Roberto Basilico, Luigi Bavagnoli, Andrea Busi, Alessandra Casini, Claudia Ninni, Andrea Thum, Davide Padovan, Gianluca Padovan, Maurizio Radacich, Klaus Peter Wilke, Domenico Zanon

To Jasmine and Thomas,
loving and beloved twin Crystals
for the next dimension

Friends, I know indeed that there is truth in these words that I now utter,
but the path to conviction, troubled and suspicious,
does not come easy to the human soul.

Empedocles, *Physical Poem (Fragments)* 101

Mariano Jacopo - Cod. Ms.a. 430, I, 295 Biblioteca Marciana of Venice (XV century).
«Si cupis capere arcem in summo monte positam, adhibeto fossores qui montem cretosum aut tufosum perfodiant ad modum scalarum donec perveniant ad fundamenta arcis quibus supponant materiam aridam oleo vel adipe suillo unctam, quibus immisso igni fundamenta arcis coruent» (*Canestrini n. d. [1945?], pg. 282).*

GENERAL INDEX

INTRODUCTION

This first volume initiates the new series of HYPOGEAN ARCHAEOLOGY (*Research and documentation of underground structures*), within the larger production of *British Archaeological Reports*. It is a study which promotes the archaeological, architectonical, speleological and historical investigation of artificial cavities.

The *Italian Cadastre of Artificial Cavities 'I'* covers the first fifty-five index cards relative to the underground works registered in the three Italian regions of Lombardy, Piedmont and Tuscany. Intended as an initial contribution to the classification and to the knowledge of artificial cavities, the cadastre includes a "Classification of artificial cavities by typology", essential for an understanding of the vast range of underground cavities.

The texts are written by Roberto Basilico, Luigi Bavagnoli, Stefano Del Lungo, Gianluca Padovan, and by Klaus Peter Wilke as editorial co-ordinator. Ivana Micheli from Edinburgh has translated the volume and Roberto Basilico, Luigi Bavagnoli, Andrea Busi, Alessandra Casini, Claudia Ninni, Andrea Thum, Davide Padovan, Gianluca Padovan, Maurizio Radacich, Klaus Peter Wilke and Domenico Zanon are the authors of the photographies.

We would like to thank all our collaborators, not only for their contribution to this volume but also for their field research. We would also like to thank those who conducted the investigations over the years, the results of which served in the acquisition and development of our knowledge of artificial cavities, thus permitting the creation of this volume.

Our special thanks go to: Chiara Aquino, Matteo Bertulessi, Sara Bianchi, Monica Bosio, Maria Antonietta Breda, Rino Bregani, Alberto Buzio, Paola Carità, Carlo Ferrero, Antonello Floris, Celestino Ghezzi, Umberto Gibertini, Guglielmo Esposito, Fabrizio Frignani, Franco Gherlizza, Gianusso Roberto, Matteo Grimoldi, Stefano Masserini, Davide Mengoli, Ivana Micheli, Chiara Nesti, Gianni Niccolai, Claudia Ninni, Davide Padovan, Alessandro Pesaro, Stefania Piccoli, Marco Placidi, Maurizio Radacich, Alberto Recanatini, Andrea Thum, Venturini Italo, Alessandro Verdiani, Cecilia Villao, Domenico Zanon and Roberto Zorzin.

Roberto Basilico, Luigi Bavagnoli, Stefano Del Lungo, Gianluca Padovan and Klaus Peter Wilke

Italian Federation Artificial Cavities (Federazione Nazionale Cavità Artificiali)

Well Cascina Torretta at Sesto San Giovanni (Milan) (photo Archive Ass. S.C.A.M.).

CHAPTER I

RESEARCH, STUDY AND CATALOGUING OF ARTIFICIAL CAVITIES

I.1 - *Archaeology and artificial cavities*

Man is instinctively drawn to the observation and exploration of structures from the past. From this he acquires knowledge. On a utilitarian front, he puts such structures to use by living in them or restoring them. In mediaeval times and during the Renaissance period, man restores and reactivates certain Roman aqueducts. In modern times, he ensures the operation of ancient sewers and canals. Tombs and rock dwelling become stables, warehouses and garages. Abandoned mines are located and brought back into operation or revitalized as tourist attractions and the surface ecosystem is often re-established. Everything comes full circle. Many cave places of worship are still in use today. The advent of the modern era and the advent of archaeology have on the one hand led to structures being abandoned or destroyed while on the other hand, structures are actively sought out and protected. These brief and succinct examples explain how artificial cavities can be restored and researched for different and often divergent purposes.

Artificial cavities are strictly of interest to the archaeological field. The reasons for this are both clear and simple. Artificial cavities are either man-made structures, created by the hand of man, or caves (natural cavities), which have been transformed or adapted either manually or by means of machinery. Burials generally fall under this category. For instance, studies of the catacombs led to the development of Christian Archaeology (Bovini 1952). Certain sites with underground areas are subject to archaeological investigation. One striking example is provided by the underground Hal-Saflieni Hypogeum (3300-2500 BC), on the island of Malta. This particular structure is built on three levels and its main chamber has been designed as if it were an external facade, in the style of buildings on the surface (Benevolo, Albrecht 2002, pg. 94).

Archaeological excavations of urban centres often reveal underground structures such as cisterns, granary pits and tombs (figs. I.1 and I.2). Of studies conducted on Italian rock settlements, the one of Vitozza in Tuscany is worthy of notice. The settlement consists of approximately 200 artificial cavities, classified according to their shape and typology (Parenti 1980). Numerous cave excavation campaigns have also taken place. Archeo-stratigraphic excavations were recently carried out in the Lazaret cave in Niece, on the slopes of the Boron Mountain (de Lumley 2004). However these relate to studies of deposits formed within natural cave environments. Of the so-called “rock shelters”, the “cliff dwellings” of south western America, inhabited by pre-Columbian cultures, should be noted (Gleria 1995, pgs. 110-113).

The Brandes en Oisant site in Haute Savoie (France) is a mining settlement known as “Argentaria de La Branda”. Operational between the XII and XIV centuries, this is a large mining and metallurgic complex. Nearby lie the ancient silver mines, during the exploration of which, speleological as well as other techniques were adopted: «L’archéologie minière est essentiellement souterraine. Elle demande des competénces qui allient celles de tout archéologue à une solide formation spéléologique. L’étude des réseaux miniers rencontre un certain nombre de contraintes. Il faut, d’abord, savoir localiser d’éventuelles entrées, le plus souvent éboulées, voire totalement colmatées. Les réouvertures et plus encore les désobstructions sont longues et pénibles, difficilement comparables à la progression d’un chantier traditionnel» (Bailly-Maitre, Bruno Dupraz 1994, pg. 13).

In principle, archaeological investigation consists of archive and surface research, its digs uncovering the entrances to underground structures. The internal areas of easily accessible structures are explored directly and stratigraphic excavations are carried out in natural cavities. It can generally be said that explorations are carried out almost exclusively in those cavities, which are easily accessible and easily explored. Or at any rate, in environments where soil can be moved with ease and stratigraphies for the phases of use can be easily carried out.

I.1.1 - *Artificial cavities as a heritage*

In the latter part of the XX century, the concept of “World Heritage” was established by UNESCO (United Nations Educational Scientific and Cultural Organization), a UN (United Nations) agency, with the Convention for the Protection of the Worlds’ Cultural and National Heritage, held in Paris in 1972.

Fig. I.1. *Structures uncovered in the Piazza Meda excavations, in Milan in 2006 (photo A. Thum).*

Fig. I.2. *Detail of one of the structures uncovered in the Piazza Meda excavations, in Milan in 2006 (photo A. Thum).*

It should be understood that artificial cavities, in the sense of man-made structures and testament to our history, are structures which should be acknowledged and protected. In relation to "heritage", Zerbi states: «Cultural Heritage consists of a heritage group, which essentially represents the traces left by man. The social conscience of a specific era considers it essential that such traces be transmitted to future generations. The notion of cultural heritage has literally taken off in the last century, grouping together an ever-increasing number of diverse elements. It should be added that a non-cultural heritage may become such through the voluntary intervention of an individual person, a body, an association, a group of people or a population» (Zerbi 2005, pg. 18).

The technological surge of the XIX and XX centuries has undoubtedly rendered a certain type of civil structure, falling within the typology of artificial cavity, such as icehouse storage rooms, uncompetitive. Socio-economic and industrial development has rapidly supplanted wells, hand-dug cisterns and aqueducts utilising natural forces with new collection-storage-distribution systems. The use of concrete in reinforced concrete structures, of perforated bricks, of metals and their alloys have rendered almost all man-made buildings financially uncompetitive and not in keeping with developing technologies. These are the immediate examples which shed light on how, even recent structures have fallen into disuse and have sunk into oblivion when not cancelled out together with the life systems and traditions to which they are linked.

I.2 - *Artificial cavities and speleology*

The application of speleological and underwater speleology methodologies means that a wider range of artificial cavities can be explored (fig. I.3). The adoption of modern single-cord descent and scaling techniques, physical and mental training, the use of underwater speleology equipment as well as an understanding of the risks and adequate compliance to safety regulations have undoubtedly opened new horizons. This has led to the discovery and exploration of a considerable number of previously unknown or simply overlooked underground structures.

Underground studies of both natural and artificial cavities were commenced many years ago; however only in the XIX century did speleological exploration take its first decisive steps. «Le véritable sens de la découverte souterraine est dans l'étude du détail, la documentation, l'expérimentation» (Bouillon 1972, pg. 83). The introduction of the speleological discipline has also greatly benefited artificial cavity research. However, the creation of a true investigation methodology has been undoubtedly far slower and far more difficult. There may be several reasons for this, however one thing is certain: maturation required its own time.

This new type of speleology was initially known as "urban speleology", as the bulk of explorations were conducted underneath the cities. In Italy, with the passing of time and with explorations being primarily conducted outwith urban centres, it has become more aptly known as "Speleology of artificial cavities". The Italian Federation Artificial Cavities (Federazione Nazionale Cavità Artificiali*)* was established in 2004 with the aim of combining speleology and archaeology in the exploration of underground structures. An internet site has been set up (www.archeologiadelsottosuolo.it) and a computerised multimedia land registry with card and bibliography management functionality is in the process of being set up for the disclosure and classification of artificial cavity research.

The «Archeologia del sottosuolo. Lettura e studio delle cavità artificiali» (Padovan 2005 a) volume was published in 2005, and provided a new name for the subject. The "First Congress on Artificial Cavities – Underground Archaeology: methodologies compared", provided a direct comparison of the common theme of artificial cavities in Archaeology, Speleology, Underwater Speleology and Biospeleology. The congress was held in Bolsena (Viterbo) from 8-11 December 2005 and was organised by the Italian Federation Artificial Cavities and the Lake Diving School Association of Bolsena (Associazione Scuola Sub Lago di Bolsena). The presentation included thirty-four studies, numerous films and slide shows, including 3D slide shows, as well as thematic exhibitions. Representatives from twenty-four speleology, underwater speleology, and diving associations as well as individual researchers, attended the Congress. There were speleologists from eleven Italian regions.

I.3 - *Underground Archaeology*

Underground Archaeology is the result of an activity, carried out through the development of original criteria where multiple aspects of other disciplines converge. This multidiscipline is based on underground permanence in man-made structures and on the ability to gather analytical data from the structure. The next step is the processing of such data. At this point, other aspects of the study such as architecture, geology, geomorphology, topography etc., contribute to our understanding of the structure in question.

Italian cadastre of artificial cavities

The research and study of artificial cavities is not incidental and is not the by-product of surface investigations or stratigraphic excavations. Just like classic speleology, Underground Archaeology has developed its own methodology. Without confusing the matter in abstractionism or particulars, it must be understood that this is also archaeology as the study relates to man-made structures. However, the aspects of data collection, graphic return, documentation, analysis and synthesis differ. The undertaking of this type research essentially entails the "documenting the underground".

I.4 - *Artificial cavities and classification*

In the first part of the XX century, Del Pelo Pardi was one of the first people to attempt to classify artificial cavities, providing his own interpretation regarding certain cunicular structures for drainage purposes (Del Pelo Pardi 1943). Artificial cavities were subsequently sub-divided within both the speleological and archaeological fields, although no specific classification was made.

The first "Bibliografia delle Cavità Artificiali Italiane" was presented at the "XVII Speleology Congress" in 1994, and divides artificial cavities according to their typology and purpose (Floris, Padovan 1997, pgs. 79-174). The first draft for the cataloguing of artificial cavities was exposed in 1999, at the "International Meeting of Studies on Methodologies for research on ancient hydraulic science" (Padovan 2002 a, pgs. 327-352). At the "XV Congress of Lombard Speleology" (Padovan 2000, pgs. 11-54), which also took place in 1999, it was proposed that artificial cavities be classified under the speleological field.

This classification was re-proposed at the "V Congress on Artificial Cavities", held at Osoppo (Udine) in 2002 (Casini, Padovan 2002, pgs. 155-184). The classification was also published in "Civita di Tarquinia: indagini speleologiche" (Padovan 2002 b). On an international level, the classification was published in 2005 in the aforementioned "Archeologia del sottosuolo. Lettura e studio delle cavità artificiali" (Padovan 2005 a).

I.5 - *The seven primary typologies*

The study of artificial cavities has resulted in the identification of a certain number of typologies and sub-typologies. Some sub-typologies may in turn present underground characteristics, which shall only be mentioned within this book and not individually covered, except in exceptional cases. The continuation of works and the development of the discipline shall hopefully lead to the broadening and integration of this list, which is intended as a simple yet solid starting point.

1. EXTRACTION WORKS
quarry, mine

2. HYDRAULIC WORKS

2 a. WATER SUPPLY AND TRANSPORT
aqueduct, artificial underground canal, artificial vaulted canal, drainage channel, natural vaulted water course, underground effluent, filtering gallery, connecting shaft

2 b. VERTICAL PERFORATIONS
artesian shaft, graduated shaft, ordinary shaft, ordinary radial shaft

2 c. STORAGE
cistern, icehouse, snowstore

2 d. WASTE DISPOSAL
septic pit, sewer, clarification (or biological) well, drainage well, cesspit, sump

3. RELIGIOUS STRUCTURES
crypt, rock hermitage, underground hermitage, "favissa", rocky place of worship, underground place of worship, mithraeum, holy well

4. FUNERARY STRUCTURES
catacomb, cemetery, columbarium, "domus de janas", "foiba", morgue, necropolis, ossuary, tomb

5. STRUCTURES FOR CIVIL USE
rocky dwelling, underground dwelling, rock apiary, "butto" (waste disposal pit), cellar, "camera dello scirocco" (sirocco chamber), columbarium, crypt, cryptoportico, underground oil mill, mushroom cultivation rooms, railway tunnel, pedestrian tunnel, road tunnel, granary pit, artificial cave, rock settlement, underground settlement, warehouse, nymphaeum, underground wine-making plant, gunpowder magazine, vault, road in cutting

6. MILITARY STRUCTURES
bastion, battery, castle, caponier, casemate, pillbox, countermine, demolition tunnel, cupola, fort, tunnel, counterscarp tunnel, demolition gallery, road tunnel, war cave, fortified cave, mine, cave structure, gunpowder magazine, postem, redoubt, reduit, shelter, artillery magazine, ravelin, vault, defensive tambour, "traditore", trench

7. UNIDENTIFIED STRUCTURES
structures, the function of which is unknown.

Fig. I.3. *Exploration and study of a well in Moncrivello (Vercelli), with a depth of 85.48 m (photo R. Basilico).*

CHAPTER II

CADASTRE OF ARTIFICIAL CAVITIES: DATA MANAGEMENT

II.1 - *Census and cataloguing of artificial cavities*

In the Italian archaeological field there have been various studies of artificial cavities yet a census of such cavities has never been carried out and no catalogue exists. However, the Ministry for Cultural Heritage and Activities has promoted the national census of those "artificial cavities", to be found in Italian parks and gardens (Scazzosi 2005, pg. 192. Please refer to: Cazzato, Fagiolo, Giusti 2001 and 2002).

In the speleological field, the census and cataloguing of underground structures, according to their typology, was initiated around twenty years ago. The first part of the National Federation Artificial Cavities, National Cadastre of Artificial Cavities is included within this volume. The typologies and sub-typologies are re-presented within this volume, together with a short description of the instrumentation used.

The study of a cavity allows the acquisition of data, to be used in the identification and description of the various aspects. Such a collection of data can be used in the implementations of a "cadastre". Within the Italian speleological field, "cadastre" refers to the census of natural cavities, which have been both identified and explored, and includes the data collected in respect of each individual cavity.

A similar yet distinct system has been adopted for artificial cavities. A "cadastral form", linked to the relevant "cadastre" is compiled for each artificial cavity. This basic form contains the fields necessary for the identification and understanding of the man-made structure and also provides information of a more general nature. Details of the planimetric survey, photographic services and any information or works relevant to the underground structure can be attached to this form. Thus copies of articles, with the results of analyses of the materials found in the cavity and the outcomes of any specific studies in the archaeological, architectonic, historical, biospeleological fields and other fields, will be available.

The "cadastre of artificial cavities" is an essential investigation tool in the promotion of strategies for the protection of the underground structure under investigation, and for those areas concerned with man-made underground structures or structures falling under the wide range of artificial cavities.

Artificial cavity census and its uses:
- location and distribution of cavities within the territory, for the compilation of a thematic map;
- exploration status;
- work status;
- status and breakdown of completed surveys;
- comparability.

II.2 - *The cadastral form and its compilation*

The cadastral form is used in the identification and classification of the underground structure under investigation. The following paragraphs detail cadastral form headings with relative annotations (figs. II.1a, II.1b, II.1c and II.1d). For improved, more rapid management this form may be transferred to electronic format.

II.2.1 - *Specifications*

- *Header:* owner/s of cadastral form data.
- *Cadastral number:* number and, where applicable, sub-number (to be assigned to the secondary environment or to be at any rate distinguished from the primary environment, through which access is normally made), Region or Province abbreviations.
- *Denomination*: name by which the underground structure is known, or alternatively, the name which it has been assigned.

II.2.2 - *Location within the territory*

- *Region-country*: tick the corresponding box.
- *Province*: Province name.
- *Municipality:* Municipality name.
- *Locality:* locality name, where applicable.
- *Location:* brief directions to the entrance of the underground structure.
- *Ownership*: provide details of the ownership of the land or building where the entrance to the underground structure is located.
- *Cartography:* cartography utilised for area identification and for subsequent identification of the point of longitudinal and latitudinal intersection.
- *Geological unit*: the territory's geological terrain and more specifically that of the area in which the underground structure extends.

II.2.3 - *Positioning*

- *Altitude*: height at which the entrance is located (specify data reliability).
- *Position*: access co-ordinates; specify data reliability in the adjacent boxes and the instrument used in the below space. For contingency reasons, these co-ordinates may be omitted.

II.2.4 - *Framework data*

- *Context*: sum of environmental (physical, geological, geomorphologic and climatic characteristics) and human factors (completion of material works, structures, infrastructures and settlement area), of which the underground structure is a part.

- *Completed operations:* the work that has been carried out (for example exploration, surveys, photographic services, etc.).
- *Work carried out by:* list the names of those who carried out the various operations.
- *Warnings:* please indicate any existing dangers or potential risks.

II.2.5 - *Classification*

- *Typology:* place a tick next to the original typology and the type of underground structure. Where the type is not indicated, details should be entered next to the relevant box, at the end of each individual list (with the exception of Typology No.7).

II.2.6 - *Notes*

- *Description*: environment description, detailing each characteristic, as well as general information, relating to the context (the surrounding area).
- *Interpretation*: brief description of the purpose, of any re-utilization as well as any possible change in use over time (with indication of the purpose for which it is intended and providing written details of the typology category followed by the specific type).
- *Dating:* chronological indication of the phases of use, starting from the creation date (where possible).
- *Notes*: space for additional information (visibility, state of vegetation, oral memory, etc.).
- *Bibliography*: please indicate any publications which deal, whether marginally or specifically, with the underground structure under investigation.
- *Data ownership:* name or names of the owners of the data contained within the cadastral form.
- *Compiler:* name of the person who physically compiled the cadastral form.

II.3 - *Notes relative to cadastral number allocation and typological classification*

An underground structure may have one or more entrances and may have one or more environments. It can be defined as a "complex" whenever it results from the unison of multiple parts which were not necessarily created at the same time and which did not or do not necessarily have the same function. An artificial cavity must immediately be assigned a primary cadastral number and a denomination (first page).

SCAM® Associazione Speleologia Cavità Artificiali Milano

Cadastre of Artificial Cavities

CADASTRAL NUMBER CA [ ][ ][ ][ ][ ] / [ ][ ] [ ][ ] [ ][ ]

CADASTRAL NUMBER — SUBORDINATE N. — REGION — PROVINCE

DENOMINATION ____________________

REGION-COUNTRY

Aosta Valley	Piedmont	Liguria	Lombardy	Trentino Alto Adige	Veneto	Friuli Venezia Giulia	Emilia Romagna	Tuscany	Marche	Umbria

Abruzzo	Molise	Latium	Sardinia	Campania	Basilicata	Apulia	Calabria	Sicily	Vatican City	Republic of S. Marino

PROVINCE ____________________

MUNICIPALITY ____________________

LOCALITY ____________________

LOCATION ____________________

OWNERSHIP ____________________

CARTOGRAPHY ____________________

GEOLOGICAL UNIT ____________________

ALTITUDE ____________ [ ] INDICATED IN THE MAP [ ]

POSITION ____________ [ ] SURE DATA [ ]

[ ] APPROXIMATE DATA [ ]

Fig. II.1.a. *Cadastral Form (Federazione Nazionale Cavità Artificiali).*

CONTEXT ____________________

OPERATIONS CONDUCTED ____________________

WORK CARRIED OUT BY ____________________

WARNINGS ____________________

TYPOLOGY

1 EXTRACTION WORKS

- ☐ quarry
- ☐ mine
- ☐

2 HYDRAULIC WORKS

2a water supply and transport
- ☐ aqueduct
- ☐ artificial underground canal
- ☐ artificial vaulted canal
- ☐ underground effluent
- ☐

2b verticals perforations
- ☐ artesian shaft
- ☐ ordinary shaft
- ☐

2c storage
- ☐ cistern
- ☐ icehouse
- ☐ snowstore
- ☐

2d waste disposal
- ☐ septic pit
- ☐ biological well
- ☐ drainage well
- ☐ cesspit
- ☐ sump
- ☐

3 RELIGIOUS STRUCTURES

- ☐ crypt
- ☐ rock hermitage
- ☐ underground hermitage
- ☐ favissa
- ☐ rocky place of worship
- ☐ underground place of worship
- ☐ mithraeum
- ☐

4 FUNERARY STRUCTURES

- ☐ catacomb
- ☐ cemetery
- ☐ columbarium
- ☐ domus de janas
- ☐ foiba
- ☐ morgue
- ☐ necropolis
- ☐ ossuary
- ☐

5 STRUCTURES FOR CIVIL USE

- ☐ rocky dwelling
- ☐ underground dwelling
- ☐ waste disposal pit
- ☐ cellar
- ☐ prison
- ☐ sirocco chamber
- ☐ rock city
- ☐ underground city
- ☐ columbarium
- ☐ crypt
- ☐ cryptoportico
- ☐ mushroom cultivation rooms
- ☐ railway tunnel
- ☐ pedestrian tunnel
- ☐ road tunnel
- ☐ rock settlement
- ☐ underground settlement
- ☐ warehouse
- ☐ road in cutting
- ☐ souterrain
- ☐ trench road
- ☐

6 MILITARY STRUCTURES

- ☐ bastion
- ☐ casemate
- ☐ countermine
- ☐ tunnel
- ☐ counterscarp tunnel
- ☐ fortified cave
- ☐ mine
- ☐ cave structure
- ☐ shelter
- ☐ ravelin
- ☐ defensive tambour
- ☐

7 UNIDENTIFIED STRUCTURES

- ☐

Fig. II.1.b. *Cadastral Form (Federazione Nazionale Cavità Artificiali).*

DESCRIPTION

INTERPRETATION

Fig. II.1.c. *Cadastral Form (Federazione Nazionale Cavità Artificiali).*

DATING

NOTES

BIBLIOGRAPHY

DATA OWNERSHIP

COMPILED BY

Fig. II.1.d. *Cadastral Form (Federazione Nazionale Cavità Artificiali).*

Having reviewed the complex and carried out the survey, a cadastral sub-number can be allocated to the various environments. For data management purposes or for a more detailed definition, a sub-number can also be allocated to each element of the same cavity, such as secondary supply conduits, natural cavities intercepted by excavation works, modest lateral spaces etc.

II.3.1 - *Typological subdivision notes*

In its typological identification (page two), the original purpose of the underground structure must be considered. Changes to the original purpose shall be taken into account, and where possible, subsequent purposes will be detailed under the *Description* field (page three). Where the purpose of the underground structure cannot be determined, the structure will be classified under Typology No. 7. "unidentified structure". Sometimes only the subsequent purpose, and not the original purpose, can be found: for contingent needs the cavity may be catalogued under the unidentifiable typology, so long as its details are specified under *Description.* Data can be more easily compared and interactions for the desirable development of Underground Archaeology can take place from a common base.

During the cataloguing phase, the following points should be taken into account:
- Where the original purpose of certain artificial cavities cannot be established and their future re-use is not evident, the cavity is always assigned to Typology No. 7.
- Where the original purpose is not evident, but its re-use is, the artificial cavity will be assigned to the typology of the latter, with a note on the cadastral form indicating that its original purpose is unknown.
- A catacomb, clearly originating from the excavation of an underground quarry and which resulted from simple adaptation of the extraction areas, will be catalogued, under Typology No.1, as an "extraction work", and the cadastral form should then provide details of its subsequent re-use under Typology No. 4.
- A catacomb excavated from scratch from underground extraction works or from an aqueduct, will be assigned to Typology No. 4.
- A rock tomb (clearly identifiable as such), subsequently re-used as a stable, will still fall under Typology No. 4.
- Where there is insufficient data to determine the funerary use of an underground structure and on a practical level its purpose is that of a stable, the structure should be classified under Typology No. 5.
- An ordinary well can be situated either in a town square or within a castle but will always be catalogued as a hydraulic structure, classifiable under Typology No. 2 b.
- When found within a city, an underground passage" will fall under Typology No. 5, whereas the underground passage of a rock hermitage will fall under Typology No. 3. (unless it is the re-used branch of an aqueduct, in which case it shall fall under Typology No. 2 a).
An underground tunnel connecting two casemates excavated into the rock is classified as a military structure (Typology No. 6) and not as a pedestrian tunnel (Typology No. 5).
- A well situated on the arched crown of an aqueduct's underground conduit is an integral part of the aqueduct itself.

Moreover: within the necropolis at Cerveteri (Rome) there are several underground conduits, excavated in tuff: one of these is situated next to the later chamber tombs. These are drainage systems, for the outflow of rain-water and although situated in a burial area these are classified as hydraulic structures for the transport of water (Typology No. 2 a). A cistern situated on the lower level of tower will always be considered a hydraulic conservation structure and is catalogued under Typology No. 2 c.

Thus, the purpose of an underground structure is not always clear and does not always fall within a specific context. Additionally, historical or archive documents, which clarify the time of excavation and its purpose, may not always be available. For typology assignation, reference must be made to the gathered documentation.

II.4 - *Bibliography and the bibliographic form*

Having conducted the survey of an underground structure, the next phase relates to its interpretation, to analysis of the structure's territorial context and to the study of any similarities and discrepancies. This provides further clarification on various aspects, such as its functionality, chronology, contacts and relations between various cultures and particularly on the diffusion of ideas. Such studies cannot be separated from bibliographic research, primarily on account of the knowledge and development of the themes covered by the investigation. The bibliographic fields of an artificial cavity's bibliographic form must also be subdivided according to their typology (Padovan 2005 e, pg. 353).

II.5 - *Data archiving*

A census of underground environments involves the recovery of acceptable documentation attesting their existence, the internal examination of their structure, the collection of relevant data and the organisation of a cadastre and a bibliographic file. This is the basis of data archiving. Technology and information technology may prove to be indispensable. Work which began with the research and exploration of an underground structure should not terminate upon rendering of the survey and its inclusion in a traditional paper-based cadastral file. The popularity of computers and modern, user-friendly programmes allow a growing number of users to reach excellent computer solutions (Bavagnoli 2005, pgs. 355-356). Data collected during the study and analysis phase can thus be arranged into a single digital container known as a database (fig. II.2).

II.5.1. - *Technology and computer science*

Diverse data, collected for study documentation purposes, may be digitalized, irrespective of its nature and origin. This means that a computerised operation takes place in which the data is converted and broken down into binary code (0 and 1). Binary code is the only code that computers are able to understand and manage. Modern technology allows us to administer data, numbers, photographs, images, cartographies, technical surveys, GPS positioning, reports, audio-clips and full-length videos in this way. Digital conversion (discrete system) of an analogue measure is nothing more than an approximation technique, however the levels obtained today through information technology do not constitute a problem.

II.5.2 - *The advantages and disadvantages of database utilization*

Large data volumes can be difficult to manage and keep in order while at the same time maintaining their efficiency. Above all, rapid data retrieval can be complex. A computerised archive has many practical advantages so long as it is adequately planned during the analysis phase. Vast quantities of information can be stored in a single CD-ROM, replacing large, dusty paper files, while the requested forms can be printed as and when required. The entire archive duplication process is simplified, requiring just a few minutes at negligible cost. Copies can be quickly sent while backup copies are retained.

The advantages of use are equally significant. In relation to research and data comparison: once research criteria have been input, the result immediately appears on the computer screen. Retrieval and printing of the indexes of a specific province rather than those of a locality is another operation which takes just a few seconds. In addition, information technology allows us to add multimedia material to each individual index, which would otherwise have been impossible. Once all the data has been converted into files, all that remains is for the files to be linked together in order that a specific cadastral form can be automatically linked to the other documents. An error in the planning phase will undoubtedly cause the data to mix and render the work useless. The database is internally organised in tables, thanks to which we are able to archive non-uniform data.

II.5.3 - *Database creation*

There are many suitable programmes, known as databases produced, by various suppliers. A choice may be made based on personal preference, licence for use costs (although some of these are free) and the level of ability required for product utilisation. Microsoft Access is popular software used in many offices and it is likely that many users will already be acquainted with the software. In the case of complex archives, professional databases such as IBM's DB2 or Oralcle's database should be used. When correctly installed, the application creates a structure capable of containing all relevant data. The more attention is paid to its architecture, the greater the database's performance and subsequent ease of maintenance. Tips gained through experience and the numerous free online manuals should be followed.

A database is a collection of tables which are interlinked. Each table has its own name so that it can be easily identified and distinguished from other tables within the same archive. In this particular case, the cadastral table is the main table and is linked to the other five tables which make up our basic initial file. Each table has a heading, known as a "field", the sole purpose of which is the unique identification of the element to which it is associated. This field is known as the primary key and will be automatically managed by our application. Element identification fields are commonly known as records.

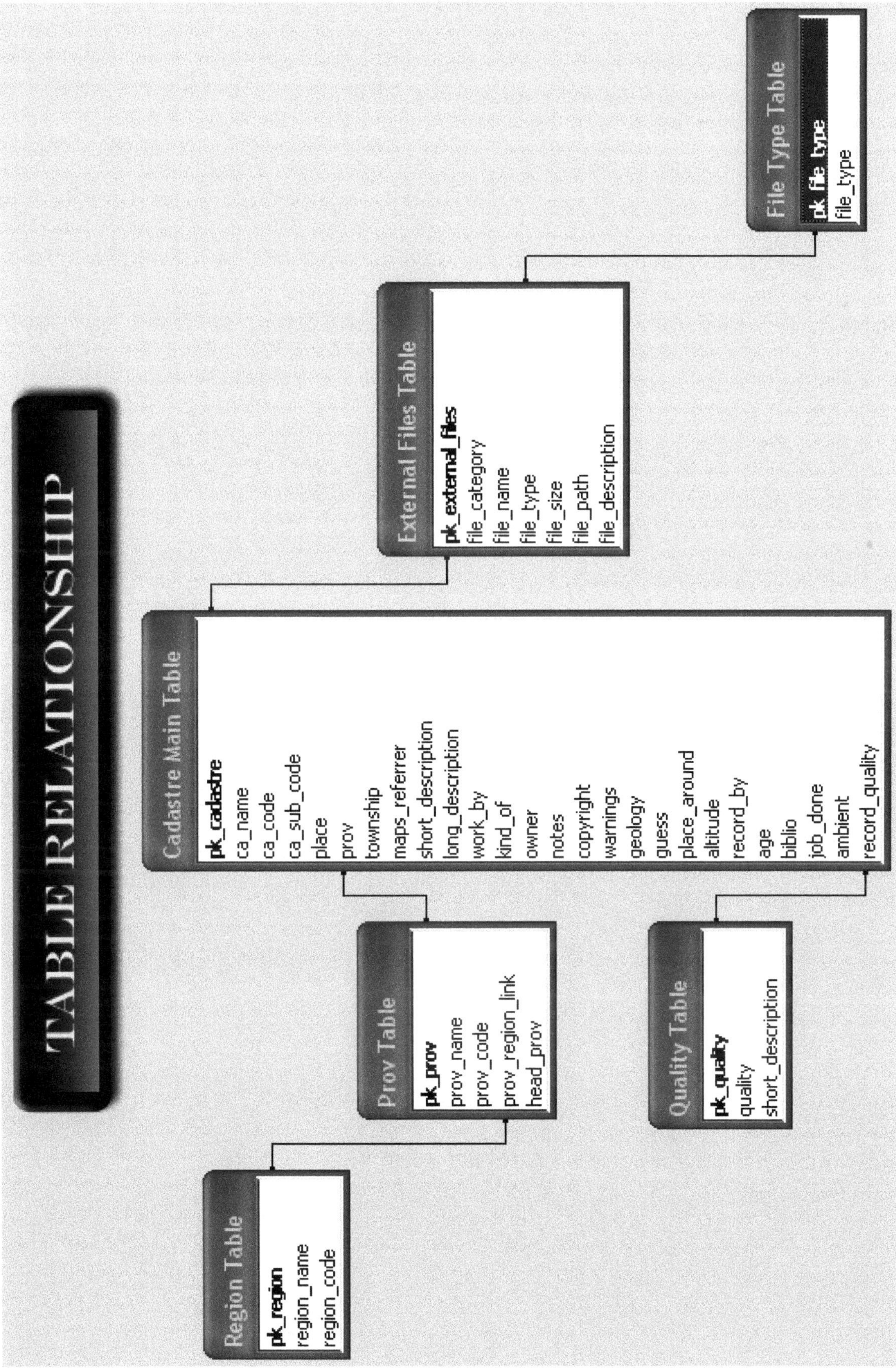

Fig. II.2. *Database creation framework (Federazione Nazionale Cavità Artificiali).*

II.5.4 - *Multimedia applications on a practical level*

Multimedia files (photographs, films, images, surveys, maps, etc.) should be kept outwith the database and stored in suitable folders. Although even large files can be added to a database, this should be avoided as it reduces database performance. The disadvantage is that not everything can be held in a single file. The benefit is in the performance, or rather in the search results. The option of updating documents without having to access the database will soon be available. The database will contain only the name and information on where to locate the file in question, as well as a suitable description for its identification. Usability should not be forgotten: that is the option of using the formats more compatible with the rest of the computer world, such as PDF (Portable Document Format) for edited text, DivX for videos and Jpeg for images.

Every aspect of the cadastre can be managed according to preference, thus preventing our efforts from being in vain. For example, the Federazione Nazionale Cavità Artificiali (National Federation Artificial Cavities) has created a computerised multimedia cadastre with cadastral form and bibliographic form management. Collected data is currently being made available to internet users on the following website: www.archeologiadelsottosuolo.it.

CHAPTER III

ITALIAN CADASTRE OF ARTIFICIAL CAVITIES: PART ONE

III.1 - *Introduction*

This is the first part of the National Federation Artificial Cavities' National Cadastre of Artificial Cavities. The underground structures registered are located on Italian territory and are situated in the following regions: Lombardy, Piedmont and Tuscany. Cadastral forms are subdivided in alphabetic order by region and by province. They are listed within the cadastre in ascending order. It should be noted that in the list of artificial cavities in Piedmont, cadastral numbers from 00001 to 00029 and from 00034 to 00036 for the province of Turin, relate to the Verrua Fortress. Cadastral numbers from 00030 to 00033, relate to Pavarolo Castle. Cadastral numbers for the province of Livorno (Tuscany) start at 01000. This is to ensure that the allocation of cadastral numbers within local areas does not overlap. The Livorno Speleological Group has in fact registered numerous mines, both old and modern, particularly in the Campiglia Marittima (Livorno) area. In order to provide additional information and thus a better understanding of each individual structure, a brief historic framework is provided for each group of cavities from a single region.

III.2 - *Lombardy*

Fig. III. 1. *Geographic framework: Lombardy Region.*

Grotta Ferrera (LC). The Meria valley, in the Grigne mountain complex, is deeply set in limestone rock (figs. III.1 and III.2). The arid environment, with its jagged landscape does not immediately present traces of past mining activity. In fact, the mineral outcrops, the discharge piles and the excavations themselves are not immediately visible, nor are any remains of industrial works. Upon first investigation it seems that there are no traces of mines in the collective memory of the area's inhabitants. One toponomastic element is in some way indicative of mining activity and relates to the name of most famous cavity in the area: the Ferrera [Ironworks] cavity. Listed in the T.C.I.'s 1:20.000 topographic map as: «Grotta del Rame (la Ferriera)» (Copper cave [Ironworks]).

The Ferrera Cave was first described by Domenico Vandelli in the 1763 work entitled "Saggio d'Istoria Naturale del lago di Como, della Valsassina, etc.". Brief references to the cave appear in subsequent works by XIX century academics. Towards the end of the 1950s, the S.E.M.-C.A.I. Milan Cave Group carried out its survey and research in collaboration with the Speleo Club Universitario Comense and the Gruppo Speleologico Ligure "A. Issel". The results, published in 1962, cover the following topics: cavity description (including cadastral data), cave speleogenesis, lithogenic study of tubular stalactites, hydrogeological, meteorological and biological observations and the results of excavation tests at the cave's entrance (Cappa, Cigna, De Michele, Parea 1962). In 1988, speleologists from Milan's Speleological Association of Artificial Cavities (S.C.A.M.), in collaboration with speleologists from Milan's S.E.M.-C.A.I. Cave Group, resume studies and explorations within the cavity.

To date, traces of mining activity in Val Meria are insufficient to permit its placing within a historical framework and pose, rather than resolve, certain issues strictly connected to the importance of mining in the valley's economy. The analysis of extraction and mining methods does not permit the accurate dating of the mining activities examined, although it can provide some information on the matter. Geological and archaeological mining surveys are therefore required. Given the continued anthropic presence, which partly destroyed ancient deposits, an archaeological dig within Grotta Ferrera may prove to be non-productive as far as the identification of pre-historic settlements is concerned. However, mining extraction through the ages could be documented.

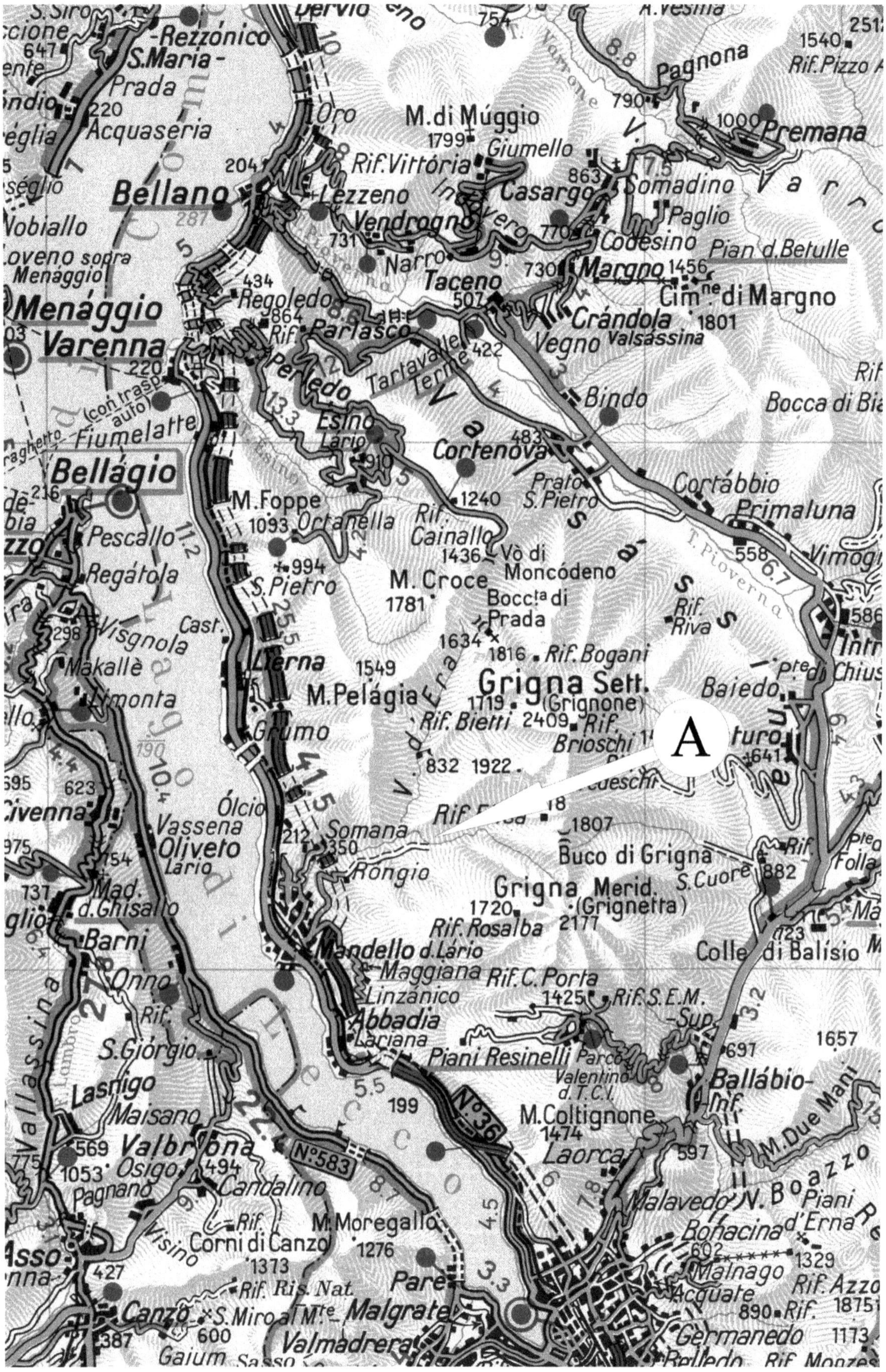

Fig. III. 2. *Geographic framework:* A. *Val Meria and Grotta Ferrera (Touring Club Italiano 2002 a, table 10).*

III.2.1 - *CA 00001 LO LC; Grotta Ferrera*

Cadastral number: CA 00001 LO LC (Cadastre of Natural Cavities: 1502 LO LC)
Denomination: Grotta Ferrera (figs. III.3, III.3.a, III.3.b and III.4)
Region-country: Lombardy, Italy
Province: Lecco
Municipality: Mandello Lario
Locality: Rongio
Location: Val Meria
Ownership: //
Cartography: I.G.M. 1: 25.000 Pasturo, F. 32
Geological unit: Esino limestone
Altitude: 590 m a.s.l.
Position: latitude N 3° 05' 54"; longitude W 45° 55' 59"
Context: situated on the right bank of the River Meria. A cart-track leads from Rongio to the Iron Bridge via Val Meria. From here a path, some parts of which consists of wide, low steps, leads from the Iron Bridge to the entrance of the Cave
Operations conducted: excavation, survey, photographic service
Work carried out by: Associazione Speleologia Cavità Artificiali Milano (S.C.A.M.) and Gruppo Grotte Milano S.E.M.-C.A.I.; 1998-2002. In 2006, a bottle-neck was widened by Ass.ne S.C.A.M. speleologists, however no significant tunnel extensions were found
Warnings: slippery sections with crevices
Typology: 1 - mine
Description: undeniably formed by natural phenomena, the Ferrera Cave was subject to extensive mining activity. The cave's short, low, external corridor is followed by an 8 m section of tunnel with average height of 2.4 m and width of 5-6 m, which leads to a large room. The oblong room has a slight bottle-neck to its centre and is 170 m long (along the main axis) and is 41 m wide at its widest point. It is situated at a depth of - 46 m and to date 516 m have been explored. At its centre, the roof slopes to the left with a 30°-40° inclination. There is evidence of water infiltration at the point where the right wall meets the roof. Speleothems are to be found almost exclusively in the final section of the cave. The floor is characterised by blocks of various shapes and sizes and by a ridge-like structure consisting of large rock segments, which have fallen from the roof. A veil of clay lines the floor deep fractures interrupt the floor's surface. The floor of the final section is basically convex and consists of stone rubble, large patches of which have been cemented together by a thin layer of calcium carbonate. The floor is crater-like in appearance: numerous large, yet shallow pits of an inverted conical shape become evident. One next to the other, the prominent detritic deposits around the edge of each pit sometimes merge with the deposits of the adjacent pit. The final section of the room is entirely man-made and is almost orthogonal to the main axis. From here, several other passages branch out and lead to small plants. There are evident signs of mining all around. Mineralization consists of red and brownish-red iron minerals (hydroxides) Mining was carried out by "following the vein". The vastness of the room undoubtedly allowed the "visible" mineralized blocks to be mined without particular difficulty. It was in fact their depletion that led the miners to follow the veins in the rock and to extend their search to the surface and collapsed blocks. Having been thus widened by mining works, the artificial cavity was used as a mineral selection and crushing area. The work areas can be divided into various sectors. Signs of excavation are immediately evident along the walls of the entrance gallery, even a few metres above the ground, a sure sign that steps and platforms were once positioned here. In some sections the ground surface is lower due to mining activity. In others, the deposits indicate that the ground surface has been raised and that this is partly due to filling material. The excavations, primarily along the west wall, were carried out below the ground. There are narrow sub-horizontal passages, two of which with semi-circular tunnel and small plants, which were either filled or partially blocked by detritus and small cave-ins. Further excavations follow the wall, coated by small to medium unworked broken stones and descend a few metres below the ground. Where the ground consisted of unconsolidated materials, dry stone masonry lining and sometimes clay (large amounts of which are to be found in the site) lining was used. It is not clear if the remaining, somewhat limited, parts of the cave were utilised for research purposes or if the now impracticable part conceals actual mining plants, although the latter is more likely. The east wall presents considerably less traces of mining. It could well be that small cave-ins and natural ground settlement have blocked and concealed plant entrances. The few visible entrances are in fact situated in the centre of large under-wall man-made depressions at a depth of between 2 and 16 m. At the time of mining activity, the Ferrera Cave would have had numerous simultaneous mining faces.

Fig. III. 3. *Grotta Ferrera plan (CA 00001 LO LC) (Ass.ne S.C.A.M. Archive).*

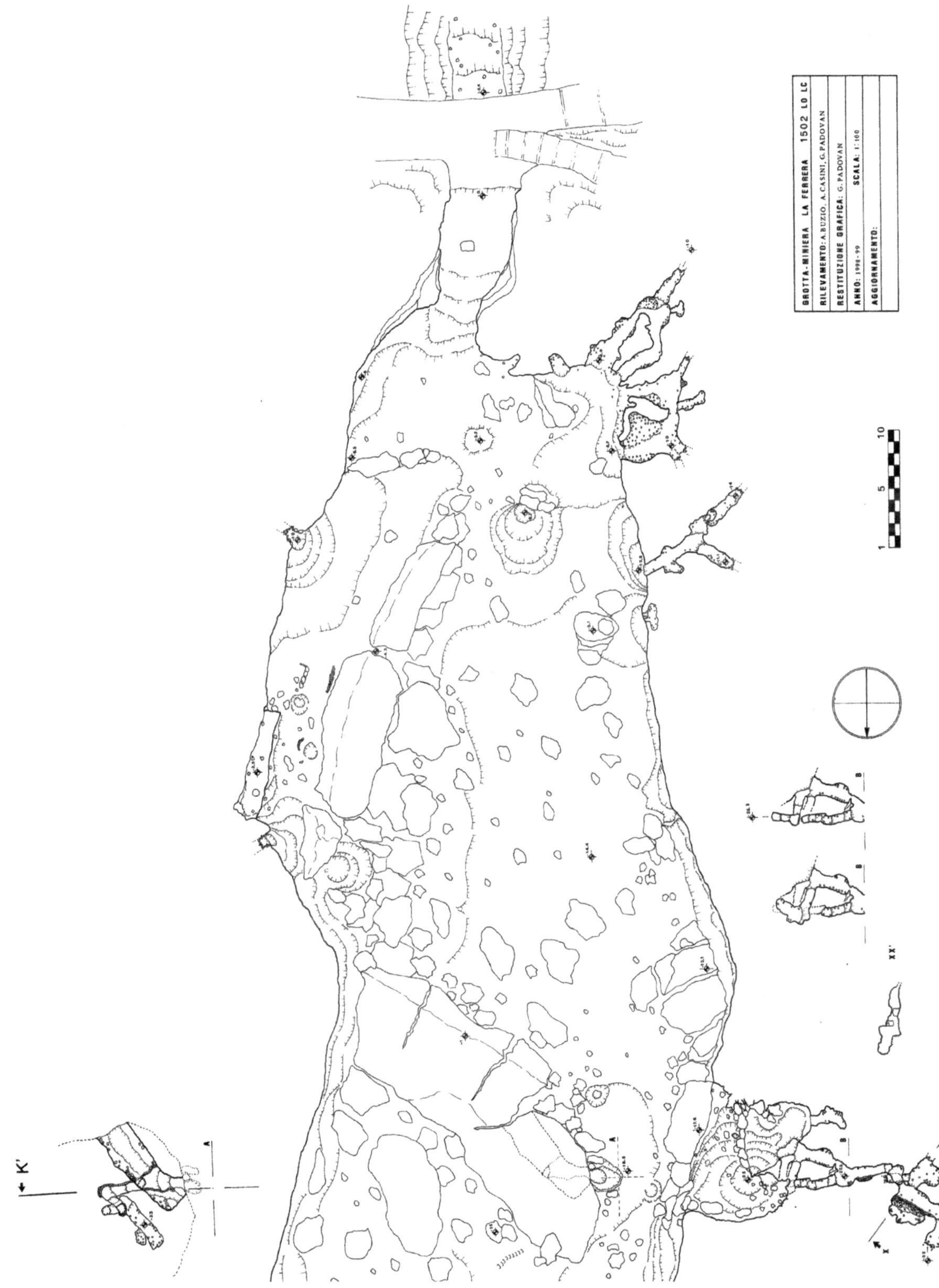

Fig. III. 3.a. *For a better view: Grotta Ferrera plan, southern part.*

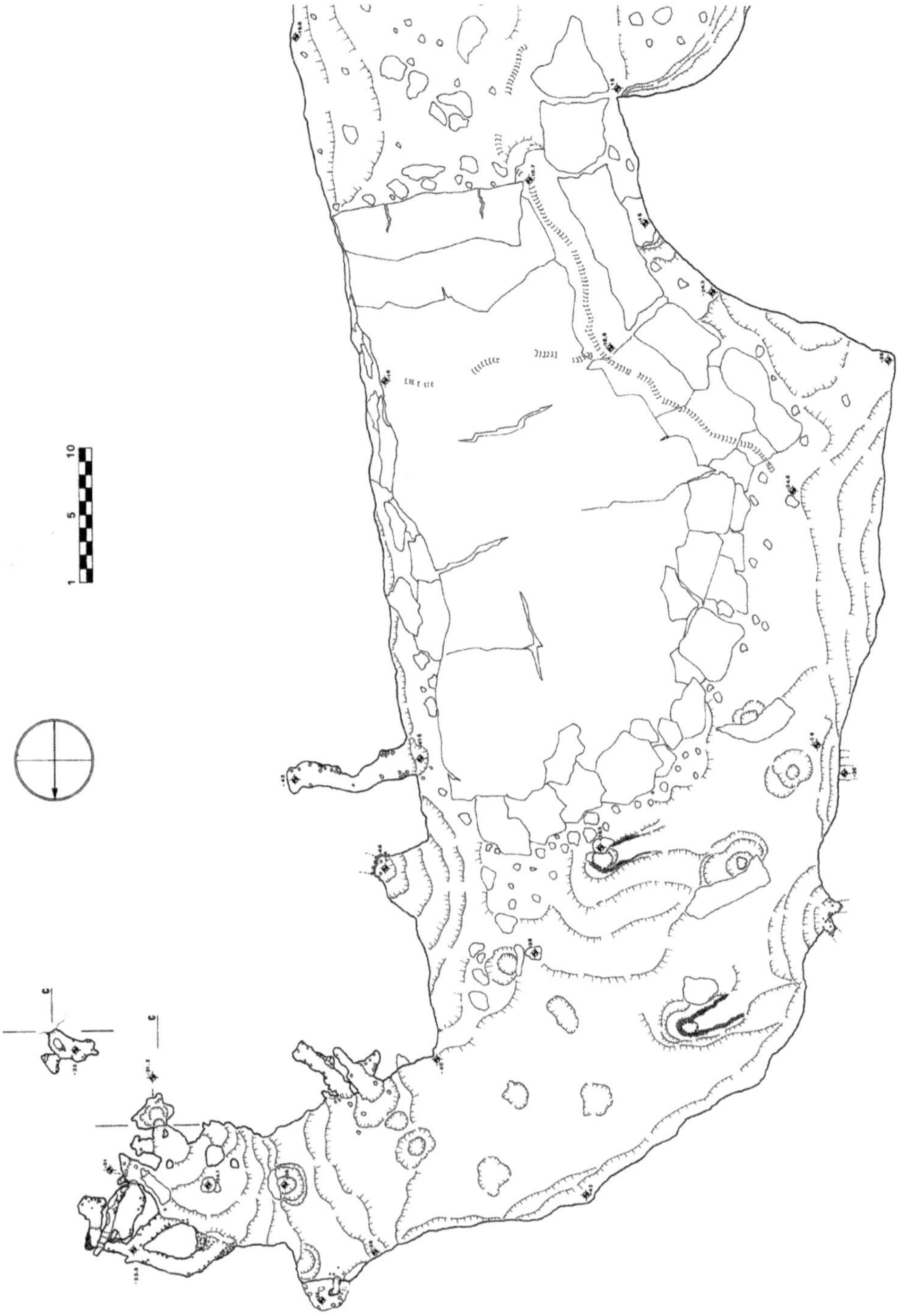

Fig. III. 3.b. *For a better view: Grotta Ferrera plan, northern part.*

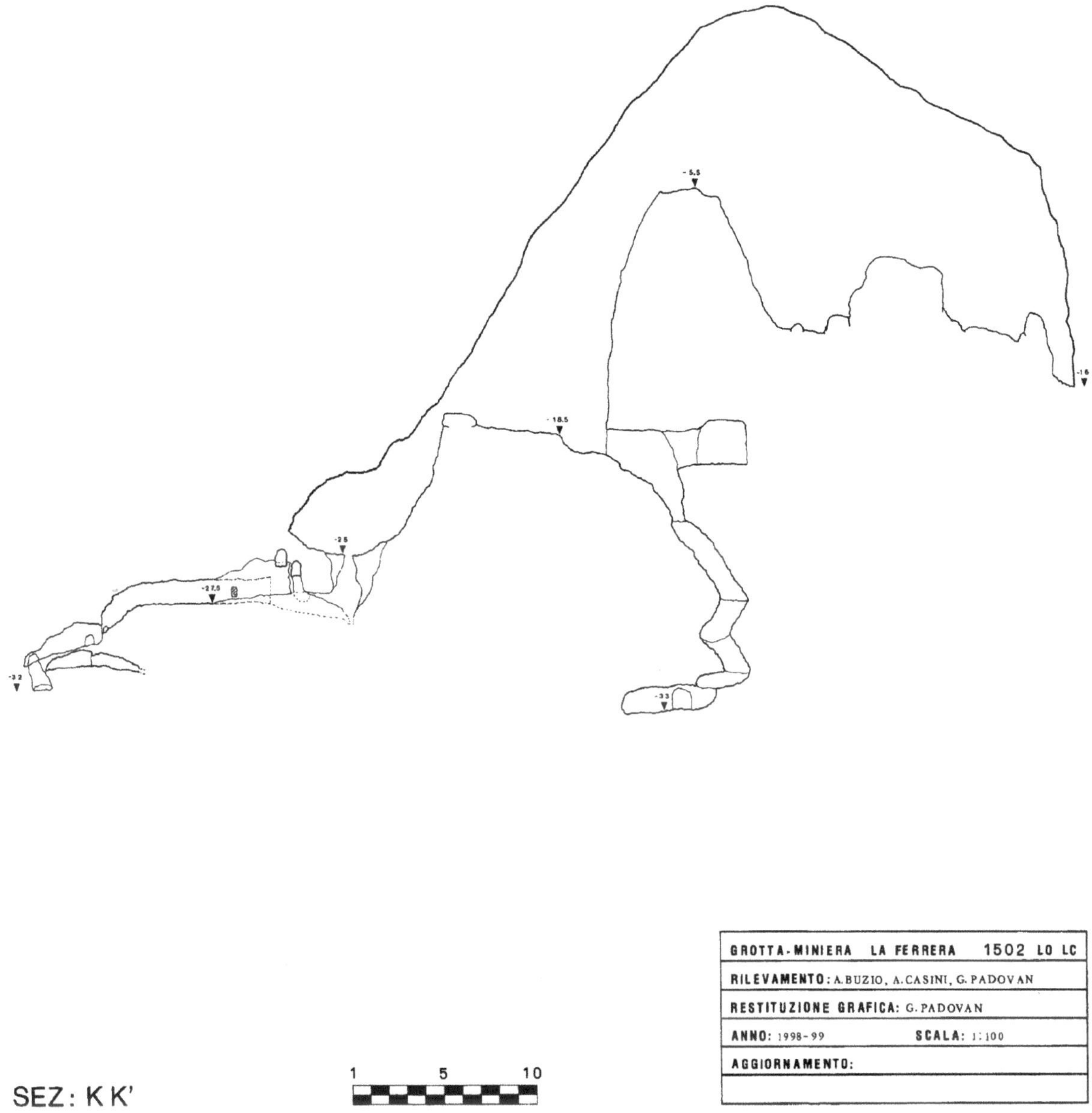

Fig. III.4. *Grotta Ferrera (CA 00001 LO LC) latitudinal section (Ass.ne S.C.A.M. Archive).*

What is far clearer is how the small right angle, which forms along the perimeter towards the fallen rocks, results from mining extraction. A ledge has been carved into the rock at a height of just over 2m. The ledge has been cut around a cylindrical rock section, which is at least 1 m in height and has a diameter of approximately 30 cm. This structure may have been the anchorage for a rope used to lower mineral containers. The same anchorage could constitute the base of a wooden frame used in wall excavations. As previously mentioned, the final section of the cave is man-made. The mineral has been removed thus leaving gaps of various dimensions. Some passages contain mining plants and what remains of the original fill material. Mining works which follow the mineralized vein were structured in an upwards direction. The base consists of unconsolidated material, which is only cemented together by formations where there is dripping water. The left passage is of some interest in that its rocky surface retains the evident marks left by the mining equipment. In addition, transparent and "milky" formations have formed on top of the fill material. All around, are the remnants of sub-vertical shaft openings surrounded by prominent and characteristic borders of prevalently fine, detritic fragments. These are quite distinct from the fallen rocks. They are primarily concentrated towards the bottom of the cavity, the ground surface of which, as previously mentioned, provides visible signs of mining plants. Piles of broken material such as residue from the mineral separation and crushing processes are present throughout. Crushing was the process, generally carried out just outside the mine's entrance, by which useful mineral was separated and sent for the following treatments: washing and metallurgic processing. Given the presence of water within the cavity, the washing process may also have taken place internally. There are small sub-vertical excavations, both around and beneath the cave-ins. These are often no longer practicable due to the collapse of containment structures and external material. The openings of the two shafts are protected by low, wide dry stone walls, which extend to form a type of entrance corridor. The low walls are in turn covered by fine detritus, which is partially cemented by a thin layer of calcium carbonate and by tiny stalagmites. Cappa & C. have interpreted the structures as follows: «There are signs, here and there, of low walls built during the 1943-1945 wartime period» (Cappa, Cigna, De Michele, Parea 1962). It should be highlighted that this information is incorrect. Furthermore, from both a tactical and defensive point of view, there is no reason for the existence of such structures as, placed on top of fallen rocks, they would be easily identifiable and attacked from above. The entrance to a sub-vertical passage lies a few metres from the west wall, on the edge of the rocky blocks and in precise correspondence with the bottle-neck. It is served by a tunnel-like structure, characterized by a series of small, neat fills. It develops under a large layer of inclined rock, which acting as a "roof" sheltered and preserved the entire plant. The passage extends for 11.5 m and is approximately 7 m deep. The last 2 m are flat, with a lateral diversion of 1.5 m, which appears to be a small, collapsed mining plant. It has numerous small containment structures which were surveyed and found to be at the limit of their stability. Consumption of the floor and of some parts of the wall edges would indicate long-term use. On the other hand, the fact that the only vague signs of excavation are to be found near the bottom of the passage and in the diversion, could indicate that this was a research structure which was later abandoned. It is plausible that mineral pockets were found along the way and removed and that mining took place far below the diversion. These walls and roofs of these underground structures show no sign of the recesses used to support wooden frameworks, which could have improved their stability, particularly during the movement of blocks. Although there are no signs of these, it cannot be excluded that beams would have been used in the most delicate and unstable areas. Further signs of intervention are to be found within the deep crevices which intersect the large blocks however multiple cave-ins and clay deposits have made these difficult to interpret. The works both within and below the blocks can be explained by a need to reach the cave's original level, now destroyed by an enormous cave-in, in order to search for further mineral deposits. Furthermore, given the irregular diffusion of the mineral within the limestone fractures, the same blocks could have held interesting mineralized sections.

Dating: hypothetically, mining activity could have taken place over various historical periods. It cannot be excluded that extraction could have been undertaken by Celtic civilisations before the Lombardy's Roman invasion. Generally speaking, mining was carried out by "following the vein", using irregular, narrow tunnels and an excavation method from ancient or mediaeval times. In a few cases there is intentional excavation of the enclosing rock (not following the mineralization), which would presuppose a different, perhaps more recent, mining method dating to the XV century. The absence of signs indicative of drill rod use excludes the use of explosives thus the mining activity appears to precede the XVI century.

Notes: the extraction method consisted in the use of tools such as pickaxes and bradawls, which were pounded with stone breaking hammers. The rocky surfaces present small bradawl marks left by a small pyramidal blade (of approximately 0.5 cm) and marks from a pickaxe with flat 2-4 cm head. There are also parallel vertical and oblique marks, which are 20-30 cm in length and were possibly made by a pickaxe or by a bradawl with pyramidal head. The last mining plant to the north presents numerous signs left by a tool with a 2.5 cm wide flat, curved head. No significant amounts of wood charcoal or roasting of the surfaces were observed; the use of fire is therefore excluded (even if the large, aerated environment would have permitted its effective use). There are no signs of bore holes (drill rods) indicating the use of explosives. Moving upwards towards the large blocks, there are a few short stepped sections, while at the very top of the collapsed material similar steps descend to the bottom of said material, almost without interruption. Continuing the descent, another brief section branches off to the right, towards the section of cave with the most dripping water, and leads to a small but permanent waterfall. The steps are made of flat layers of limestone with irregular edges, which are on average 50-60 cm wide, with tread of between 30 and 4 cm. They are of essentially the same height, of between 15 and 20 cm. The descending section is well-preserved and only a few areas have collapsed; there are 98 steps. This is undoubtedly the path used by the miners to reach the mining plants and transport the materials and to reach the spring. At the end of the XV century, Leonardo da Vinci wrote: «The highest known mountains to be found in this country are the Mandello Mountains, near the Leche and Gravidonia Mountains. Towards Bellinzona, 30 miles from Lecco are the Valchiavenna Mountains; however the highest of them all is that of Mandello, having at its foot a cave facing the lake, with 200 descending steps, here where there is constant ice and wind» (Leonardo da Vinci, Atlantic Code, f. 214 v.e.). Popular tradition would believe that «the cave facing the lake» refers to Ferrera Cave. It is irrefutably part of the Grigne mountain group (more commonly referred to as "Grigna") and identifiable as «the highest is that of Mandello», that is the Mandello Mountain. Sheet 573 of the Atlantic Code already states: «Grigna is the highest mountain in this country, and it is free of vegetation». It is situated on the lower mountain slope («having at its foot»), towards the Lecco del Lario branch (Lake Como), however the «there is constant ice and wind» does not appear to be consistent as its position is simply concealed, "in the shade". In any event, climatologic studies relative to the period in which Leonardo visited or passed through the area, would have to be conducted. On the other hand, the «cave» could refer to the following: this being a karst and inaccessible area, there are surely numerous, unknown caves. The only part indicative of the path, «with 200 descending steps», is the set of steps, in particular the latter, of which approximately 100 steps survive. Taking into account the missing sections, the descending stairway would have had at least 150 steps or thereabouts. It should however be noted that no mention of mining activity is made. It could be assumed that mining had ceased some time before or that Leonardo did not actually see the cave and that it was simply described to him. We do not believe, had he visited Ferrera, that he would not have been struck at least by its vastness. While taking into consideration the various exceptions, we can reasonably assume that the Ferrera Cave is the cavity mentioned, but that information on its existence was provided to Leonardo by third parties.

Bibliography: Cappa G., Cigna A., De Michele E., Parea G. C., *Ricerche sugli aspetti del fenomeno carsico profondo nel Gruppo delle Grigne (Lombardia). IV. La caverna Ferrera di Mandello 1502 Lo*, in Museo Civico di Storia Naturale, *Atti della Società Italiana di Scienze Naturali e del Museo Civico di Storia Naturale di Milano*, vol. CI, Milano 1962, pgs. 20-42.
Buzio A., Casini A., Padovan G., *Attività estrattive nelle Grigne. Alcune note riguardo la Grotta del Pallone e la Grotta Ferrera,* in G. Padovan and I. Riera (edited by), *Atti XV Congresso di Speleologia Lombarda. Sant'Omobono Imagna Terme (2-3 Ottobre 1999)*, Volume 3 - Speleologia in Cavità Artificiali, Milano 2000, pgs. 141-162.
Data ownership: Associazione Speleologia Cavità Artificiali Milano (S.C.A.M.); 2002.
Compiled by: Gianluca Padovan (Ass.ne S.C.A.M.).

III.3 - *Piedmont*

References to various defence works in the area now known as Piedmont date to as far back at the X century (fig. III.5). Some works survive, having undergone significant transformation. Others were built over the years to meet the changing needs of the towns in respect of defence, territory control and blockades for possible invasion directives from adjacent areas. Defensive works continued to be built until the mid XX century.

Fig. III. 5. *Geographic framework: Piedmont Region.*

Fort San Vittorio in Tortona (Alessandria) and Fort Demonte in Valle Stura (Cuneo). Between 1714 and 1770, the Savoia ordered that strategic barriers be developed for their properties in north-western Italy. This was because the previous network of fortresses erected by the Dukes of Savoia from the XVI to the XVII centuries had suffered extensive damage during the wartime operations of the Spanish Succession. The acquisition of new territories thus imposed the construction of modern fortification works. Thus, new building plants for the erection of strongholds to block the old Piedmont invasion roads were opened.

Each stronghold was functionally placed within the territory of the Kingdom. Each was positioned in such a way as to receive rapid assistance if threatened by enemy armies and to act as a logistics base in the event of offensive action. The large calibre battery cannons and the imposing walls followed in the footsteps of blockade fortifications as a deterrent to Piedmont invasion. The Brunetta di Susa, Exilles and Demonte Forts, the Fenestrelle Forts, the Citadels of Alessandria and Turin and the Fort of San Vittorio in Tortona allowed the Savoia army to hold their positions on the western alpine front and on the south-eastern front. However, during the so-called War in the Alps, the lack of blockade works in the south-western sector allowed French armies to outflank the defensive works in April 1796 and to thus defeat the King of Sardinia in the field of battle. The victorious French carried out the demilitarisation of the Strongholds which had prevented the Italian peninsula from coming under attack. Thus, the defensive Piedmont enceinte was destroyed between 1796 and 1801, by mine and pickaxe. For a number of reasons, the only works to survive were Fenestrelle Fort and the Citadels of Alessandria and Turin (Vigilino Davico 1989).

One of the military engineers who was to confront the new wartime realities and adapt Piedmont strongholds was Lorenzo Bernardino Pinto, Count of Bari (Bianzè 1704 - Turin 1789). Among other things, he dealt with the restoration of Fort San Vittorio in Tortona and Fort Demonte in Valle Stura.

The construction of Fort San Vittorio began in 1774 on the hill above Tortona (fig. III.6). the building was built on a previous fortification with mediaeval layout. The western crownworks were retained and restructured accordingly. The rear Hornwork parapet is raised and defiles Fort San Vittorio's western front. Of an irregular rectangular shape, it has angle ramparts, ditches and counterscarp works. Despite its small dimensions, the Fort is well-structured and its batteries are positioned on multiple levels. Occupied by French troops following the War in the Alps, it was besieged by Austro-Russian troops in 1799 (Comoli Mandracci, Marotta 1995). It was demolished a few years later. The current park layout to some extent preserves structures covered by soil and debris, to be restored at a later date.

Fort Demonte once blocked Valle Stura and was built on the site of the old Consolata Fort, following its attempted demolition on the part of French troops in 1744 (fig. III.7). In 1754 the construction plants, utilised for more than ten years, were supervised by Bernardino Pinto. The fort was built on multiple storeys so the enemy could be attacked with plunging fire. It was disarmed and mined following Napoleonic invasion. Erroneously regarded as a simple pile of ruins, the fort's core remained intact, carved as it was, directly into the rock. In fact, only the masonry structures were almost completely demolished and the Fort was at any rate, recoverable as a historic monument.

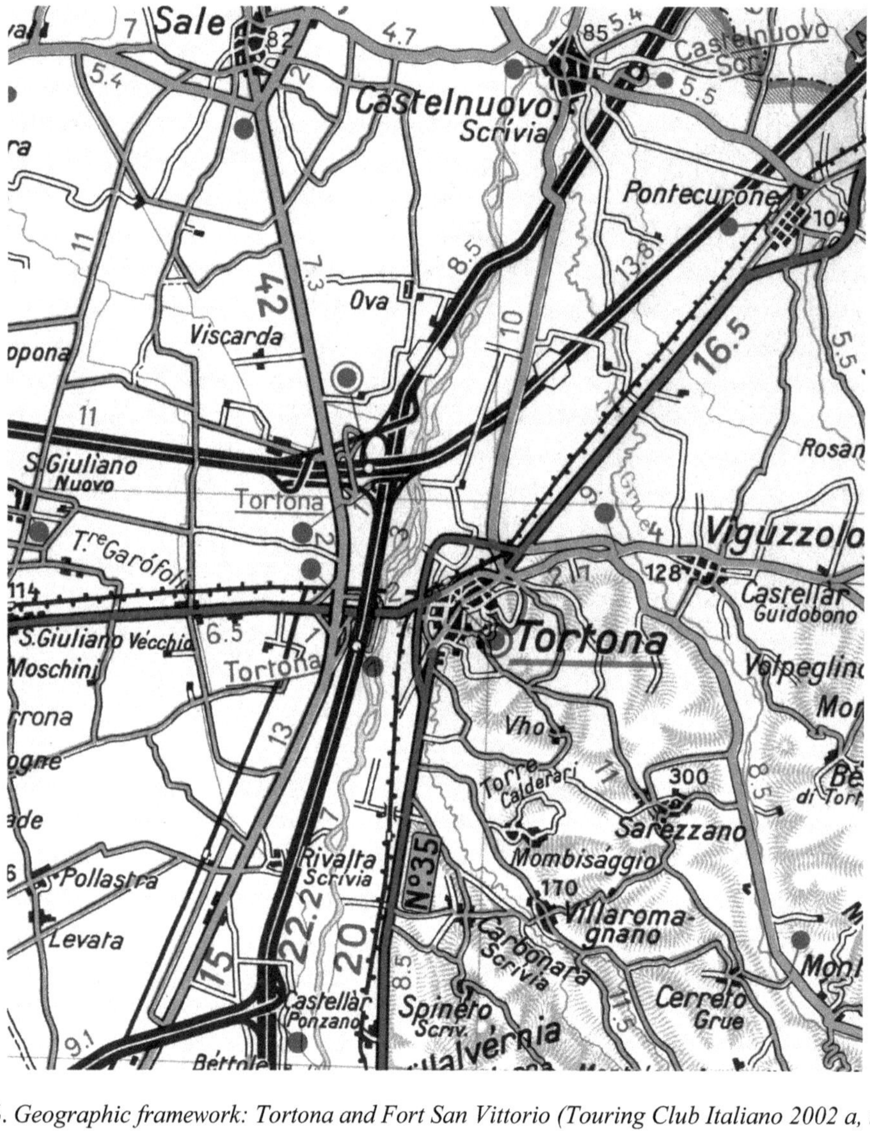

Fig. III. 6. *Geographic framework: Tortona and Fort San Vittorio (Touring Club Italiano 2002 a, table 22).*

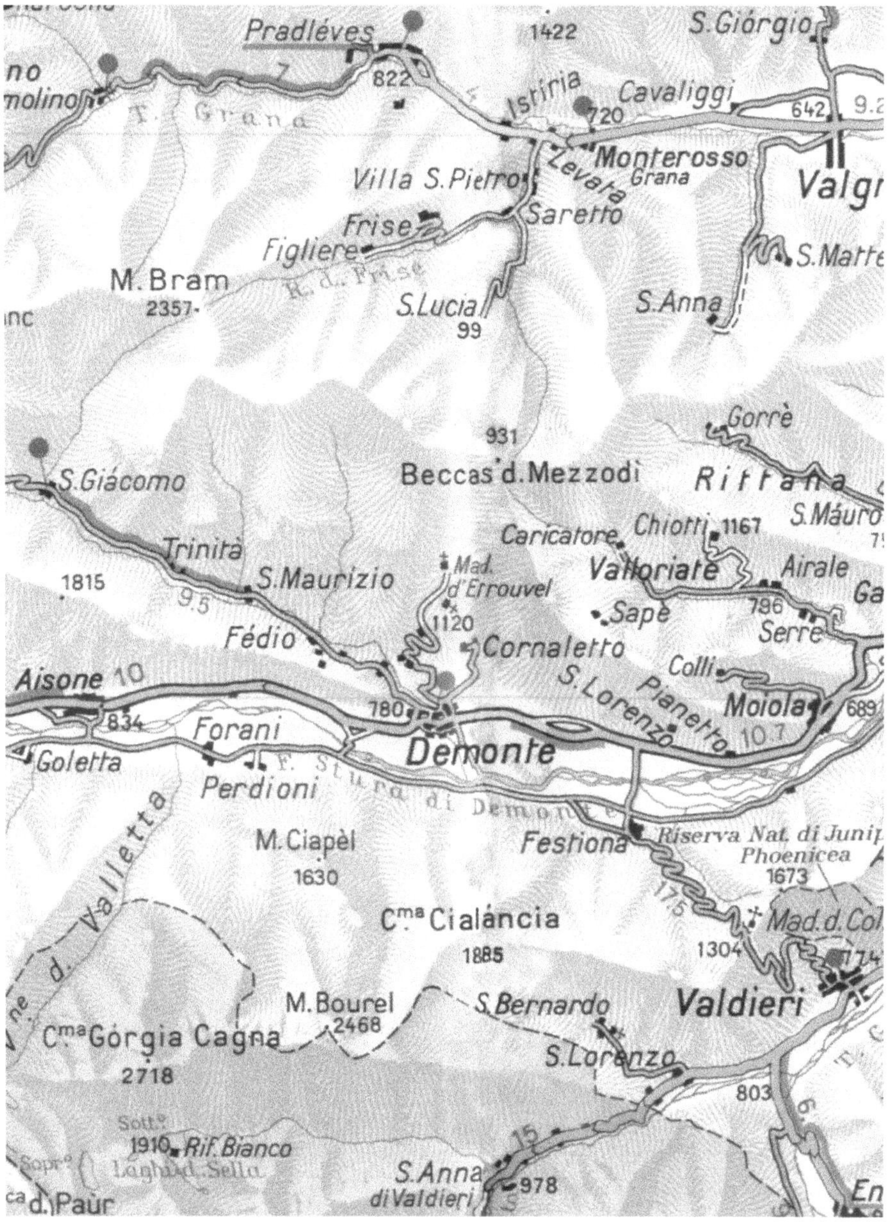

Fig. III.7. *Geographic framework: Valle Stura and Fort Demonte (Touring Club Italiano 2002 a, table 27).*

Verrua Fort (Turin). Built on a hill to the northwest of the River Po, only the heart of the Fort, consisting of the Donjon, the "bomb-proof" Barracks and a few bastioned sections, still survives (fig. III.8). The first reference to the "Verruca" settlement appears in a diploma issued in 999 by Emperor Otto III in favour of Leo, Bishop of Vercelli (Bazzi 1907, pg. 6. Signorelli A., Signorelli B. 2000, pg. 174). In 1159, Emperor Frederic Barbarossa orders that the Verrua defence works be improved and during his reign these are repaired and extended. At the end of the XVII century, the structure of Verrua Fort is that of a triple bastion fortification protecting the only flank not protected by rock. It communicates with the fortified settlement of Crescentino, situated on the opposite bank of the river, by means of boats and terrepleine works. On the opposite side, in a south-south-westerly direction, two defensive lines branch out from the fort forming a trenched field, which communicates with the Fort Royal on the nearby Carbignano hill. The sieges which would make the fort famous throughout Europe take place in 1625, against Spanish troops and in 1704-1705 against French troops. Following the second siege, the Fort was partly destroyed and only its central core remained functional. In 1955, the hill and therefore the ruins were purchased by "Cementi Victoria S.p.A." and a quarry was opened. The Fort was thus partially eliminated, with no intervention for the conservation of the structure on the part of competent authorities. With the approval of the Municipality of Verrua Savoia, Milan's Speleological Association of Artificial Cavities (Associazione Speleologia Cavità Artificiali Milano – S.C.A.M.) registered and studied the fort's underground and surface structures.

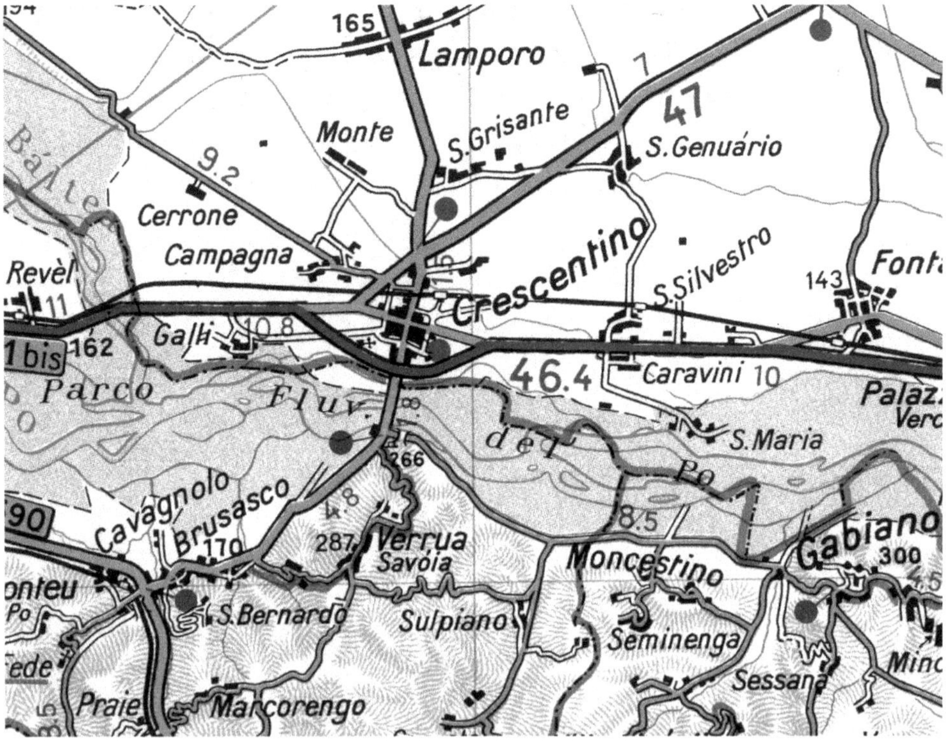

Fig. III.8. *Geographic framework: Fort Verrua (Touring Club Italiano 2002 a, table 10).*

Pavarolo Castle (Turin). Pavarolo territory is in north-western Monferrato, on the left bank of Rio Morto (fig. III.9). The fortification works which still occupy the hilltop date back to the XVIII century restoration of the XIV century castle. The castle consisted of a curtain wall protecting the keep and the residence of the yet to be identified Castle Lord. The remains of the curtain can be seen under the last road curve just before the top of the hill. Registration of the artificial cavities was possible thanks to the support of the castle's current owners, the Zavattaro Ardizzi Family.

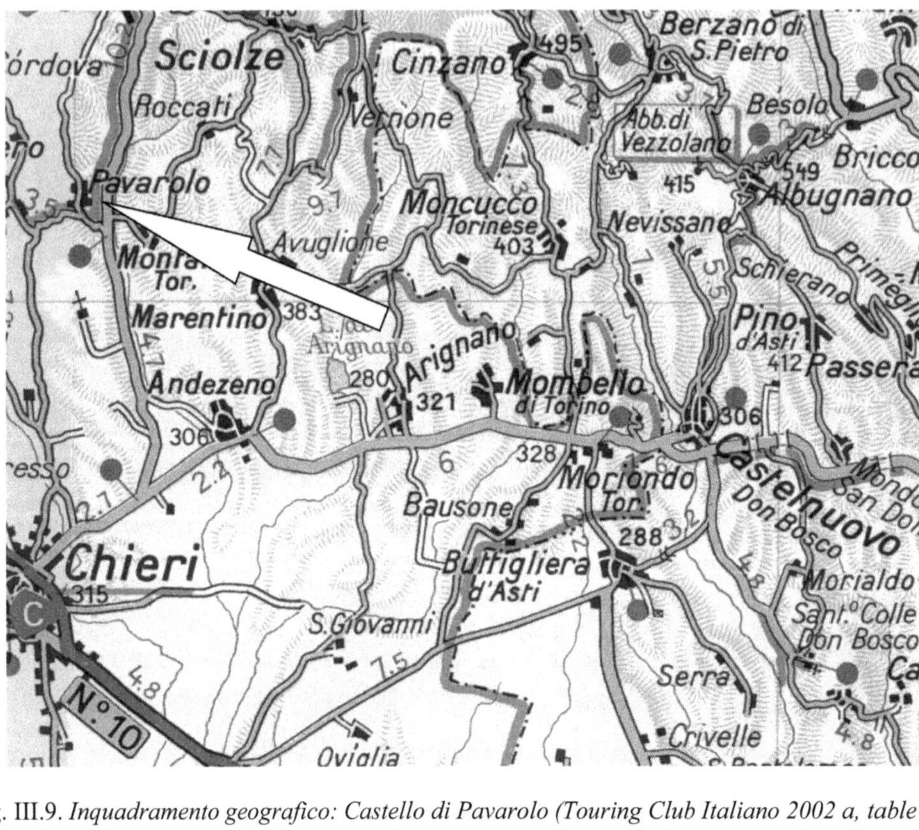

Fig. III.9. *Inquadramento geografico: Castello di Pavarolo (Touring Club Italiano 2002 a, table 22).*

III.3.1 - *CA 00001 PI AL; "Barbarossa Tomb"; Countermine Tunnel in the Counterscarp Ditch of Gate San Vittorio*

Cadastral number: CA 00001 PI AL
Denomination: "Barbarossa Tomb"; Countermine Tunnel in the Counterscarp Ditch of Gate San Vittorio (figs. III.10.a, III.10.b, III.11 and III.11.a)
Region-country: Piedmont, Italy
Province: Alessandria
Municipality: Tortona
Locality: Colle di Tortona, Fort San Vittorio
Location: Fort San Vittorio ditch
Ownership: Municipality of Tortona
Cartography: //
Geological unit: the municipality of Tortona in the province of Alessandria is situated along the central basin of the River Po, between the right bank of the Scrivia and the foothills of the Ligurian Apennines. The relief towering over the city, upon which Fort San Vittorio stands, is the Mombisaggio Formation consisting of yellow sandstones and calcarenites through to the lower, coarser sandstones with sandy Miocene marl intercalation (Serravalliano - Langhiano). From a tectonic point of view, the "Sperone di Tortona" (Tortona Spur) consists of two anticlinal structures, one passing through Berzano di Tortona and the other through Cerreto Grue, at the centre of the "Calcari di Zebedassi" (Zebedassi Limestones) emerge. These are separated by a synclinal structure essentially consisting of Monte Piano Marls, Ranzano Sandstones and Antognola Marls with axial depression, highlighting the Mombisaggio Formation
Altitude: //
Position: //
Context: Fort San Vittorio
Operations conducted: excavation, survey, photographic service
Work carried out by: Associazione Speleologia Cavità Artificiali Milano (S.C.A.M.); 2002
Warnings: presence of organic and inorganic waste in the first section and structural collapse in the final section
Typology: 6 - countermine or demolition tunnel
Description: the entrance to an underground work is visible for the counterscarp ditch protecting Gate San Vittorio. Known locally by the name of "Barbarossa Tomb", the work is currently located just above the bottom of the ditch and is partially filled due to the demolition of the defensive structure. The counterscarp masonry lining is almost completely missing therefore it has not been possible to ascertain whether or not it was excavated through a wall. The entrance was equipped with a masonry structure which served as a metal gate. This was removed, most probably when the area was transformed into a park. The underground system is carved in sandstone and its exposed roof and walls present visible signs left by early tools. Unfortunately the first section is full of debris and waste material, while the final section shows signs of structural collapse. It extends for 34.12 m and there are three sections facing different directions. The initial part of the first section has a short and worn descending stairwell consisting of eight steps, covered in detritus and waste material; the tunnel points in a 282° direction. The walls and roof are irregular in appearance. This can be attributed in part to the natural characteristics of the rock matrix, in this section presenting fissures and pockets of unconsolidated material, which has become detached over time leaving small vacuums. In certain points there are clear signs of rock extraction, which have altered its original appearance. The height varies between 2.18 m at the entrance and 1.93 m half-way along the stairwell and 1.96 at the end of the room. The width also varies, being 1.27-1.28 m at the base of the piers and 1.42-1.32 m at the top of the piers. Overall, the excavation is regular and level, with a height of between 1.96 and 1.8 m and a width of between 1.28 m and 1.02 m. Beyond the stairwell, the passage continues in a 275° direction. The roof consists of a drop arch, the height of which decreases towards the end of the first section. The walls are slightly curved and parallel one to the other. The excavation face is relatively well-finished. It is therefore unlikely that the change of direction in the second section resulted from a structural error when positioning the first section. The second section branches off following a rectilinear 200° direction. This long tract has a slightly smaller section that the latter, with a base width of between 1.02 and 1.12 m. Its roof has a semi-circular vaulting, which rises slightly in respect of the entrance. In some points the walls curve outwards, in others they simply narrow towards the base. The floor is covered in dust and detritus. Towards the second section there is a partial breach of the walls of unknown origin. The first section presents just one niche for the placing or hooking of a lamp, whereas there are ten in the second section. They are 13 cm wide by 14 cm high and 6 cm deep on average. Although indicative of a certain acquired order on the part of those who conducted the excavation and the regularization of the passage walls, the positioning of the niches is not perfectly regular. On the left wall such niches are positioned close together in couples, the distance between the second niche of the first "couple" and the first niche of the next couple being double the distance between the first and the second niche of the previous couple. On the other wall only two niches are visible, positioned half way between the first and second

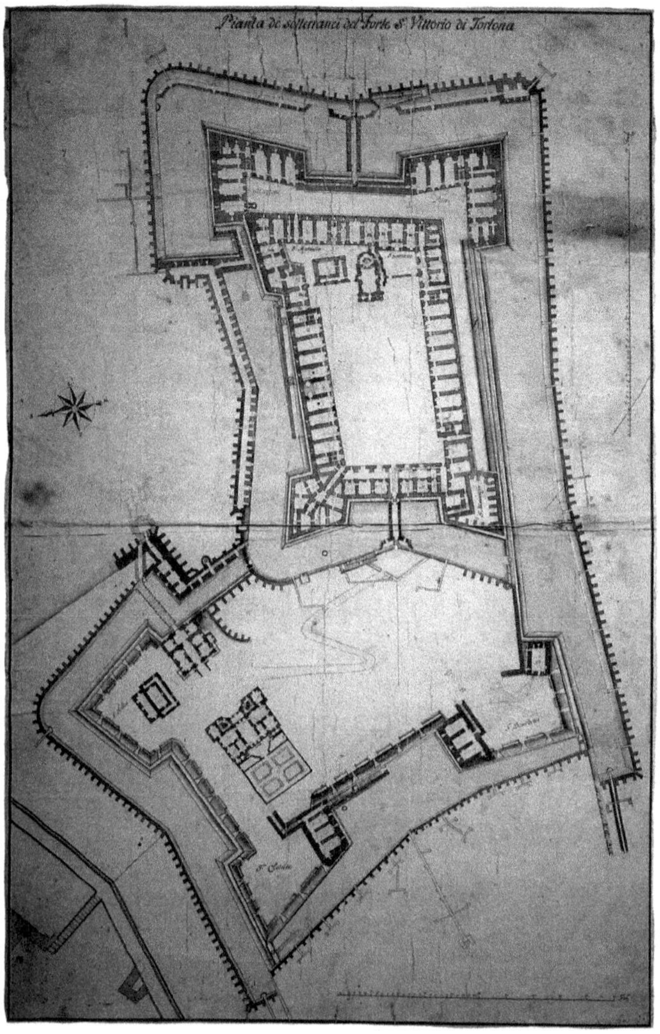

a

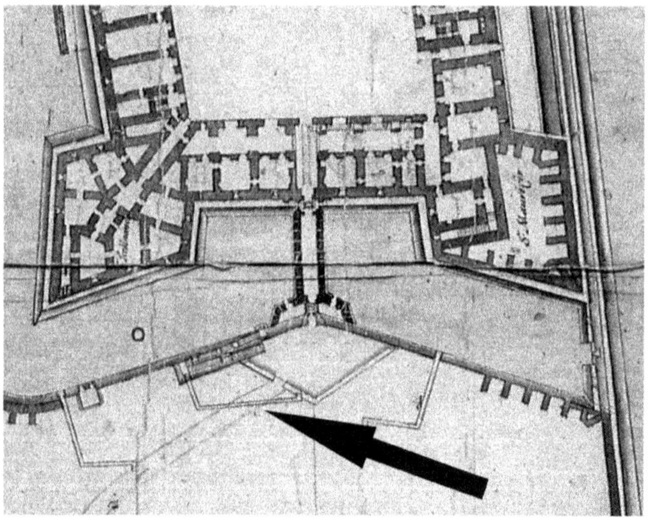

b

Fig. III.10.a. *Plan of Fort San Vittorio subterranean structures in 1799 (source: Comoli Mandracci, Marotta 1995, pg. 59).*
Fig. III.10.b. *Arrow indicates the Countermine Tunnel in the Counterscarp Ditch of Gate San Vittorio.*

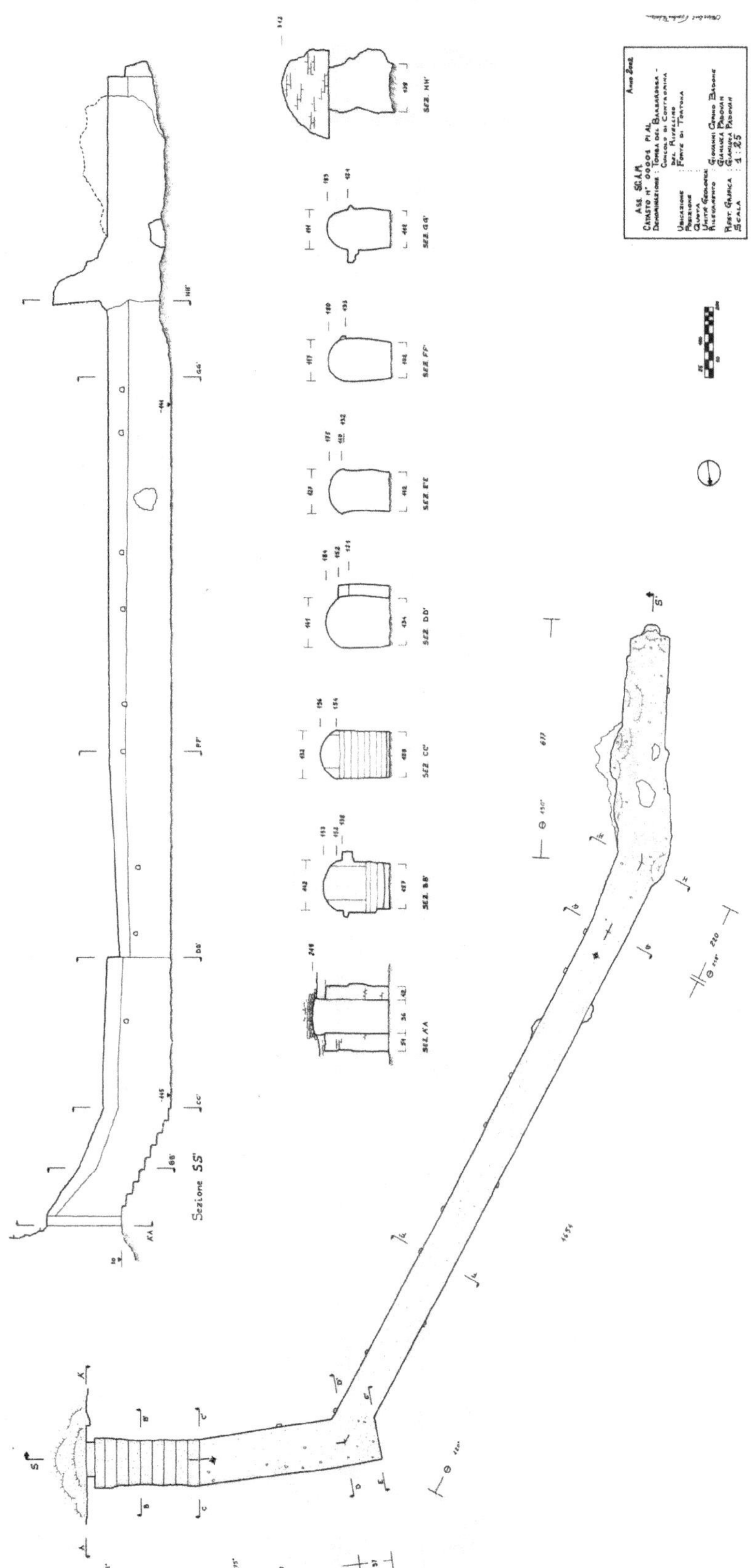

Fig. III.11. *Ravelin Countermine Tunnel Planimetry (CA 00001 PI AL), known locally as "Barbarossa Tomb". The structure extends below the ravelin protecting Gate San Vittorio (Ass.ne S.C.A.M. Archive).*

Fig. III.11. a. *Ravelin Countermine Tunnel (CA 00001 PI AL), known locally as "Barbarossa Tomb"(photo G. Padovan).*

couple and the second and third couple. There may have been a further niche in the location of the current breach in the wall. Furthermore, if the first couple is situated under the springer then the second couple is in line with the springer, the third is above it and the fourth is at an even higher level. The last section, 6.77 m long, curves to the left in a 190° direction. Towards the base of the work, there are collapsed walls and the ceiling excavation intercepts a slightly sloping brick wall, which appears to be the external wall facing of the overlying ravelin. The only perplexing aspect is that we would have expected the battered plinth to slope in the opposite direction: this is the only aspect that remains unclear. Just beyond this, the left wall has collapsed, however the structure has not been entirely compromised. Here the floor is obstructed by detritus and large blocks of rock. The bottom of the passage appears to have been lowered a few tens of centimetres by irregular excavations, which probably took place before the work was built. Only one niche is visible on the right wall, in proximity of the excavation front. A small, regular channel has been cut into the arch of the latter, which at the head of the excavation follows the wall down to the right. Two nails with metal heads would normally indicate an electric cable recess therefore this is a relatively recent work. What needs to be ascertained is whether the structure was used as an air-raid shelter during the Second World War.

Interpretation: late XVII century plans show no defence work opposite San Vittorio Gate. However, a table from 1799 confirms the existence of a ravelin (or structure identifiable as such) with three nearby underground works. Taking into account that the tunnel system in question is accessed through a broken wall, it is possible that it was created (most probably at the same time as the other two) by French troops during their short occupation of the Tortona stronghold. In any case the passage works open onto the ditch counterscarp. The first and the third tunnel are L-shaped and each has a short terminal branch, in all appearances a demolition chamber, facing outwards from the Fort's nucleus. This second tunnel first leads to the top of the ravelin's external point, almost reaching it by means of a slightly obtuse angle. As in previous examples, there is a short tunnel section at the apex, but its chamber faces directly into the ravelin, directly below the perimeter wall. The first and third tunnels are thought to be countermine tunnels, while the sole purpose of the second tunnel seems to be the demolition of the ravelin. Given its position, it may also have been used as a countermine branch excavation base. The evidence present a work that is slightly different from a standard countermine tunnel: it has no demolition chamber and the tunnel extends beyond the foundation wall. As already mentioned, the wall should slope in the direction of the chamber rather than in the opposite direction. This could indicate that the wall is something other than the anticipated left ravelin wall or it could be indicative that the tunnel comes into contact with the ravelin's right facade, passing by the other without exposing it. The latter assumption would imply that the foundation wall of the left façade is not as deep as the other. There is in any case no demolition chamber.

Dating: 1797-1798.

Notes: the evidence again indicates that following the fort's demolition little was left. The casual discovery of this basic tunnel was completely unexpected. First of all, the information emerging from the current park layout is testament to the fact that the military structures were irreparably damaged but not obliterated. Clearing the debris would therefore reveal the entire defensive perimeter. Secondly, we should ask ourselves why the tunnel was never used: when charged it could easily have blown up the salient angle of the ravelin through to the foundations and far more quickly and effectively. It should come as no surprise that the utmost of care was taken to ensure that the structures, especially those structures of intricate design were not irreparably damaged, as in the case of Fort Demonte.

Bibliography: Padovan G., *Due noci dure da rompere. I forti di Demonte e di Tortona alla fine del XVIII secolo: l'organizzazione della difesa, la rete di contromina e l'approvvigionamento idrico*, in Anzanello E., Dal Cin F., Gasparetto P., Gava S. (edited by), *Atti Montello 2002. Conglomeriamoci. 21° Incontro Internazionale di Speleologia. Nervesa della Battaglia 1-3 Novembre 2002*, Villorba 2003, pgs. 293-365.

Data ownership: Associazione Speleologia Cavità Artificiali Milano (S.C.A.M.); 2002.

Compiled by: Gianluca Padovan (Ass.ne S.C.A.M.).

III.3.2 - *CA 00001 PI CN; Counterscarp Gallery of the Bastion of Saint Ignatius*

Cadastral number: CA 00001 PI CN

Denomination: Counterscarp Gallery of the Bastion of Saint Ignatius (figs. III.12, III.13, III.14, III.15, III.15.a, and III.16)

Region-country: Piedmont, Italy

Province: Cuneo

Municipality: Demonte

Locality: Promontorio del Podio

Location: Demonte Fort, situated on the left bank of the River Stura of Demonte, a few hundred metres east of the town of Demonte

Ownership: Municipality of Demonte

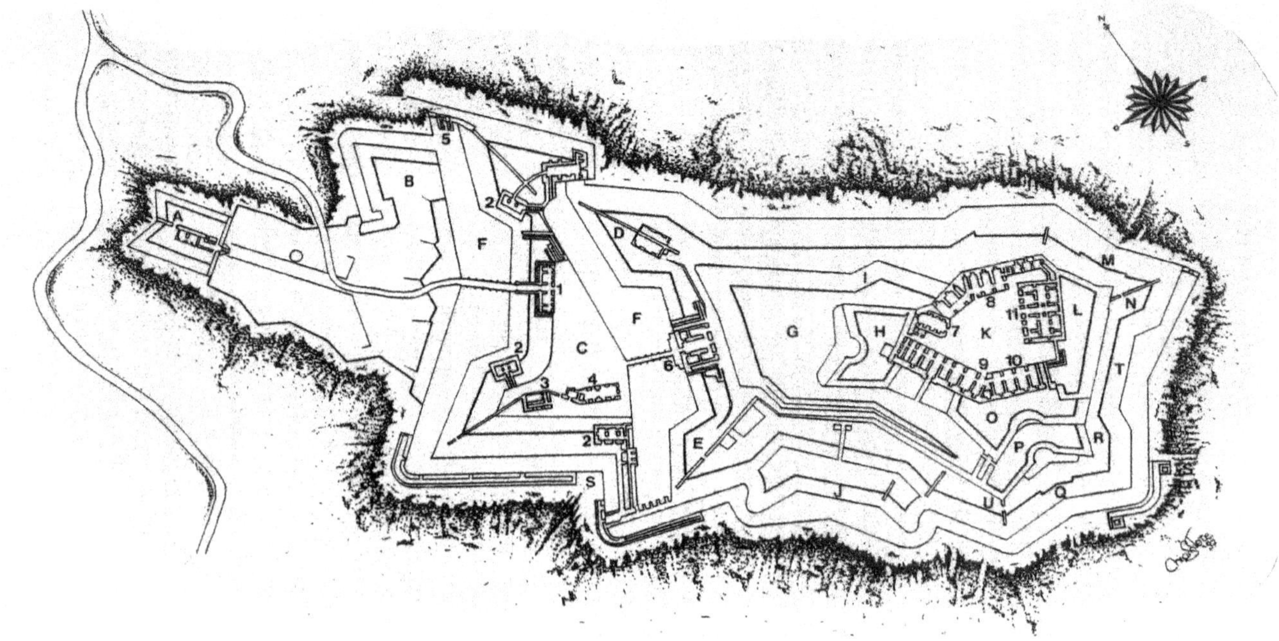

Fig. III.12. *Graphic exemplification of the Fort Demonte system during the latter part of the XVIII century, detailing the primary defensive structures (from: Gariglio 1997, pg. 195). A: San Marcellino redoubt. B: lunette. C: hornworm. D: San Paolo Bastion. E: Bastion of Saint Ignatius. F: ditch. G: Beato Amedeo battery or Royal Battery H: San Ferdinando Bastion Fortification. I: San Anna Demi-Bastion. J: Low work or Tennaillon facing Stura. K: Superior or Knight Square. L: Royal Battery. M: Tenaille of San Giuseppe. N: San Giuseppe Demi-Bastion. O: San Maurizio Bastion Fortification. P: San Lorenzo Bastion Fortification. Q: Tenaille of San Michele. R: San Michele Demi-Bastion. S: counterscarp gallery. T: front facing Podio. U: Green Demi-Bastion. 1: Royal gate and guard-house. 2: casemates. 3: gunpowder magazine. 4: laboratory and armoury. 5: possible location of San Marcellino Well (if this was in fact excavated). 6: main gate. 7: San Carlo Church. 8: Nordo or San Carlo districts with bread oven and cistern. 9: South districts. 10: cistern. 11: Governor's Palace (Ass.ne S.C.A.M. Archive).*

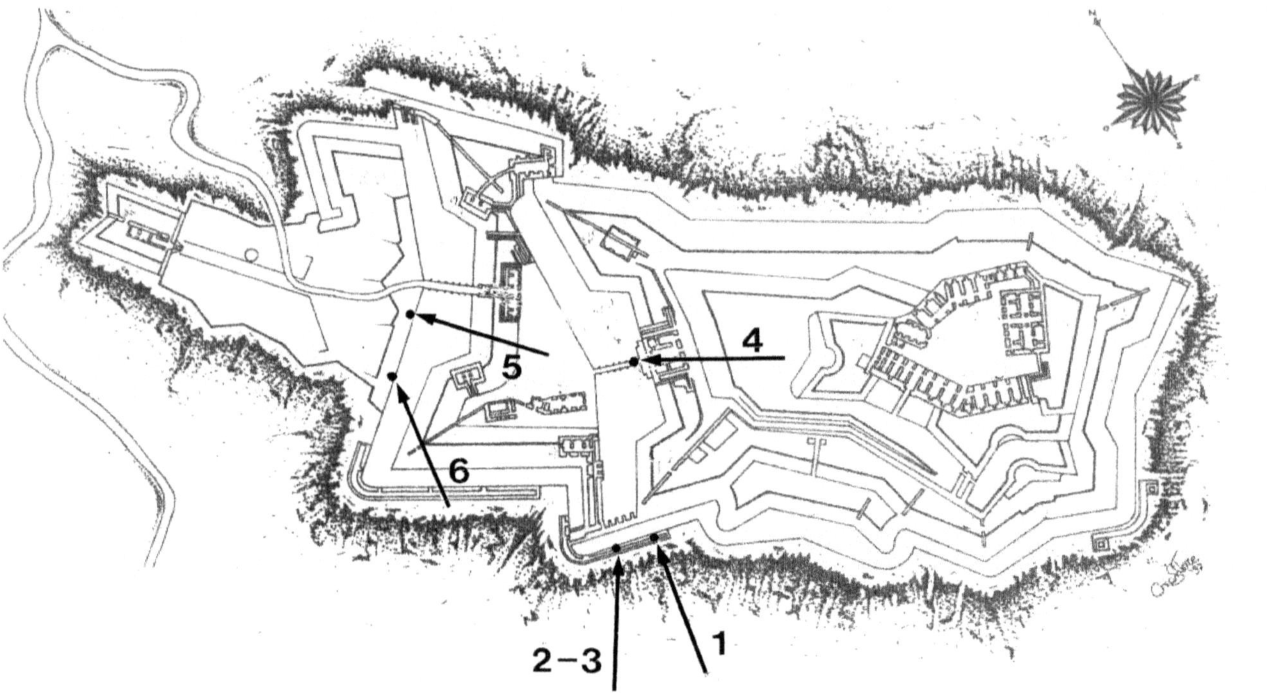

Fig. III.13. *Graphic rendition of the Fort Demonte installation during the latter part of the XVIII century (from: Gariglio 1997, pg. 195). The artificial cavities surveyed are listed below: 1: Counterscarp Gallery of the Bastion of Saint Ignatius. 2: Countermine Tunnel of the Bastion of Saint Ignatius. 3: Ditch Drainage Tunnel of the Bastion of Saint Ignatius. 4: Ditch Counterscarp Cistern of the Main Gate. 5: Primary Shelter in the Hornwork Counterscarp . 6: Secondary Shelter in the Hornwork Counterscarp (Ass.ne S.C.A.M. Archive).*

Cartography: C.T.R. 1:10.000
Geological unit: Fort Demonte rises on a low and narrow rocky spear on the left bank of the River Stura, in proximity to an alluvial cone. The substratum consists of grey arenaceous or entrochal limestone with streaks of pink or yellow-red, with flint layers and Crinoid fossils and grey or greenish calcareous Dogger-Malm slate (upper-late Jurassic). These are marginal marine environment deposits
Altitude: //
Position: //
Context: Fort Demonte
Operations conducted: excavation, survey, photographic service
Work carried out by: Associazione Speleologia Cavità Artificiali Milano (S.C.A.M.); 2002
Warnings: structural collapse of the current entrance
Typology: 6 - counterscarp gallery
Description: the Bastion of Saint Ignatius is external to the stronghold and faces a south-westerly direction, towards the River Stura. Given the less pronounced slope of the relief as compared to the other flanks, this side is the most exposed, as attested by the 1744 siege. The Bastion is protected by a ditch with counterscarp gallery, of which the final, north-east facing section with sortie is essentially intact. The other sections have been destroyed and only rubble remains. In any case, a more detailed investigation confirms that the gallery remains intact as far as the springer and only a few sections of the upper breastwork are damaged; various other sections provide evidence of the vaulted roof. All the loopholes should therefore be intact. A second postern, with its original roofing, can be seen towards the end of the north-east branch. However, the ditch is full of rubble. The counterscarp gallery is a key point in the external defence of the sector, from which it kept control over the ditch itself by loophole gunfire and the launch of sortie attacks. The north-east sector was also controlled by enfilade fire along the re-entrant flank and the south-west salient of the Tenaillon facing Stura: imposing and carved in the rock almost in its entirety, so much so that it is perfectly visible despite its demolition, it is surrounded by a well-defined ditch, which despite being full of rubble, measures 12 m in depth. The section of gallery which has been uncovered is rectilinear and measures 11.36 m in length. Only the first 2.2 m no longer have the original vaulted covering. It is partially buried and the some sections of the wall facings show evidence of demolition attempts. A short tract of gallery, now completely buried and which once lead to the postern, branches out three metres from the bottom. The main gallery is 1.82 m wide; its current maximum height is 3.22 m, and 2.4 m from the springer. The punch-dressed masonry and stone walls are lined. The external wall (facing south-east) is more than 1.72 m deep, as can be calculated from the demolition chamber created through a wall, without entirely perforating it and in front of the postern. This side also has a niche, which was presumably used for lighting purposes. It has a lined semi-circular brick barrel vault, rising slightly beyond the lateral branch, the cross vault of which breaks its uniformity. In an almost entirely vault-free section, the breach almost perpendicularly cuts the structure thus revealing its construction style: three rows of bricks placed one next to other, one on top of the other and head to head form the design of the barrel arch. Further up, regular stone rows cover an area of approximately half a metre and come into contact with at least one line of stone bricks sloping outwards towards the parapet. There are further rows above this and although these are difficult to interpret, they are needlessly regular and are almost all made of stone. The roof is therefore both thick and solid and is undoubtedly bomb-proof. The part overlooking the ditch has three brick-framed loopholes, which are scratched and exposed by demolition attempt. These are partially buried by detritus and are currently situated at a depth of just over 2 metres. Inside the structure, at least the pillar arch and the walls are made of exposed brick. Strangely, one is different from the other. The latter is wider, the central one is narrower and distinctly higher than the latter and the third, near the current entrance, is slightly wider than the central loophole while its height cannot be determined on account of the soil which covers it. Despite the fact that the entire complex is in a state of ruin, the precision and care involved in the construction of the structure are evident. Furthermore, if the loopholes are not all the same, there is surely a valid reason for this. The walls of the other short tunnel have been partially demolished together with a short section of roof lining although the tunnel itself is fairly solid and presents no danger. Demolition chambers constructed by breaking through walls can be seen on both sides of the tunnel. The base of the tunnel holds the entrance to the postern as well as the collapsed pillars and part of the vaulted arch. The visible part of said arch consists of perfectly square and perfectly positioned stones. On the left of the room's granite frame is a hole, which would have held one of the two door closure hinges. The postern would thus have been accessed by means of a flight of stairs or a sloping floor.
Interpretation: surviving section of counterscarp gallery with ditch sortie.
Dating: XVIII century.
Notes: an adequate dig would completely uncover the structure. The recovery of the entire ditch is also recommended.
Bibliography: Padovan G., *Due noci dure da rompere. I forti di Demonte e di Tortona alla fine del XVIII secolo: l'organizzazione della difesa, la rete di contromina e l'approvvigionamento idrico*, in Anzanello E., Dal Cin F.,

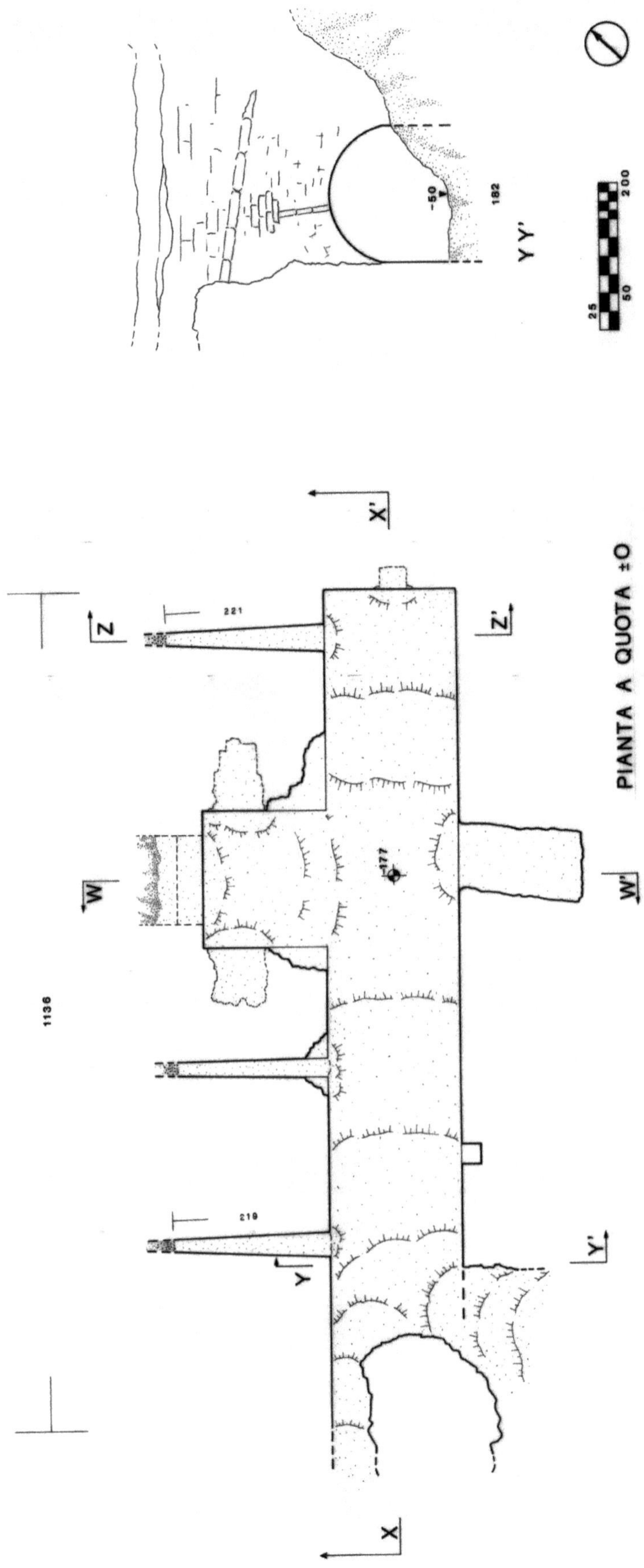

Fig. III.14. *Plan of the Counterscarp Gallery of the Bastion of Saint Ignatius (CA 00001 PI CN) (Ass.ne S.C.A.M. Archive).*

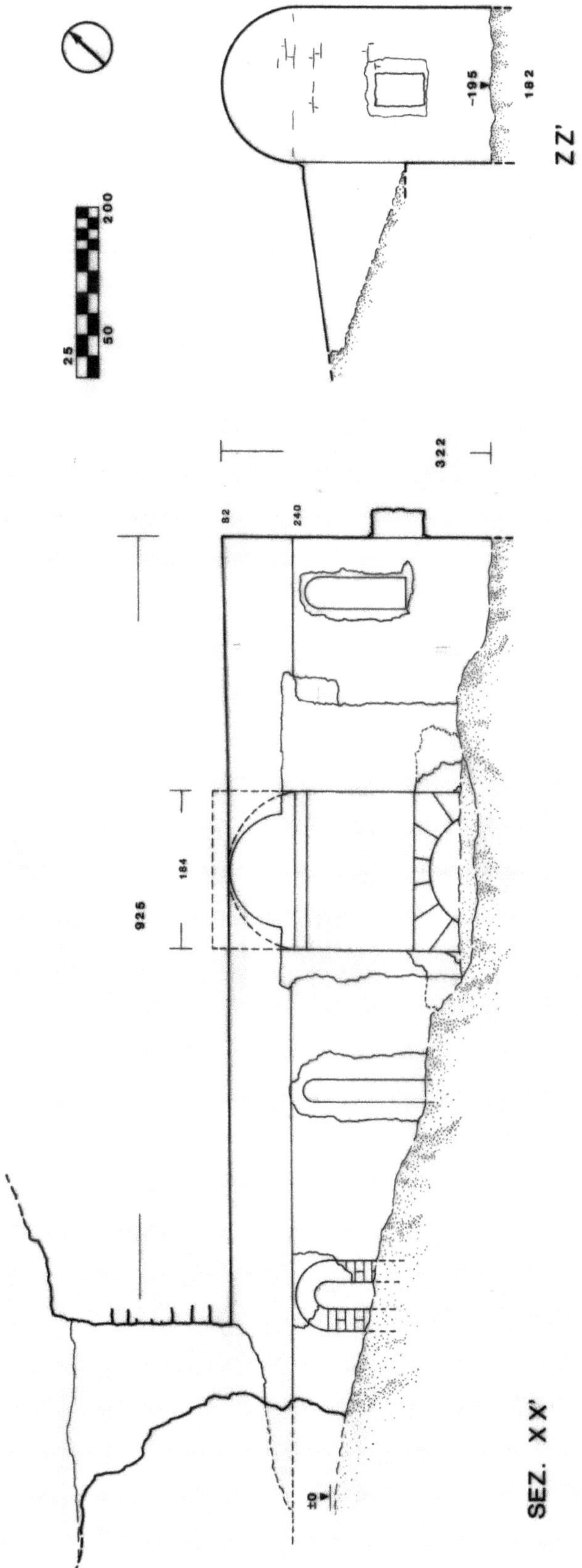

Fig. III.15. *Longitudinal section of the Counterscarp Gallery of the Bastion of Saint Ignatius (CA 00001 PI CN) (Ass.ne S.C.A.M. Archive).*

Fig. III.15. a. *Counterscarp Gallery of the Bastion of Saint Ignatius (CA 00001 PI CN) (photo G. Padovan).*

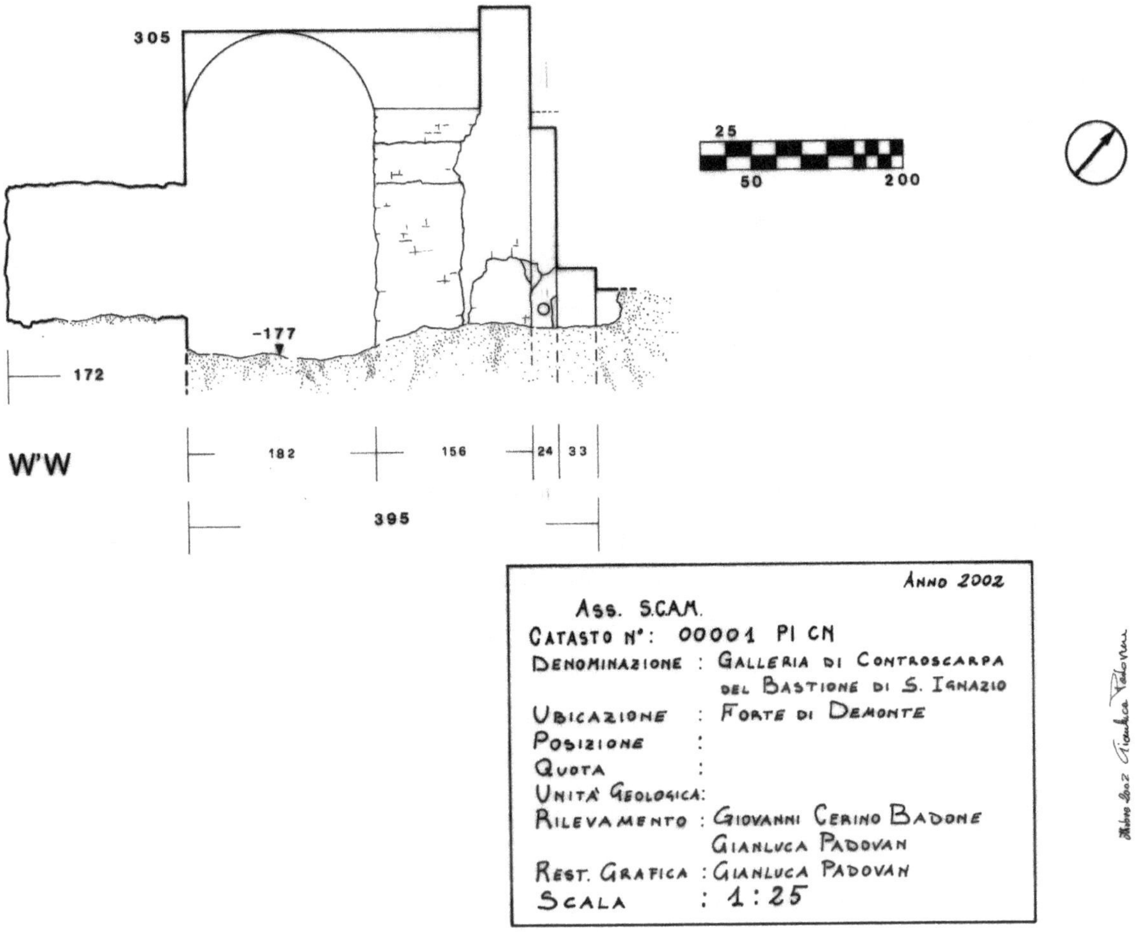

Fig. III.16. *Transversal section of the Counterscarp Gallery of the Bastion of Saint Ignatius (CA 00001 PI CN) (Ass.ne S.C.A.M. Archive).*

Gasparetto P., Gava S. (edited by), *Atti Montello 2002. Conglomeriamoci. 21° Incontro Internazionale di Speleologia. Nervesa della Battaglia 1-3 Novembre 2002*, Villorba 2003, pgs. 293-365.
Data ownership: Associazione Speleologia Cavità Artificiali Milano (S.C.A.M.); 2002.
Compiled by: Gianluca Padovan (Ass.ne S.C.A.M.).

III.3.3 - *CA 00002 PI CN; Countermine Tunnel of the Counterscarp Gallery of the Bastion of Saint Ignatius*

Cadastral number: CA 00002 PI CN
Denomination: Countermine Tunnel of the Counterscarp Gallery of the Bastion of Saint Ignatius (figs. III.17, III.18 and III.18.a)
Region-country: Piedmont, Italy
Province: Cuneo
Municipality: Demonte
Locality: Promontorio del Podio
Location: Fort Demonte, situated on the left bank of the River Stura in Demonte, a few hundred metres east of the town of Demonte
Ownership: Municipality of Demonte
Cartography: C.T.R. 1:10.000
Geological unit: Fort Demonte rises on a low and narrow rocky spear on the left bank of the River Stura, in proximity to an alluvial cone. The substratum consists of grey arenaceous or entrochal limestone with streaks of pink or yellow-red, with flint layers and Crinoid fossils and grey or greenish calcareous Dogger-Malm slate (upper-late Jurassic). These consist of marginal marine environment deposits
Altitude: //
Position: //

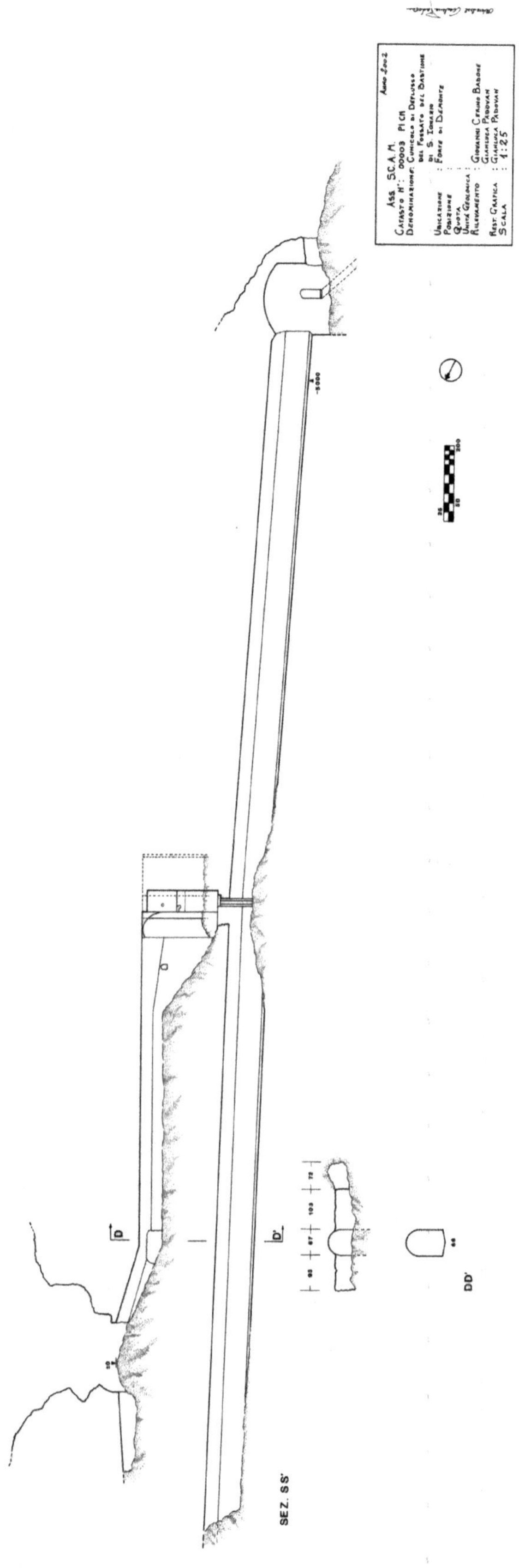

Fig. III.17. *The section shows the upper section (altitude 0) of the uncovered Counterscarp Gallery of the Bastion of Saint Ignatius, its loophole overlooking the ditch (to the left). On the right is the entrance to the Counterscarp Gallery of the Bastion of Saint Ignatius (CA 00002 PI CN), which provides access to the underlying Ditch Drainage Tunnel of the Bastion of Saint Ignatius (CA 00003 PI CN). Section DD' shows the two demolition tunnels which would have contained unutilised demolition chambers (Ass.ne S.C.A.M. Archive).*

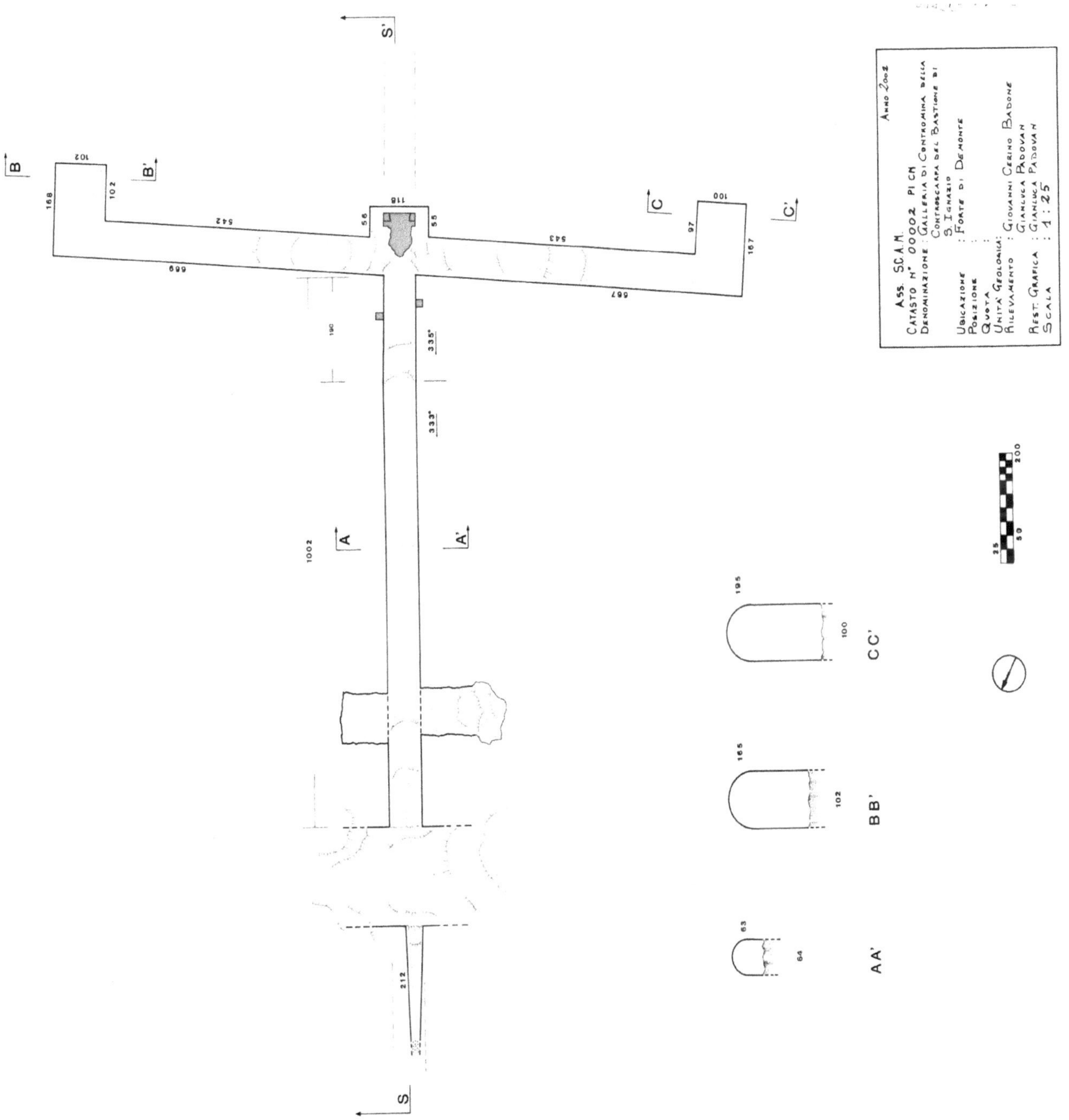

Fig. III.18. *Plan of the Countermine Tunnel of the Counterscarp Gallery of the Bastion of Saint Ignatius (CA 00002 PI CN). From the left are the loophole overlooking the ditch, the uncovered Counterscarp Gallery of the Bastion of Saint Ignatius and the Countermine Tunnel with its two demolition tunnels; at the front is the guide for the sluice gate of the underlying Drainage Tunnel from which, the countermine branches and their relative chambers laterally branch (Ass.ne S.C.A.M. Archive).*

Fig. III.18.a. *"Tenaglione verso Stura" cut into the limestone, situated east of the Bastion of Saint Ignatius (photo G. Padovan).*

Context: Fort Demonte
Operations conducted: excavation, survey, photographic service
Work carried out by: Associazione Speleologia Cavità Artificiali Milano (S.C.A.M.); 2002
Warnings: the tunnel is currently accessed by means of an opening excavated in collapsed material
Typology: 6 - countermine tunnel
Description: the entrance to the Countermine Tunnel is situated along the Counterscarp Tunnel branch of the Bastion of Saint Ignatius. It is approximately 30 m from the vaulted counterscarp gallery. It is buried to almost the full height of the piers. The uncovering has salvaged a short section of masonry vaulted arch and a deep loophole opposite the countermine, partially blocked by detritus, through which the pillar arch and the exposed bricks can be seen. The countermine branch is in the classic T-shape, with an unusual (and perfectly aligned) underlying hydraulic conduit. The length of the branch access varies between 0.64 m and 0.67 m; it has a lined semi-circular barrel arch and debris fills 5/6 of the tunnel, almost as far as the springer. The tunnel has a short, descending first section and then proceeds on a 153°-333° axis almost to its end where it slopes slightly to the right by 2°. Here, despite the fact that the arch maintains the same height the tunnel gives the impression of descending via a set of steps. The short, final tract of tunnel presents a niche on each side, once used for lighting purposes, as attested by the weak halos of carbon black on the small arches of each niches. It should be mentioned that breaking down a wall in the point where the first descending tact and the rectilinear tract meet, revealed two demolition chambers, one opposite the other. Despite the scrap contained within them, they show that the masonry structure was made of stone bricks and stone material. After the first 1.03 m, the one on the right continues a further 0.72 m, indicative that the tunnel was created using the cut and cover method and that its lateral wall is at any rate one metre deep. The demolition chambers are fully brick-lined and are sufficiently large to be charged with "globes of compression". Returning to the entrance branch, this has a small room at its opening, to provide room to control a sluice gate for the underlying hydraulic conduit. Two horizontal granite inserts are grooved in such a way as to support the winch. The granite guides for the sluice gate are on the floor, which has now subsided.
Interpretation: countermine system with two demolition chambers, which were probably used with "globes of compression".
Dating: XVIII century.
Notes: an adequate dig would completely uncover the structure. Restoration of the entire ditch and counterscarp gallery is desirable. Similar works, that is countermines with underlying drainage systems, would also have been created within Fort San Vittorio in Tortona (Alessandria). A multi-colour watercolour painting of Fort San Vittorio, dated 14 September 1799, depicts one serving the ditch opposite the Hornwork (Comoli Mandracci, Marotta 1995, pg. 51 and pg. 167).
Bibliography: Comoli Mandracci V., Marotta A., *Tortona e il suo castello. Dal dominio spagnolo al periodo postunitario*, Alessandria 1995.
Padovan G., *Due noci dure da rompere. I forti di Demonte e di Tortona alla fine del XVIII secolo: l'organizzazione della difesa, la rete di contromina e l'approvvigionamento idrico*, in Anzanello E., Dal Cin F., Gasparetto P., Gava S. (edited by), *Atti Montello 2002. Conglomeriamoci. 21° Incontro Internazionale di Speleologia. Nervesa della Battaglia 1-3 Novembre 2002*, Villorba 2003, pgs. 293-365.
Data ownership: Associazione Speleologia Cavità Artificiali Milano (S.C.A.M.); 2002.
Compiled by: Gianluca Padovan (Ass.ne S.C.A.M.).

III.3.4 - *CA 00003 PI CN; Ditch Drainage Tunnel of the Bastion of Saint Ignatius*

Cadastral number: CA 00003 PI CN
Denomination: *Ditch Drainage Tunnel of the Bastion of Saint Ignatius* (figs. III.19 and III.20)
Region-country: Piedmont, Italy
Province: Cuneo
Municipality: Demonte
Locality: Promontorio del Podio
Location: Fort Demonte, situated on the left bank of the River Stura in Demonte, a few hundred metres east of the town of Demonte
Ownership: Municipality of Demonte
Cartography: C.T.R. 1:10.000
Geological unit: Fort Demonte rises on a low and narrow rocky spear on the left bank of the River Stura, in proximity to an alluvial cone. The substratum consists of grey arenaceous or entrochal limestone with streaks of pink or yellow-red, with flint layers and Crinoid fossils and grey or greenish calcareous Dogger-Malm slate (upper-late Jurassic). These consist of marginal marine environment deposits
Altitude: //

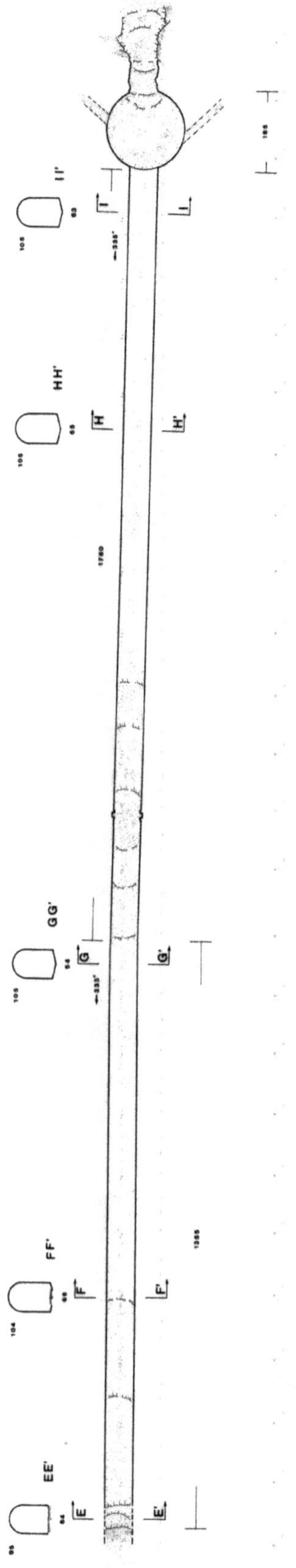

Fig. III.19. *Transversal section plan of the Ditch Drainage Tunnel of the Bastion of Saint Ignatius (CA 00003 PI CN) (Ass.ne S.C.A.M. Archive).*

Fig. III.20. *Ditch Drainage Tunnel of the Bastion of Saint Ignatius (CA 00003 PI CN) (photo G. Padovan).*

Position: //
Context: Fort Demonte
Operations conducted: excavation, survey, photographic service
Work carried out by: Associazione Speleologia Cavità Artificiali Milano (S.C.A.M.); 2002
Warnings: the Fort's external exit was created by breaking a wall and has partially collapsed
Typology: 2d - drainage tunnel
Description: this structure was used for the discharge of rainwater and icewater, which would otherwise stagnate in the ditch (the system was designed in such a way that it would remain dry and would not be subject to flooding). Almost the entire channel survives and despite its entrance now being buried, water still filters into the channel. The hydraulic system is level with the overlying tunnel which leads to the mine branches: it starts off in a 153° direction and again slopes by 2° and follows the 155°-335° axis. The first section from the ditch follows a -3° inclination as far as the granite guides, at which point it follows a -5° as far as the circular chamber. The centre of the passage is 1.05 m high and its length is of 0.63 m to 0.66 m; just over 30 m are practicable. It has mortar-lined punch-dressed masonry and stone piers. The semi-circular barrel vault is in exposed brick. The passage bottom has a slight V-shape for improved runoff. It is made of large, flat, coupled schistous stones, each one converging towards the centre; each couple overhangs the next thus forming a step-like structure with 0.4 m tread and height of 4-7 cm. This inclined floor was covered by a layer of coarse, pale yellow, hydraulic mortar, the bulk of which has been eroded by water, thus revealing the layout of the underlying stones. The tunnel terminates in a circular chamber at slightly higher altitude, almost entirely in exposed brick with occasional blocks and unworked stone. The chamber has a conch vault and its base is covered by debris and detritus. Its diameter is of 1.85 m. The XVIII century planimetry (Viglino Davico 1989, pg. 192), table confirms that the room has three branches, the lateral branches (the sizes of which have been calculated from survey information) could be approximately 10 m long, while the central branch is at least 15 m long. The system appears to have various sections. There are two conduits similar to loopholes along the walls. In respect of the continuation of tunnel axis, they are not exactly symmetrical one to the other. One faces east while the other faces south. The first is slightly larger than the second and both are brick-lined, while their bases consist of an inward-protruding single large, flat stone. Both distinctively slope downwards and both are blocked by detritus. Opposite the tunnel, in a trench-type structure the walls of which are made of large dry stone blocks is a breach in the surface wall. Originally there would also have been a conduit of a similar type to the two previously mentioned. White and/or black water drainage system outlets are notorious for being the weak points of fortified perimeters. An optimal solution to obviate the danger posed by the eventual transit of enemy soldiers within the discharge conduit was created. We shall now attempt to explain said solution. First of all, the system has been designed in such a way that it does not lead directly to the surface. At its head is the circular chamber which probably acts as a settling basin and prevents detritus and twigs from being dragged inside and blocking the three narrow conduits. The conduits, in turn, communicated with three tunnels, significantly below the chamber, which discharge the waters externally. It is not thought possible that 10 m to 15 m long inaccessible conduits were created for the simple reason that in the event of obstruction or collapse of one of these conduits, the only solution would have been to demolish the entire conduit until the section to be cleared could be located. For the very same reason, it is thought that the three terminal branches would have been accessible from the surface. Periodic maintenance would thus have been possible. The purpose of the connecting branches, that is the three sloping conduits, is therefore that of preventing direct access to the system while at the same time allowing the rapid discharge of water. An enemy could have easily accessed the terminal branches, possibly by means of excavation to the underlying conduit, but this would have taken a long time and would have been easily thwarted by the simple act of throwing a hand grenade from the chamber into the conduits. Loading the conduits with explosives would have been of little use to the besieger, both on account of the depth and lower altitude and on account of the distance between the conduits and the fort's external defences. For instance, the countermine chambers are more than 15 m away from the circular chamber. No advantage would have been gained by blocking the conduit other than flooding of the ditch.
Interpretation: ditch water drainage tunnel.
Dating: XVIII century.
Notes: removing the debris would permit the system to be fully restored.
Bibliography: Padovan G., *Due noci dure da rompere. I forti di Demonte e di Tortona alla fine del XVIII secolo: l'organizzazione della difesa, la rete di contromina e l'approvvigionamento idrico*, in Anzanello E., Dal Cin F., Gasparetto P., Gava S. (edited by), *Atti Montello 2002. Conglomeriamoci. 21° Incontro Internazionale di Speleologia. Nervesa della Battaglia 1-3 Novembre 2002*, Villorba 2003, pgs. 293-365.
Viglino Davico M., *Fortezze sulle Alpi. Difese dei Savoia nella Valle Stura di Demonte*, Cuneo 1989.
Data ownership: Associazione Speleologia Cavità Artificiali Milano (S.C.A.M.); 2002.
Compiled by: Gianluca Padovan (Ass.ne S.C.A.M.).

III.3.5 - *CA 00004 PI CN; Ditch Counterscarp Cistern of the Main Gate*

Cadastral number: CA 00004 PI CN
Denomination: *Ditch Counterscarp Cistern of the Main Gate* (fig. III.21)
Region-country: Piedmont, Italy
Province: Cuneo
Municipality: Demonte
Locality: Promontorio del Podio
Location: Fort Demonte, situated on the left bank of the River Stura in Demonte, a few hundred metres east of the town of Demonte
Ownership: Municipality of Demonte
Cartography: C.T.R. 1:10.000
Geological unit: Fort Demonte rises on a low and narrow rocky spear on the left bank of the River Stura, in proximity to an alluvial cone. The substratum consists of grey arenaceous or entrochal limestone with streaks of pink or yellow-red, with flint layers and Crinoid fossils and grey or greenish calcareous Dogger-Malm slate (upper-late Jurassic). These consist of marginal marine environment deposits
Altitude: //
Position: //
Context: Fort Demonte
Operations conducted: excavation, survey, photographic service
Work carried out by: Associazione Speleologia Cavità Artificiali Milano (S.C.A.M.); 2002
Warnings: the cistern is being restored by the Municipality
Typology: 2c - cistern
Description: a cistern for the storage of meteoric water is situated under the grade plane in the counterscarp ditch wall, which protected the Main Gate. Current excavations have removed debris and detritus from the ditch and uncovered the entrance to a small room, the floor of which contains the opening of a cylindrical well providing access to the underlying storage chamber. The well's shaft is brick-lined and its internal diameter is 1.53 m. The water level reaches to just under the cistern's edge and provides a glimpse of the underlying room, which may have a conch vault, at a depth of almost 3 m; the halyard, on the other hand, reaches the base of the cistern, which possibly consists of a detritic cone at a depth of 6.38 m. The room, lined at least on the outside with large stone blocks, was undoubtedly well protected.
Interpretation: cistern for the storage of meteoric water.
Dating: presumably XVIII century.
Notes: the cistern and the ditch should be emptied for a full examination and recuperation of the entire area.
Bibliography: Padovan G., *Due noci dure da rompere. I forti di Demonte e di Tortona alla fine del XVIII secolo: l'organizzazione della difesa, la rete di contromina e l'approvvigionamento idrico*, in Anzanello E., Dal Cin F., Gasparetto P., Gava S. (edited by), *Atti Montello 2002. Conglomeriamoci. 21° Incontro Internazionale di Speleologia. Nervesa della Battaglia 1-3 Novembre 2002*, Villorba 2003, pgs. 293-365.
Data ownership: Associazione Speleologia Cavità Artificiali Milano (S.C.A.M.); 2002.
Compiled by: Gianluca Padovan (Ass.ne S.C.A.M.).

III.3.6 - *CA 00005 PI CN; Primary Shelter on the Hornwork Counterscarp*

Cadastral number: CA 00005 PI CN
Denomination: Primary Shelter on the Hornwork Counterscarp (fig. III.22 and III.22.a)
Region-country: Piedmont, Italy
Province: Cuneo
Municipality: Demonte
Locality: Promontorio del Podio
Location: Fort Demonte, situated on the left bank of the River Stura in Demonte, a few hundred metres east of the town of Demonte
Ownership: Municipality of Demonte
Cartography: C.T.R. 1:10.000
Geological unit: Fort Demonte rises on a low and narrow rocky spear on the left bank of the River Stura, in proximity to an alluvial cone. The substratum consists of grey arenaceous or entrochal limestone with streaks of pink or yellow-red, with flint layers and Crinoid fossils and grey or greenish calcareous Dogger-Malm slate (upper-late Jurassic). These consist of marginal marine environment deposits
Altitude: //

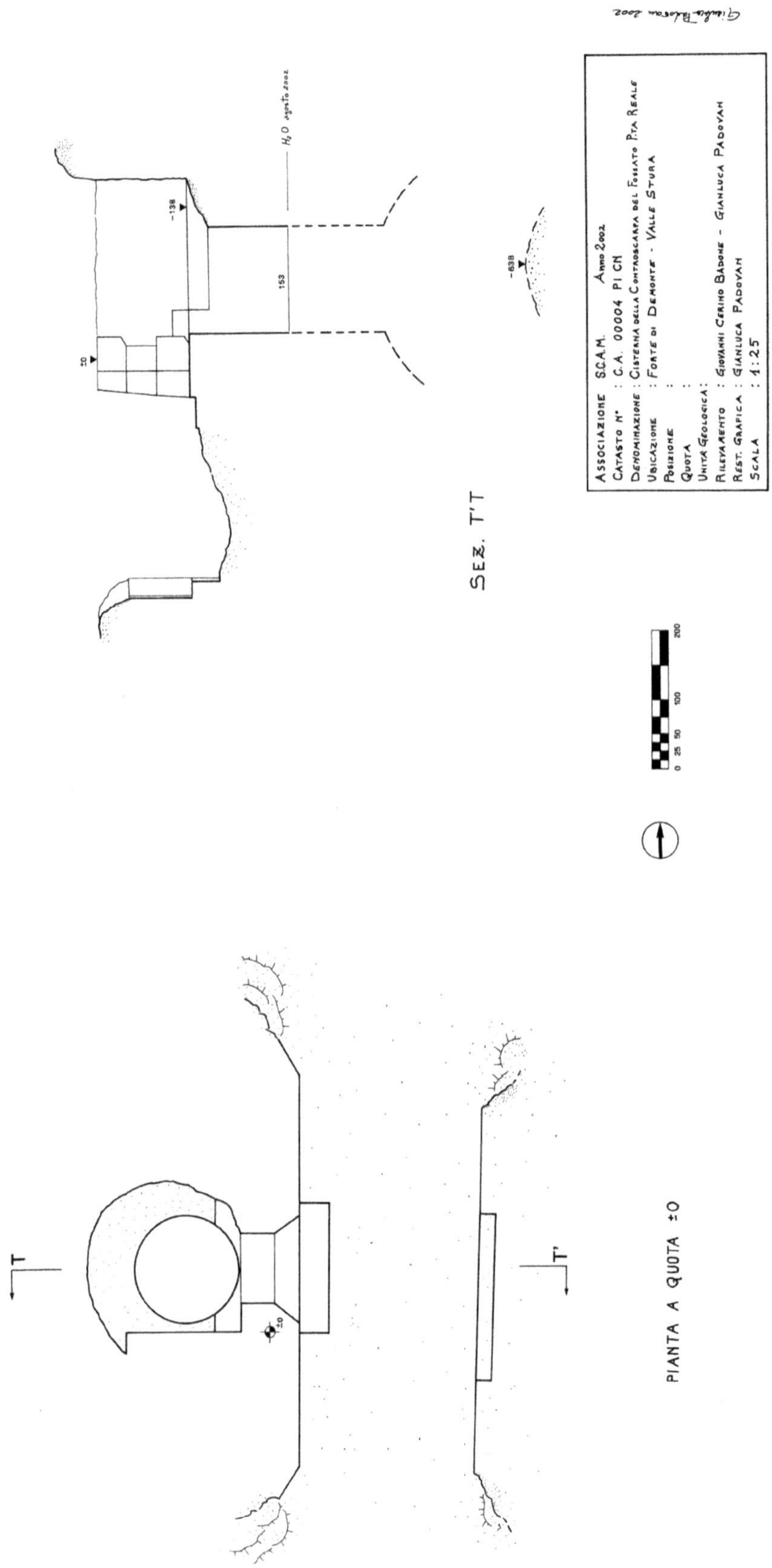

Fig. III.21. *Counterscarp Cistern of the Royal Gate Ditch (CA 00004 PI CN) (Ass.ne S.C.A.M. Archive).*

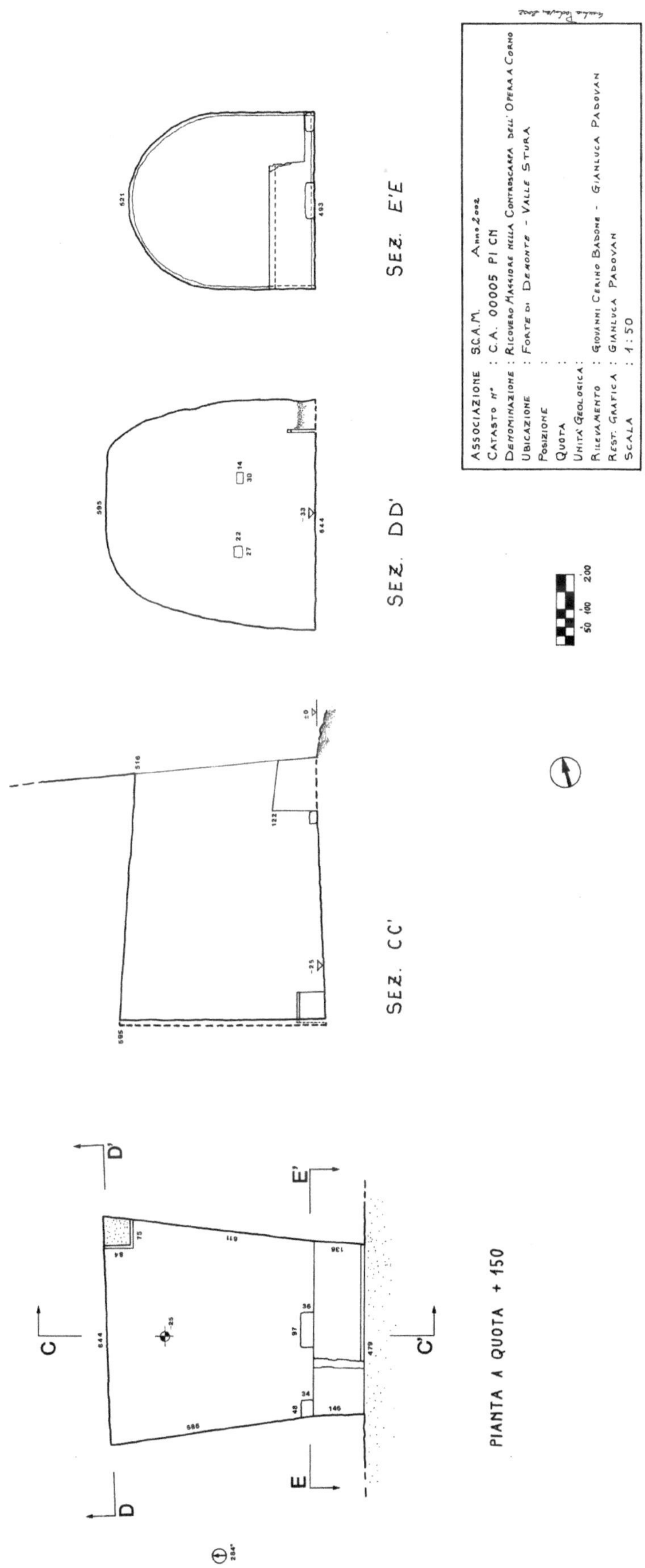

Fig. III.22. *Primary Shelter on the Hornwork Counterscarp (CA 00005 PI CN) (Ass.ne S.C.A.M. Archive).*

Fig. III.22.a. *Primary Shelter on the Hornwork Counterscarp (CA 00005 PI CN) (photo G. Padovan).*

Position: //
Context: Fort Demonte
Operations conducted: excavation, survey, photographic service
Work carried out by: Associazione Speleologia Cavità Artificiali Milano (S.C.A.M.); 2002
Warnings: none
Typology: 6 - military structure
Description: two stone-carved rooms are to be found in the counterscarp of the cyclopean ditch protecting the Hornwork, which is carved into the rock. The first, known as the Primary Shelter, is situated next to the stairwell which leads from the counterscarp to the bottom of the ditch. The large, trapezoidal room (4.79 x 7.32 x 6.44 x 7.47 m) is easily accessed and contains no debris. Inside there are currently tables and chairs as well as a fireplace in the inside left corner. At the entrance, the semi-circular vaulted arch measures 5.16 m whereas it is squashed towards the base. It is 5.95 m high. The entrance is protected by a low but thick masonry wall, the upper part of which gently slopes outwards. It could easily be the original boundary rather than a later structure built during the restoration of the covered area, despite the traces of fairly recent hydraulic mortar, thought to have been used during maintenance works. The rock floor, which gently slopes inwards, cannot be dated due to the fine layer of detritus and hay, which cover it; the exposed walls show clear evidence of excavation tools. The bottom part has two recesses (of 27 x 22 cm and 30 x 14 cm), believed to have been used for the placement of wooden supports. There are non recesses for hinges or bolts, therefore the does not appear to have had a door closure. In the event of siege, the room would certainly not have been considered safe.
Interpretation: its purpose is assumed to be that of storage area for wood or fodder for draught, pack and saddle animals. Under enemy fire, this material could clearly have caught fire, however on the whole this is the best place for such an eventuality as little or no damage would be caused.
Dating: thought to be XVIII century.
Notes: the large ditch could easily be cleared of debris and restored to its original, imposing form, with little effort or expense.
Bibliography: Padovan G., *Due noci dure da rompere. I forti di Demonte e di Tortona alla fine del XVIII secolo: l'organizzazione della difesa, la rete di contromina e l'approvvigionamento idrico*, in *Atti Montello 2002. Conglomeriamoci*, 21° Incontro Internazionale di Speleologia. *Nervesa della Battaglia 1-3 Novembre 2002*, Villorba 2003, pgs. 293-365.
Data ownership: Associazione Speleologia Cavità Artificiali Milano (S.C.A.M.); 2002.
Compiled by: Gianluca Padovan (Ass.ne S.C.A.M.).

III.3.7 - *CA 00006 PI CN; Secondary Shelter on the Hornwork Counterscarp*

Cadastral number: CA 00006 PI CN
Denomination: Secondary Shelter on the Hornwork Counterscarp (fig. III.23)
Region-country: Piedmont, Italy
Province: Cuneo
Municipality: Demonte
Locality: Promontorio del Podio
Location: Demonte Fort, situated on the left bank of the River Stura in Demonte, a few hundred metres east of the town of Demonte
Ownership: Municipality of Demonte
Cartography: C.T.R. 1:10.000
Geological unit: Fort Demonte rises on a low and narrow rocky spear on the left bank of the River Stura, in proximity to an alluvial cone. The substratum consists of grey arenaceous or entrochal limestone with streaks of pink or yellow-red, with flint layers and Crinoid fossils and grey or greenish calcareous Dogger-Malm slate (upper-late Jurassic). These consist of marginal marine environment deposits
Altitude: //
Position: //
Context: Fort Demonte
Operations conducted: excavation, survey, photographic service
Work carried out by: Associazione Speleologia Cavità Artificiali Milano (S.C.A.M.); 2002
Warnings: care should be taken at the entrance, which is subject to falling material from the top of the ditch
Typology: 6 - military work
Description: in all appearances the same as the Primary Shelter on the Hornwork Counterscarp (CA 00005 PI CN), but situated on the south side of the ditch at a slightly lower altitude; difficult to interpret as it is almost completely

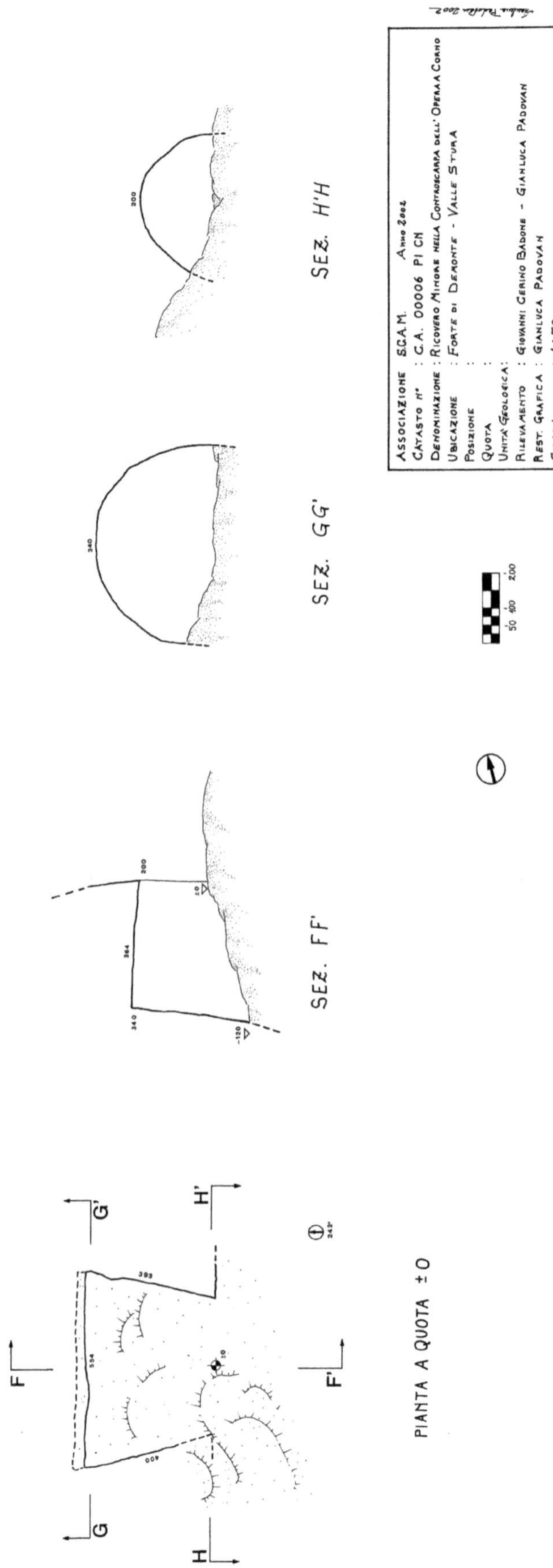

Fig. III.23. *Secondary Shelter on the Hornwork Counterscarp (CA 00006 PI CN) (Ass.ne S.C.A.M. Archive).*

buried. It is carved into the rock and is trapezoidal in shape (1.93 x 4 x 5.54 x 3.93 m) with a height measures 2 m at the entrance and 3.4 m high at the base. The semi-circular entrance arch leans slightly towards the base. The buried section does not allow for a clear interpretation of the entrance and it has not been possible to determine whether the shelter is protected by a low wall in similar fashion to the other shelter. It appears to have been carved less accurately however this could be due from the fact that material has become detached from the roof, resulting in uneven surfaces.
Interpretation: its purpose is assumed to be that of storage area for wood or fodder for draught, pack and saddle animals. Under enemy fire, this material could clearly have caught fire, however on the whole this is the best place for such an eventuality as little or no damage would be caused.
Dating: thought to be XVIII century.
Notes: the room and rear-lying ditch could be restored.
Bibliography: Padovan G., *Due noci dure da rompere. I forti di Demonte e di Tortona alla fine del XVIII secolo: l'organizzazione della difesa, la rete di contromina e l'approvvigionamento idrico*, in Anzanello E., Dal Cin F., Gasparetto P., Gava S. (edited by), *Atti Montello 2002. Conglomeriamoci. 21° Incontro Internazionale di Speleologia. Nervesa della Battaglia 1-3 Novembre 2002*, Villorba 2003, pgs. 293-365.
Data ownership: Associazione Speleologia Cavità Artificiali Milano (S.C.A.M.); 2002.
Compiled by: Gianluca Padovan (Ass.ne S.C.A.M.).

III.3.8 - *CA 00001 PI TO; First Cannon Room*

Cadastral number: CA 00001 PI TO
Denomination: First Cannon Room (figs. III.24, III.25, III.26, III.27, III.28, III.29 and III.30)
Region-country: Piedmont, Italy
Province: Turin
Municipality: Verrua Savoia
Locality: Verrua Hill
Location: "Bomb-proof" Barracks
Ownership: private
Cartography: CTR 1:10.000
Geological unit: marls, sandstones and Pliocene limestone resting on an Eocene Clay Shale Formation
Altitude: (ellipsoid) 304.07 m
Position: U.T.M. 429359.7506 E; 5002668.4365 N
Context: within the Donjon structure, upper area
Operations conducted: excavation, survey, photographic service
Work carried out by: Associazione Speleologia Cavità Artificiali Milano (S.C.A.M.); 2002
Warnings: structural collapse at entrance, poor air circulation
Typology: 6 - shelter
Description: the structure underlies the semi-circular parapet of the Donjon and is on the radial axis. The chamber is composite in shape and has a barrel vault, formed by a trapeze, the base of which supports a section of circle, the curved side of which follows the internal profile of the semi-donjon while the lower base faces the centre of the structure. Its base is significantly raised due to the soil which fills the room. The entrance consists of the collapsed entrance archway. From the entrance, the structure measures 1.91 x 4.17 x 2.97 x 4.29 m; the maximum height of the buried section is 1.64 m. As the first section is buried to above the piers, the room is believed to be slightly wider and longer. The piers and the vaulted roof are made of brick, while the infill wall is made of regular sections of pebbles, broken rocks and stones and occasional stone fragments, alternating with regular sections of bricks.
Interpretation: room under the parapet, which may have been used to shelter a piece of artillery and/or to keep munitions in the event of an attack.
Dating: built at the same time as the Donjon, it dates to between the third and fourth quarters of the XVII century.
Notes: the probe descends into the buried section by more than one metre and the room is thought to be more than 2 m in height. Near the entrance, the walls may slightly converge towards the centre, further enclosing the entrance, in the same fashion as the structure of the Fourth Cannon Room. Except for the collapse at the entrance, the structure presents no damage. There are several small formations on the roof. In order to conduct a complete investigation, the structure and the entire upper section of the Donjon should be cleared from within.
Bibliography: Padovan D., Padovan G., Bordignon L., Ottino M. 1997, *La Fortezza di Verrua Savoia*, in Club Alpinistico Triestino - Gruppo Grotte Sezione Ricerche e Studi su Cavità Artificiali, *Atti del IV Convegno Nazionale sulle Cavità Artificiali – Osoppo 30/31 Mggio - 1 Giugno 1997*, Trieste 1997, pgs. 187-208.

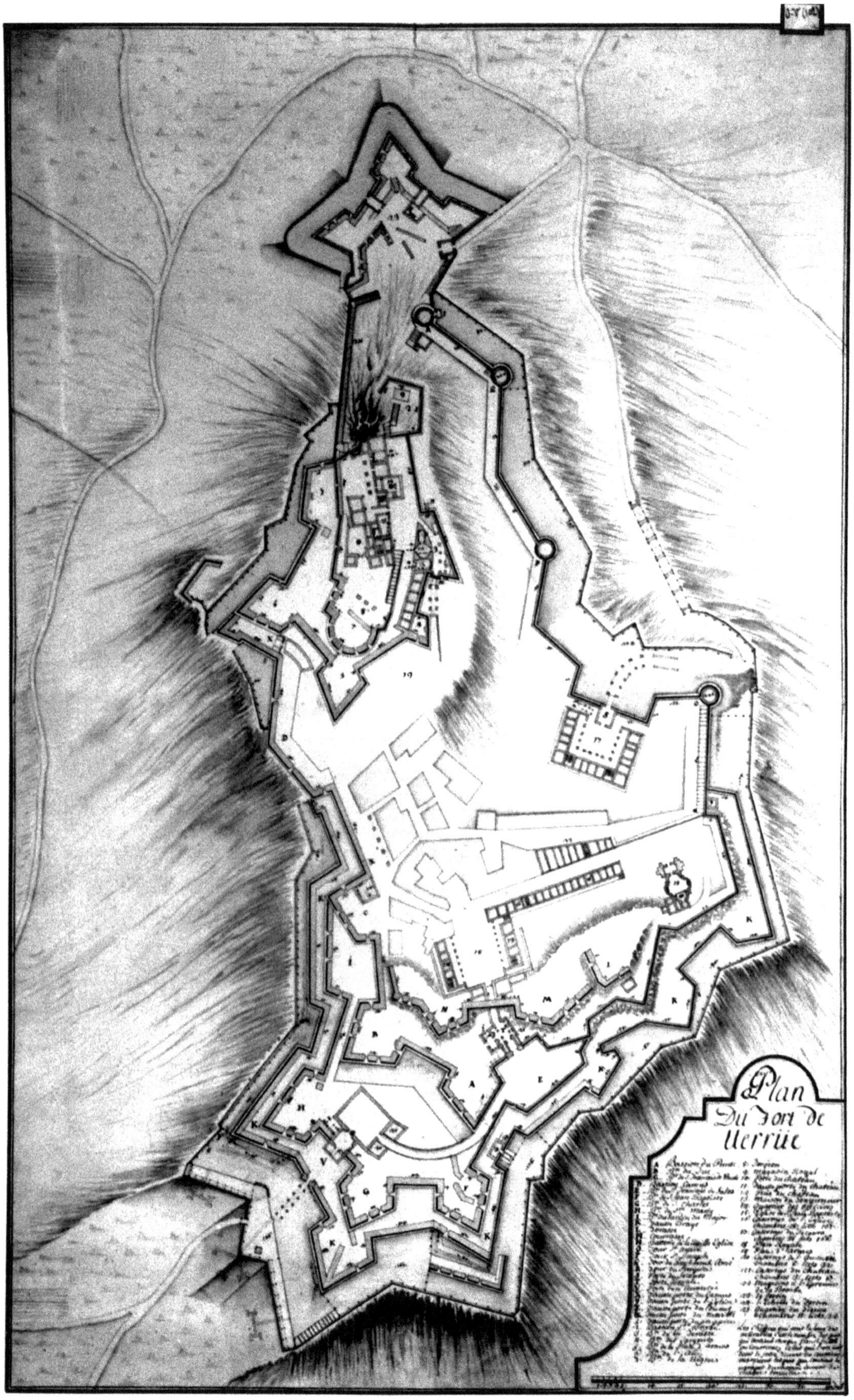

Fig. III.24. *Anonymous, circa 1704, Fortress Plan (700x450 mm watercoloured pen drawing; Royal Library, Turin O.V. 102).*

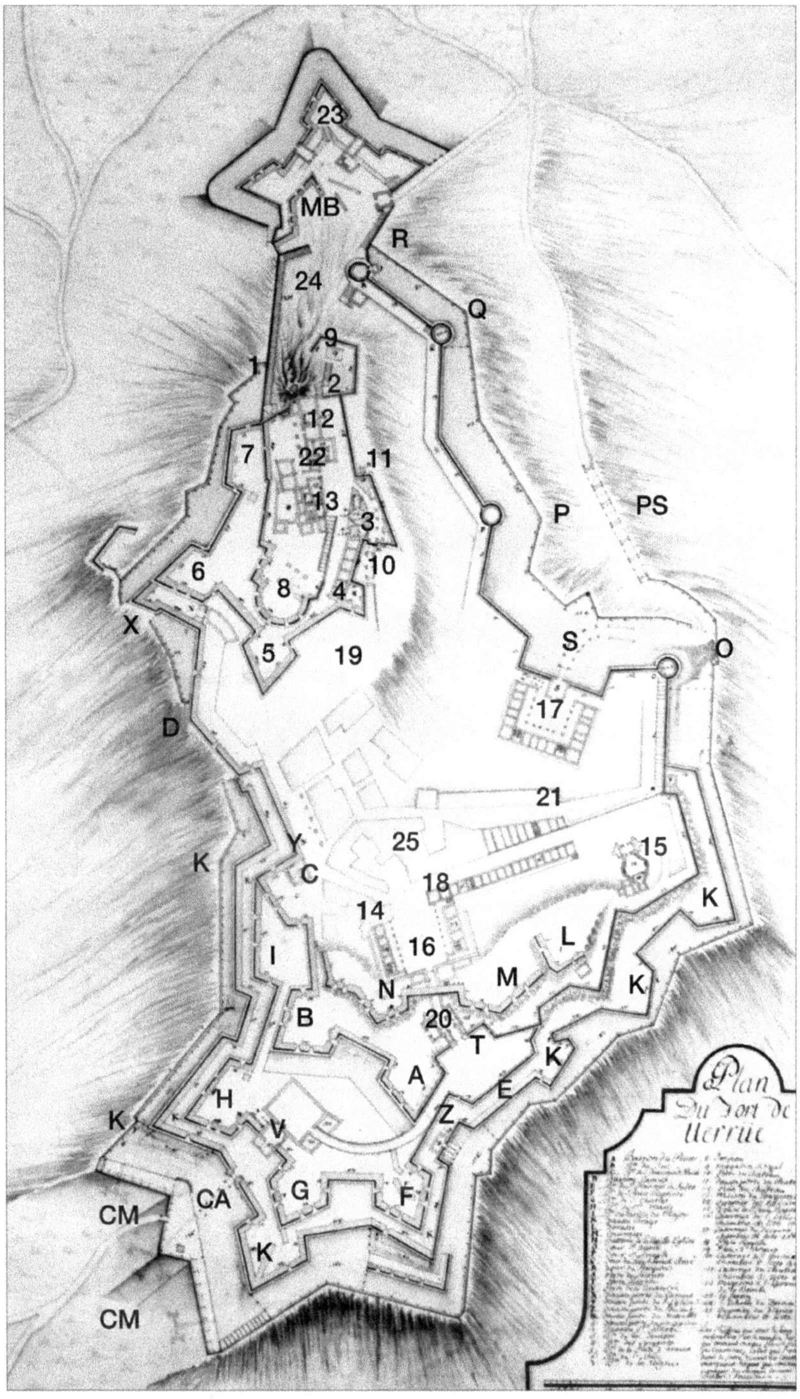

Fig. III.25. *Graphic elaboration of the Fortress Plan by Anonymous, circa 1704. The sections of the fortified complex are highlighted, complete with original key. «Plan Du Fort de Uerrue. A: Bastion du Prince. B: Bastion du Duc. C: B.^{n} de St Francois de Paule. D: Bastion Camus. E: B.^{n} de St Francois de Sales. F: B.^{n} de St Jean Baptiste. G: B.^{n} de St Charles. H: B.^{n} de Ste Marie. I: B.^{n} du Jardin du Maior. K: Fausse Braye. L: Torasse. M: Courtasse. N: Batterie della Vieille Eglise. O: Tour St André. P: Tour St Joseph. Q: Tour du Bienhereux. R: Tour du Precipice. S: Porte du Secours. T: Porte Royale. V: Porte del Auancée. X: Fausse porte du Camus. Y: Fausse porte de l'Eglise. Z: Fausse porte du Lieutenant &. Fausse porte du Moretti. 1: Fausse porte du precipice. 2: Bastion de Ste Barbe. 3: B.^{n} dè la Terasse. 4: B.^{n} des Sergeants. 5: B.^{n} de la Place d'Armes. 6: B.^{n} de l'Alle. 7: B.^{n} de la Uigne. 8: Donjeon. 9: Magasin Royal. 10: Porte du chateau. 11: Fausse porte du chateau. 12: Puis du chateau. 13: Maison du Gouverneur. 14: Quartier des Officiers. 15: Eglise de St Baptiste. 16: Casermes de l'Eglise. Chambres 28: licts 107. 17: Casermes du Secours – Chambres 38: licts 166. 18: Place Royale. 19: Place d'Armes. 20: Casermes de l'Auancée – Chambres 8: licts 32. 21: Casermes du Chateau – chambres 3: licts 17. 22: Magasins à l'Epreuve de la Bombe. 23: Le fortin. 24: l'echelle du Fortin. 25: Quartier du Gierico – Chambres 11: licts 32. Les Chiffres qui sont le long des murailles c'est le nombre des pas qui contient chaque flanchfaus ou courtines, celles que l'on nas dans le fosse devant les courtines marquent les pas que contient le parapet du chemin couvert de chaque tenaille».*
Furthermore there are: MB: Half Bastion. PS: First Aid Bridge. CA: caponier. CM: countermine (Ass.ne S.C.A.M. Archive).

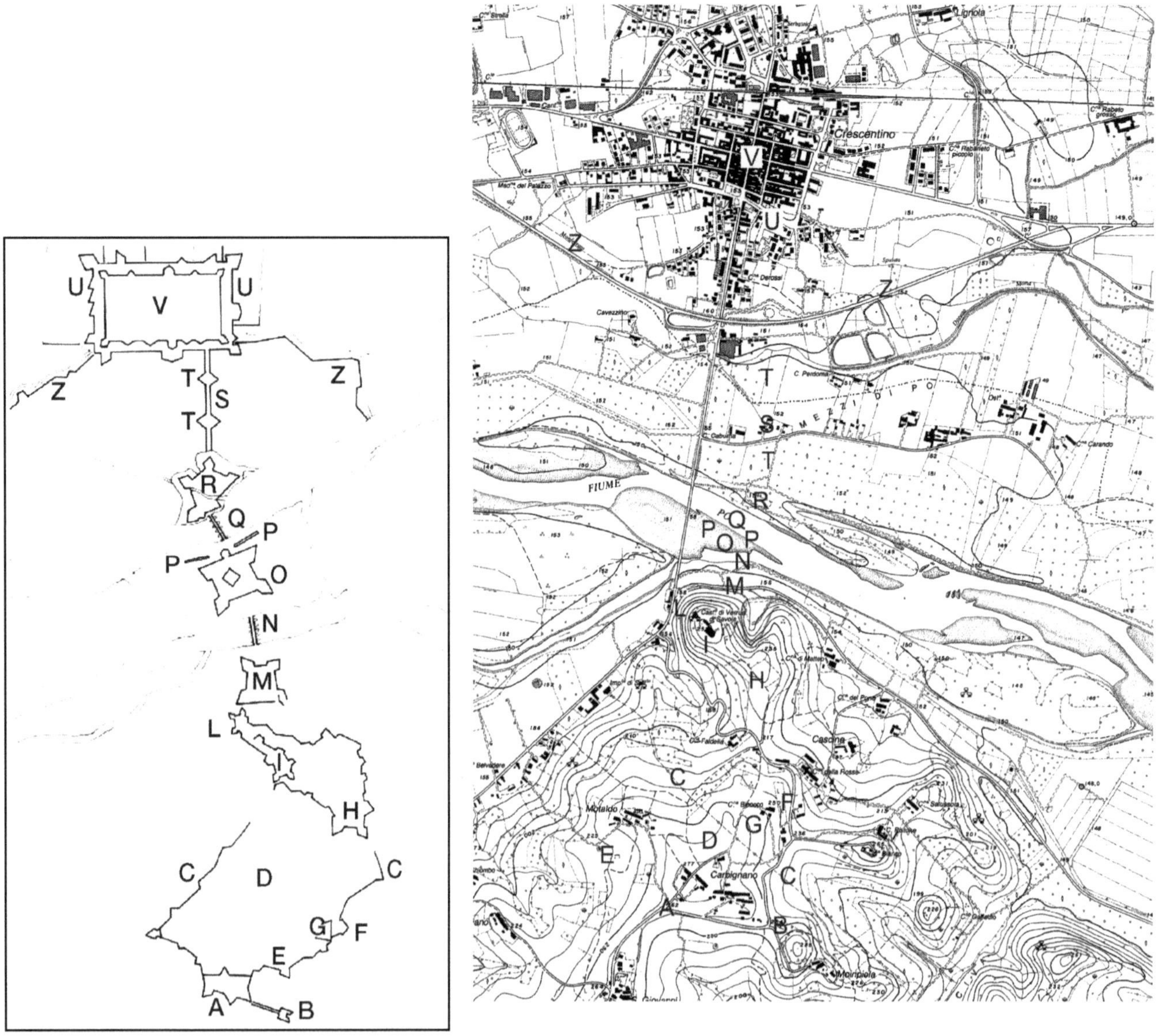

Fig. III.26. *Reconstruction of the fortification complex of which Verrua was a part. The importance of the Carbignano-Verrua-Crescentino defensive system lies in the fact that reinforcements from Turin could reach Crescentino and therefore Verrua by crossing the Dora Baltea. Furthermore, the trenched field between Verrua Fortress and Fort Royale could accommodate a few thousand soldiers who could march to Crescentino should the town be under threat. A: Fort Royale (Carbignano). B: Hornwork. C: Connecting works. D: Trenched Field Area. E: Grande Cascina Redoubt. F: Bicocca Redoubt. G: Redoubt within the Cascina Bicocca. I: Verrua Fortress. L: Lower Fort. M: Fort Wallis. N: Pontoon Bridge. O: Ognissanti Fort. P: Traverses. Q: Pontoon Bridge. R: Bridge Fort. S: Covered road. T: Fortlets. U: External Crescentino walls. V: Crescentino. Z: Entrenchments (Ass.ne S.C.A.M. Archive).*

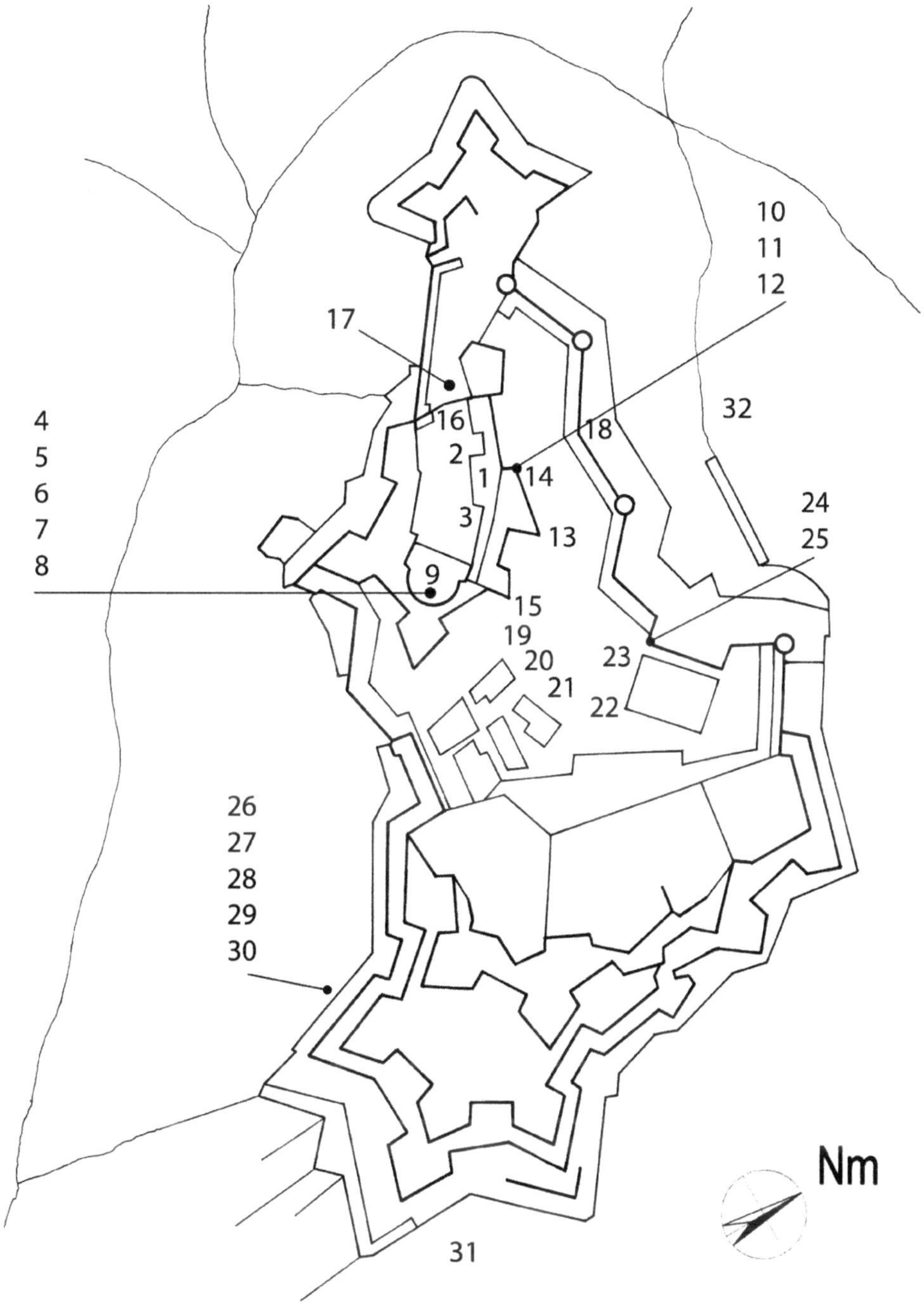

Fig. III.27. *General plan of Verrua Fortress indicating the location of registered artificial cavities. 1: Small Hypogeum Well (CA 00005 PI TO). 2: Wasps' Static Tank (CA 00025 PI TO). 3: Stairway Static Tank (CA 00026 PI TO). 4: First Cannon Room (CA 00001 PI TO). 5: Second Cannon Room (CA 00002 PI TO). 6: Third Cannon Room (CA 00003 PI TO). 7: Fourth Cannon Room (CA 00006 PI TO). 8: Fifth Cannon Room or Merlon Room (CA 00008 PI TO). 9: Donjon Cistern (CA 00004 PI TO). 10: Sortie Tunnel (CA 00010 PI TO). 11: Drawbridge Cistern (CA 00007 PI TO). 12: False Cistern or Graffiti Room (CA 00009 PI TO). 13: Second Demolition Tunnel (CA 00021 PI TO). 14: Third Demolition or Cell Tunnel (CA 00022 PI TO). 15: First Demolition Tunnel (CA 00020 PI TO). 16: TE.S.E.S. Room (CA 00034 PI TO). 17: Postern Room or Arcadio Corner (CA 00029 PI TO). 18: Mini-nymphaeum (CA 00028 PI TO). 19: Filled Well (CA 00016 PI TO). 20: Chlorine Cistern (CA 00015 PI TO). 21: Celestino Tunnel (CA 00011 PI TO). 22: First Tunnel Section (CA 00013 PI TO). 23: Second Tunnel Section (CA 00014 PI TO). 24: Stanza del Foro (CA 00012 PI TO). 25: TE.S.E.S. Tunnel (CA 00027 PI TO). 26: First Countermine Tunnel (CA 00017 PI TO). 27: Second Countermine Tunnel (CA 00018 PI TO). 28: Third Countermine Tunnel (CA 00019 PI TO). 29: Small Quadrivium Room (CA 00035 PI TO). 30: Fourth Countermine Tunnel (CA 00036 PI TO). 31: Small Rampart Cistern (CA 00023 PI TO). 32: Po Quarry (CA 00024 PI TO) (Ass.ne S.C.A.M. Archive).*

Fig. III.28. *Aerial photograph of Verrua Hill. Close-up of the semi-circular Donjon to the left of which is the building built on the Sergeants' Bastion. The "bomb-proof" Barracks is behind this. The River Po and the bridge can be seen in the background (photo G. Padovan).*

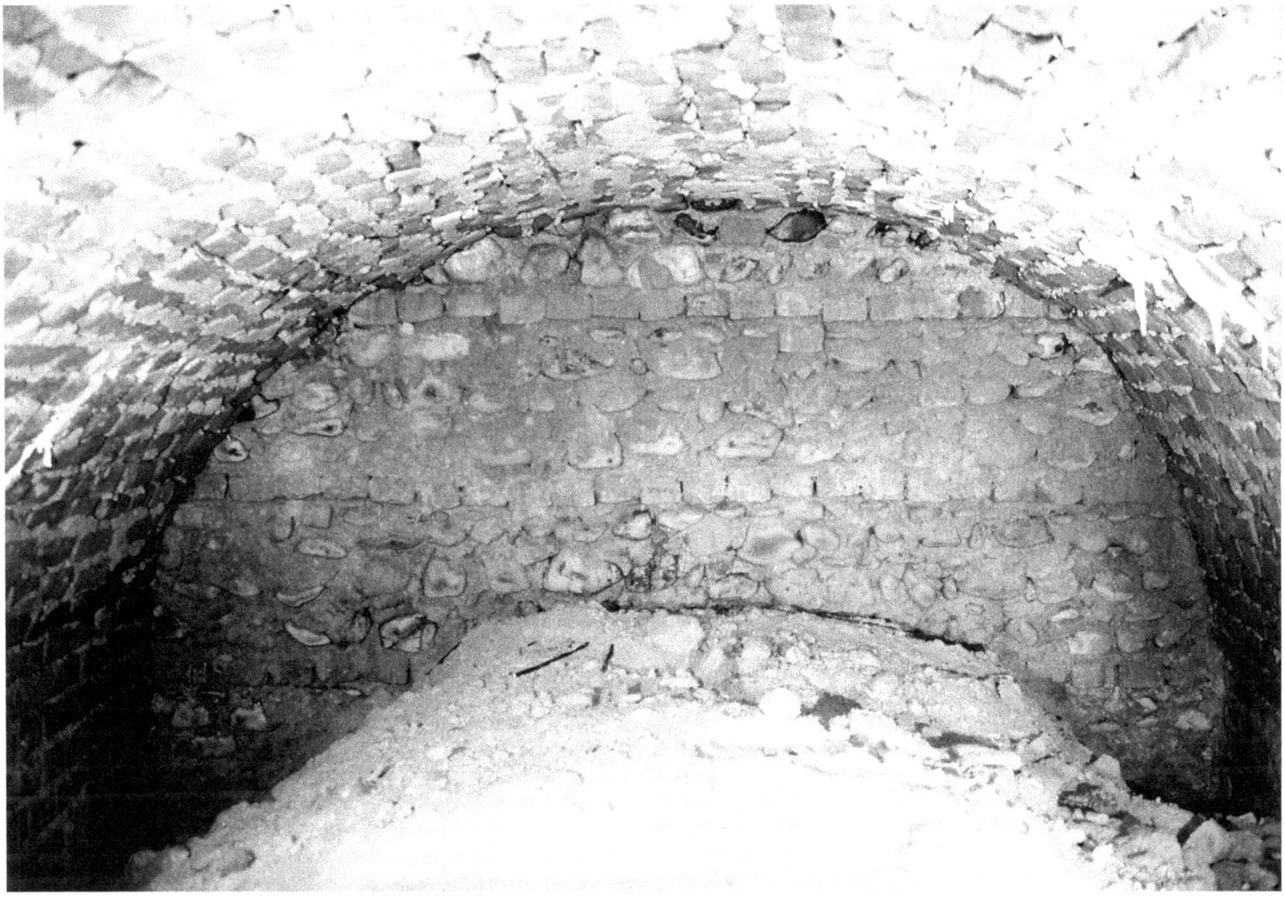

Fig. III.29. *The rear wall of the First Cannon Room (CA 00001 PI TO) (photo G. Padovan).*

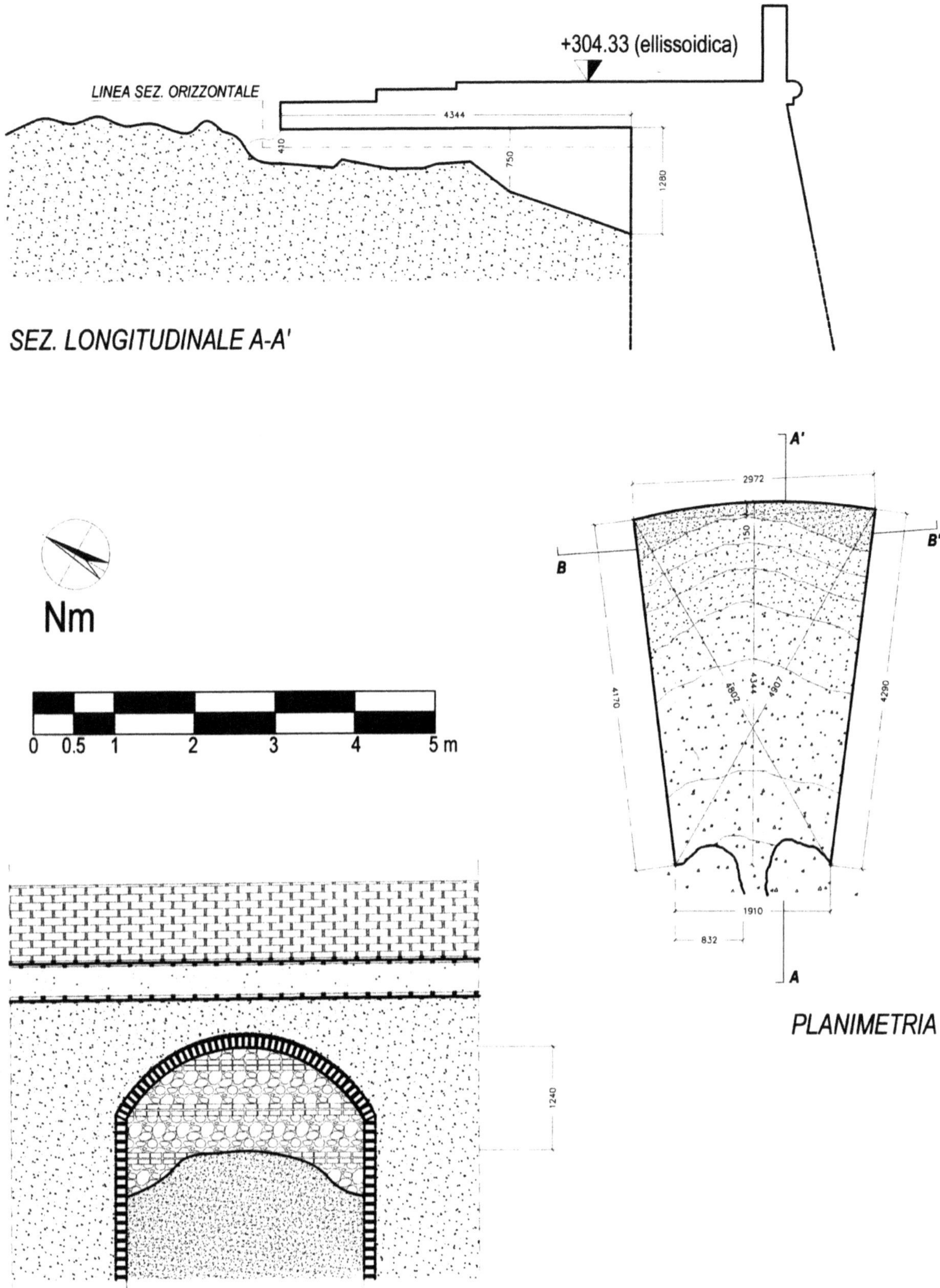

Fig. III.30. *First Cannon Room (CA 00001 PI TO) (Ass.ne S.C.A.M. Archive).*

Padovan G., *Verrua Savoia: indagini sotterranee*, in Associazione Culturale "Amici della Biblioteca" di Crescentino, *Atti del Convegno Storico "Terre sul Po dal Medioevo alla Resistenza" – Crescentino 2-3 Ottobre 1988*, Crescentino 2002, pgs. 213-243.
Data ownership: Associazione Speleologia Cavità Artificiali Milano (S.C.A.M.).
Compiled by: Gianluca Padovan (Ass.ne S.C.A.M.).

III.3.9 - *CA 00002 PI TO; Second Cannon Room*

Cadastral number: CA 00002 PI TO
Denomination: Second Cannon Room (fig. III.31)
Region-country: Piedmont, Italy
Province: Turin
Municipality: Verrua Savoia
Locality: Verrua Hill
Location: "Bomb-proof" Barracks
Ownership: private
Cartography: CTR 1:10.000
Geological unit: marls, sandstones and Pliocene limestone resting on an Eocene Clay Shale Formation
Altitude: (ellipsoid) 304.04 m
Position: U.T.M. 429361.6662 E; 5002665.1952 N
Context: within the Donjon structure, upper area
Operations conducted: exploration, planimetric survey, photographic service
Work carried out by: Associazione Speleologia Cavità Artificiali Milano (S.C.A.M.); 2002, 2003
Warnings: structural collapse at entrance, poor air circulation
Typology: 6 - shelter
Description: the structure underlies the rectilinear parapet, as does the Fifth Cannon Room. It is trapezoidal in shape with a barrel vault, of which only a small part is visible in that the base is significantly raised due to soil which fills the room. The roof is made of brick, while the infill wall is made of regular sections of pebbles, broken rocks and stones and occasional stone fragments, alternating with regular sections of bricks. The piers lie below the soil fill. The entrance consists of the collapsed entrance archway. From the entrance, the structure measures 0.87 x 4.9 x 3.9 x 3.92 m; the maximum height of the buried section is 0.55 m. As the room is buried to above the piers, the room is believed to be slightly wider and longer.
Interpretation: room under the parapet, which may have been used to shelter a piece of artillery and/or to keep munitions in the event of an attack.
Dating: built at the same time as the Donjon, it dates to between the third and fourth quarters of the XVII century.
Notes: due to the fact that the bulk of the structure is buried, few details have been uncovered; in any case, the environment should be similar to the other Cannon Rooms. There are several small formations on the roof. Except for the collapse at the entrance, the structure presents no damage. In order to conduct a complete investigation, the structure and the entire upper section of the Donjon should be cleared from within.
Bibliography: Padovan D., Padovan G., Bordignon L., Ottino M. 1997, *La Fortezza di Verrua Savoia*, in Club Alpinistico Triestino - Gruppo Grotte Sezione Ricerche e Studi su Cavità Artificiali, *Atti del IV Convegno Nazionale sulle Cavità Artificiali – Osoppo 30/31 Mggio - 1 Giugno 1997*, Trieste 1997, pgs. 187-208.
Padovan G., *Verrua Savoia: indagini sotterranee*, in Associazione Culturale "Amici della Biblioteca" di Crescentino, *Atti del Convegno Storico "Terre sul Po dal Medioevo alla Resistenza" – Crescentino 2-3 Ottobre 1988*, Crescentino 2002, pgs. 213-243.
Data ownership: Associazione Speleologia Cavità Artificiali Milano (S.C.A.M.).
Compiled by: Gianluca Padovan (Ass.ne S.C.A.M.).

III.3.10 - *CA 00003 PI TO; Third Cannon Room*

Cadastral number: CA 00003 PI TO
Denomination: Third Cannon Room (fig. III.32)
Region-country: Piedmont, Italy
Province: Turin
Municipality: Verrua Savoia
Locality: Verrua Hill
Location: "Bomb-proof" Barracks
Ownership: private

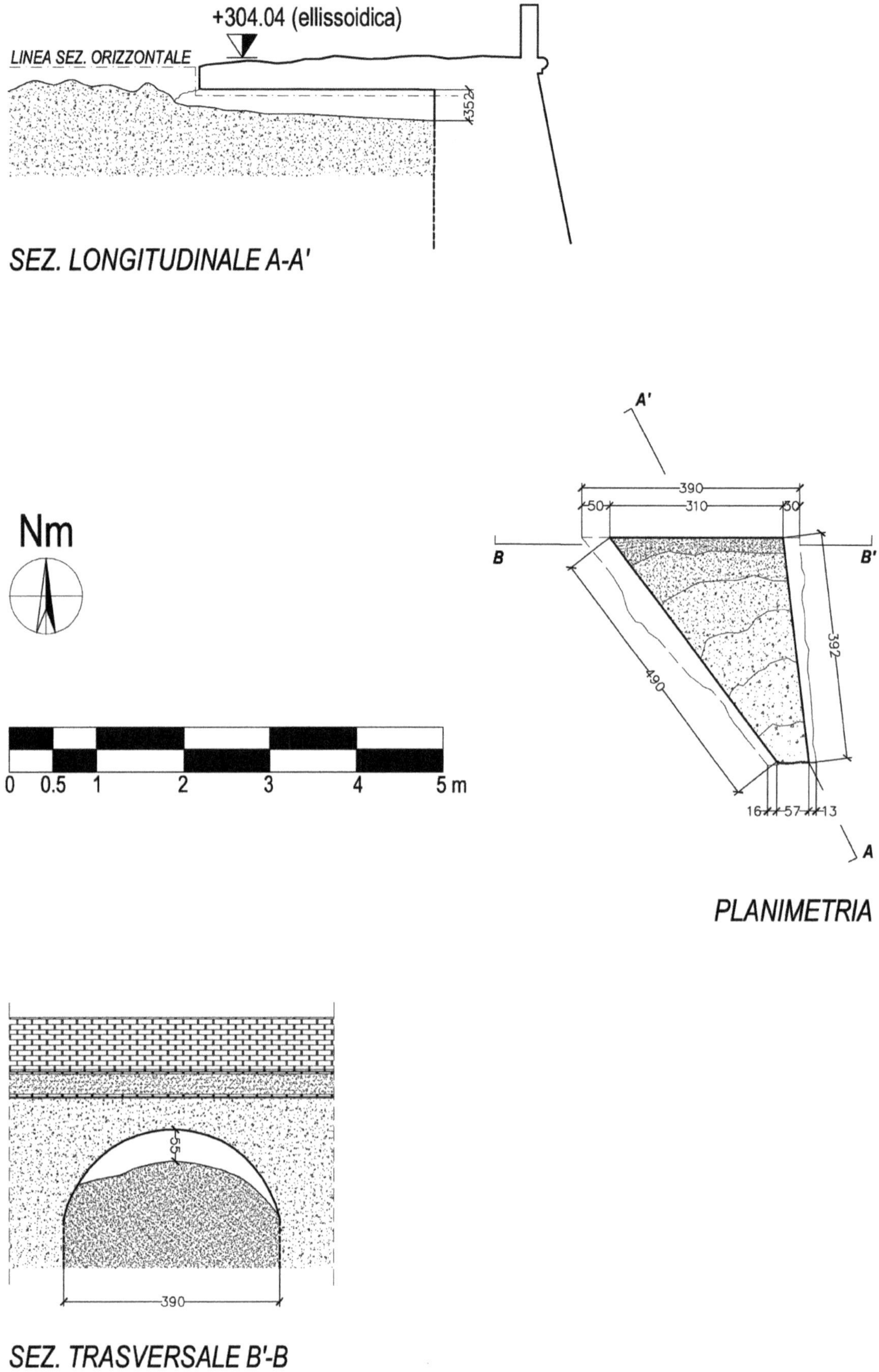

Fig. III.31. *Second Cannon Room (CA 00002 PI TO) (Ass.ne S.C.A.M. Archive).*

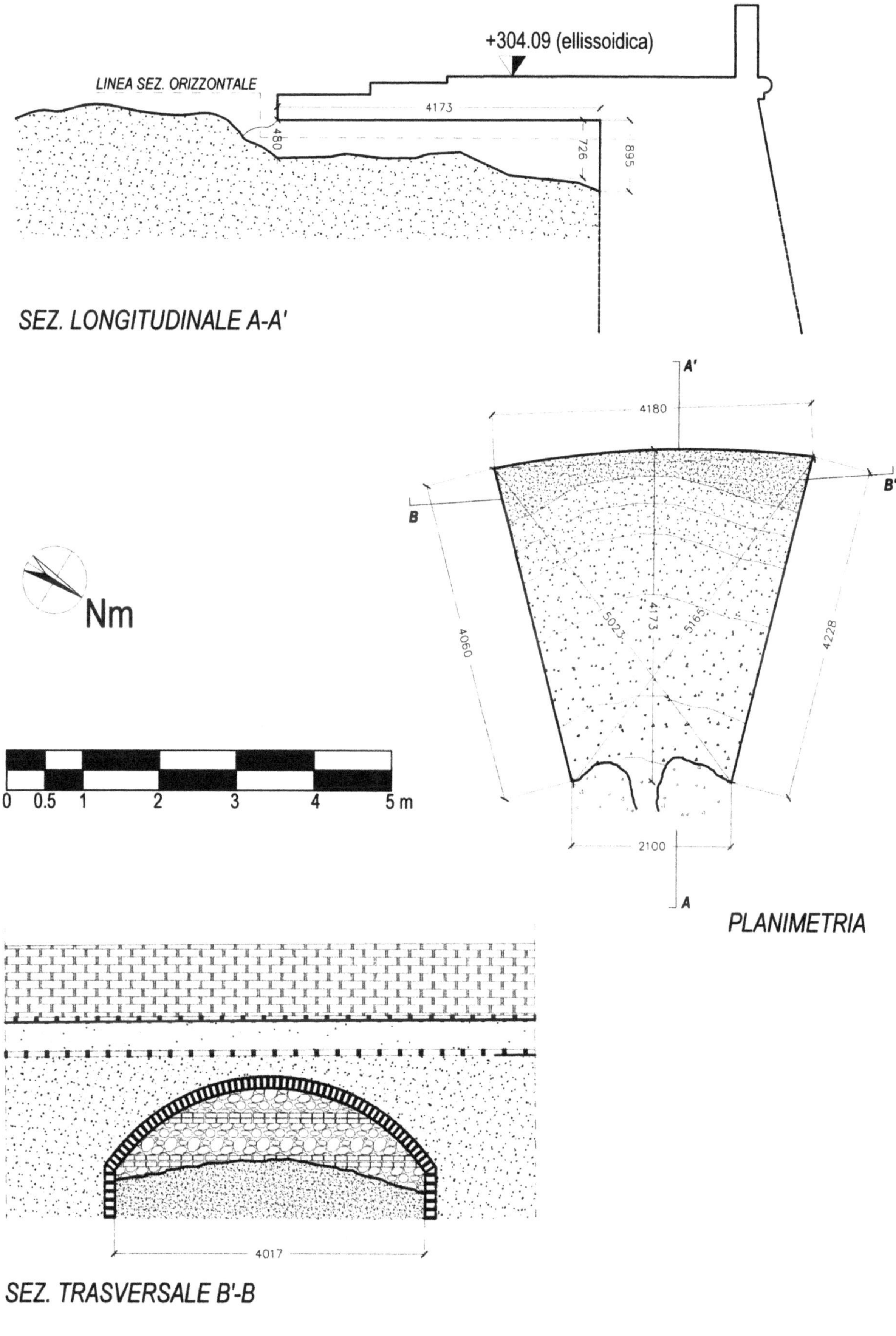

Fig. III.32. *Third Cannon Room (CA 00003 PI TO) (Ass.ne S.C.A.M. Archive).*

Cartography: CTR 1:10.000
Geological unit: marls, sandstones and Pliocene limestone resting on an Eocene Clay Shale Formation
Altitude: (ellipsoid) 304.07 m
Position: U.T.M. 429359.7506 E; 5002668.4365 N
Context: within the Donjon structure, upper area
Operations conducted: exploration, planimetric survey, photographic service
Work carried out by: Associazione Speleologia Cavità Artificiali Milano (S.C.A.M.); 2002, 2003
Warnings: structural collapse at entrance, poor air circulation
Typology: 6 - shelter
Description: the structure underlies the semi-circular parapet of the Donjon and is on the radial axis. The chamber is composite in shape and has barrel vault, formed by a trapeze, the base of which supports a section of circle, the curved side of which follows the internal profile of the semi-donjon while the lower base faces the centre of the structure. Its base is significantly raised due to the soil which fills the room. The entrance consists of the collapsed entrance archway. From the entrance, the structure measures 2.1 x 4.06 x 4.18 x 4.22 m; the maximum height of the buried section is 0.91 m. As the first section is buried to above the piers, the room is believed to be slightly wider and longer. The piers and the vaulted roof are made of brick, while the infill wall is made of regular sections of pebbles, broken rocks and stones and occasional stone fragments, alternating with regular sections of bricks.
Interpretation: room under the parapet, which may have been used to shelter a piece of artillery and/or to keep munitions in the event of an attack.
Dating: built at the same time as the Donjon, it dates to between the third and fourth quarters of the XVII century.
Notes: due to the fact that the bulk of the structure is now buried, few details have been uncovered; in any case, the environment should be similar, if larger, than the other Cannon Rooms. There are several small formations on the roof. Except for the collapse at the entrance, the structure presents no damage. In order to conduct a complete investigation, the structure and the entire upper section of the Donjon should be cleared from within.
Bibliography: Padovan D., Padovan G., Bordignon L., Ottino M. 1997, *La Fortezza di Verrua Savoia*, in Club Alpinistico Triestino - Gruppo Grotte Sezione Ricerche e Studi su Cavità Artificiali, *Atti del IV Convegno Nazionale sulle Cavità Artificiali – Osoppo 30/31 Mggio - 1 Giugno 1997*, Trieste 1997, pgs. 187-208.
Padovan G., *Verrua Savoia: indagini sotterranee*, in Associazione Culturale "Amici della Biblioteca" di Crescentino, *Atti del Convegno Storico "Terre sul Po dal Medioevo alla Resistenza" – Crescentino 2-3 Ottobre 1988*, Crescentino 2002, pgs. 213-243.
Data ownership: Associazione Speleologia Cavità Artificiali Milano (S.C.A.M.).
Compiled by: Gianluca Padovan (Ass.ne S.C.A.M.).

III.3.11 - *CA 00004 PI TO; Donjon Cistern*

Cadastral number: CA 00004 PI TO
Denomination: Donjon Cistern (fig. III.33)
Region-country: Piedmont, Italy
Province: Turin
Municipality: Verrua Savoia
Locality: Verrua Hill
Location: "Bomb-proof" Barracks
Ownership: private
Cartography: CTR 1:10.000
Geological unit: marls, sandstones and Pliocene limestone resting on an Eocene Clay Shale Formation
Altitude: (ellipsoid) 304.8498 m a.s.l.
Position: (topographic GPS) 45°10'26.46486" N; 8°06'03.77959" E
Context: upper part of the XVII century Donjon
Operations conducted: exploration, planimetric survey, photographic service
Work carried out by: Associazione Speleologia Cavità Artificiali Milano (S.C.A.M.); 2002, 2003
Warnings: stagnant water at the bottom of the cistern; in the event of removal of the cement cover, the environment should be aired for a few days before being entered
Typology: 6 - military work, reutilised as a typology 2 c - cistern
Description: at the top of the Donjon, on the overhanging east side, a square opening provides access to an irregularly-shaped underground chamber. The entrance consists of a rectangular well in the roof, measuring 0.67 x 0.87 m. Dimensions*:* the structure measures 4.3 x 4.58 x 3 x 0.52 x 2 x 3.71 m. It is 3.41 m at the springer and 5.18 m at the top of the, with an overall depth of 5.42 m. A layer of fine cement renders the walls impermeable from the springer (3.41 m) to 4.12 m to the fill-walls, above which the facing consists of badly made bricks placed in fairly

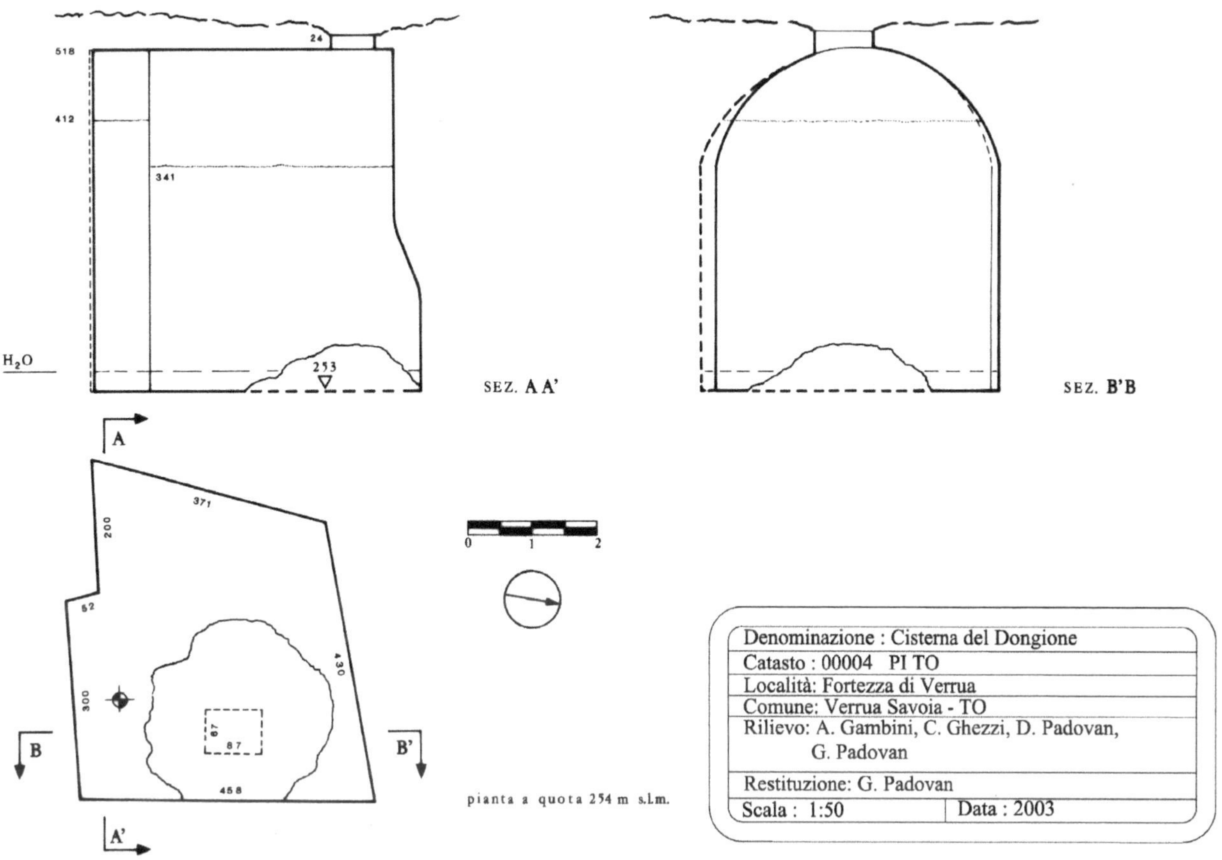

Fig. III.33. *Donjon Cistern (CA 00004 PI TO) (Ass.ne S.C.A.M. Archive).*

regular rows but with large gaps; the base is also lined with cement. The barrel vault is also made of brick. At the entrance, the floor is partly blocked by a detritic cone and there was water to a height of 0.3 m. Circular tubing overhangs the centre of the east wall, just under the roof; this is of recent construction. It is well preserved. The room served the Donjon rose garden for water storage purposes; it is now in disuse and has been closed off with cement.
Interpretation: the wall facing visible above the waterproofed area confirms that the structure was once a room within the Donjon and that it communicated with other rooms. To date the rooms and the original entrance cannot be located. However the survey of fill material would assist in locating such rooms. The cistern was therefore created from the removal of several underground rooms, which leads to the conclusion that the internal parts of the structure preceding the Donjon (part of the previous stronghold) may have been retained and reutilised in the later building to be then destroyed when the monument came to serve the rose.
Dating: in all probability, the original structure was built at the same time as the Donjon, while its transformation into a storage tank for meteoric water is undoubtedly recent and possibly dates to when the area was used as a rose garden in the first part of the XX century.
Notes: there should be a rainwater collection surface, now covered by soil and agrestal vegetation.
Bibliography: Padovan D., Padovan G., Bordignon L., Ottino M. 1997, *La Fortezza di Verrua Savoia*, in Club Alpinistico Triestino - Gruppo Grotte Sezione Ricerche e Studi su Cavità Artificiali, *Atti del IV Convegno Nazionale sulle Cavità Artificiali – Osoppo 30/31 Mggio - 1 Giugno 1997*, Trieste 1997, pgs. 187-208.
Padovan G., *Verrua Savoia: indagini sotterranee*, in Associazione Culturale "Amici della Biblioteca" di Crescentino, *Atti del Convegno Storico "Terre sul Po dal Medioevo alla Resistenza" – Crescentino 2-3 Ottobre 1988*, Crescentino 2002, pgs. 213-243.
Padovan G., *Note per la catalogazione e la comprensione delle opere idrauliche sotterranee*, in Giorgetti D. e Riera I. (edited by), In binos actus lumina. *Rivista di studi e ricerche sull'idraulica storica*, La Spezia 2002, pgs. 327-352.
Data ownership: Associazione Speleologia Cavità Artificiali Milano (S.C.A.M.).
Compiled by: Gianluca Padovan (Ass.ne S.C.A.M.).

III.3.12 - *CA 00005 PI TO; Small Hypogeum Well*

Cadastral number: CA 00005 PI TO
Denomination: Small Hypogeum Well (fig. III.34)
Region-country: Piedmont, Italy
Province: Turin
Municipality: Verrua Savoia
Locality: Verrua Hill
Location: "Bomb-proof" Barracks
Ownership: private
Cartography: CTR 1:10.000
Geological unit: marls, sandstones and Pliocene limestone resting on an Eocene Clay Shale Formation
Altitude: (ellipsoid) 295.27 m
Position: UTM 5002714.7999 N; 429325.6605 E
Context: connects to the "bomb-proof" Barracks
Operations conducted: exploration, planimetric survey, photographic service
Work carried out by: Associazione Speleologia Cavità Artificiali Milano (S.C.A.M.); 2003
Warnings: poor air circulation
Typology: 7 - unidentified structure
Description: ground floor of the "bomb-proof" Shelter, accessible via the wall breach of a small room next to the rock face. This trapezoidal chamber has a short entrance corridor, situated in a non-central position, intercepted by the "bomb-proof" Barracks and later demolished; a small quadrangular well has been excavated in the west corner. The entrance consists of a brick room which is 1.18 x 1.15 m and 1.05 m deep. A breach in the central area provides access to the hypogeum. The abutments present evidence of shelf corbels and it is probable that the room served as a wall cupboard. The overall length of the room is 10.85 m; the far wall is 5.4 m wide and 2.75 m in height at the centre of the vaulted roof. The walls, which are slightly curved and narrow towards the base, are unlined and still present clear signs left by excavation tools. Other than the removal (or the collapse) of the east angle of the entrance corridor, there are no signs of subsequent extension of the hypogeum. The elliptic vault presents a few cracks, but shows no sign of structural collapse. The floor is covered by soil and debris and inclines inwards. The original ground surface was clearly lower than that of the Barracks room and was presumably flat. Overall, the excavation is fairly regular and there appear to be no partition walls. The small well is fairly standard, measuring 0.4 x 0.48 x 0.38 x 0.51 m with a depth of 0.85 m; the base consists of soil, bricks, rubble and stones, a clear sign that it was filled at some point. On several occasions, following prolonged rainfall, it was found to be almost completely full of water. It is unlikely that it was connected to an underlying room.
Interpretation: it is most probably a cellar, an icehouse or a "bomb- proof" room.
Dating: the hypogeum predates the "bomb-proof" Barracks.
Notes: the presence of debris and the conversion of the entrance to a shelved room indicate that the hypogeum was first used for discharge purposes and demolished during the restructure of the Barracks following the last siege, when its defensive function was no longer required. The small well was possibly used to collect infiltration water and an attempt to determine whether it was made at the same time as the cavity or later, should be made. For a full investigation, the debris must first be removed. The marks left by excavation tools could also be studied. It should not be excluded that there may once have been other similar rooms, now destroyed or eliminated by building transformations or by the collapse of part of the Barracks in 1957.
Bibliography: Padovan D., Padovan G., Bordignon L., Ottino M. 1997, *La Fortezza di Verrua Savoia*, in Club Alpinistico Triestino - Gruppo Grotte Sezione Ricerche e Studi su Cavità Artificiali, *Atti del IV Convegno Nazionale sulle Cavità Artificiali – Osoppo 30/31 Mggio - 1 Giugno 1997*, Trieste 1997, pgs. 187-208.
Padovan G., *Verrua Savoia: indagini sotterranee*, in Associazione Culturale "Amici della Biblioteca" di Crescentino, *Atti del Convegno Storico "Terre sul Po dal Medioevo alla Resistenza" – Crescentino 2-3 Ottobre 1988*, Crescentino 2002, pgs. 213-243.
Padovan G., *Note per la catalogazione e la comprensione delle opere idrauliche sotterranee*, in Giorgetti D. e Riera I. (edited by), In binos actus lumina. *Rivista di studi e ricerche sull'idraulica storica*, La Spezia 2002, pgs. 327-352.
Data ownership: Associazione Speleologia Cavità Artificiali Milano (S.C.A.M.).
Compiled by: Gianluca Padovan (Ass.ne S.C.A.M.).

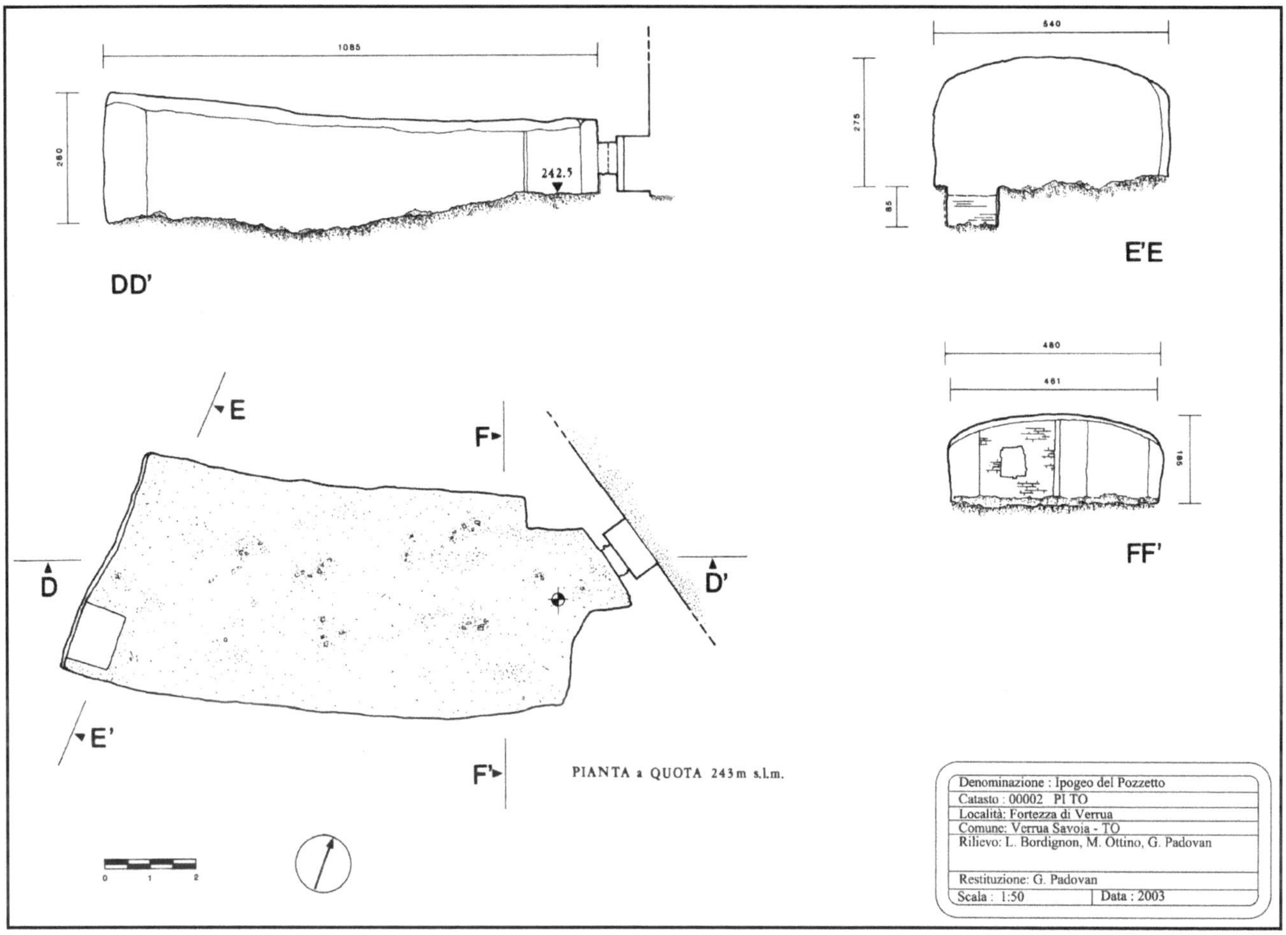

Fig. III.34. *Small Hypogeum Well (CA 00005 PI TO) (Ass.ne S.C.A.M. Archive).*

III.3.13 - *CA 00006 PI TO; Fourth Cannon Room*

Cadastral number: CA 00006 PI TO
Denomination: Fourth Cannon Room (fig. III.35)
Region-country: Piedmont, Italy
Province: Turin
Municipality: Verrua Savoia
Locality: Verrua Hill
Location: "Bomb-proof" Barracks
Ownership: private
Cartography: CTR 1:10.000
Geological unit: marls, sandstones and Pliocene limestone resting on an Eocene Clay Shale Formation
Altitude: (ellipsoid) 303.77 m
Position: U.T.M. 429359.0710 E; 5002659.4688 N
Context: within the Donjon structure, upper area
Operations conducted: exploration, planimetric survey, photographic service
Work carried out by: Associazione Speleologia Cavità Artificiali Milano (S.C.A.M.); 2002
Warnings: structural collapse at entrance, poor air circulation
Typology: 6 - shelter

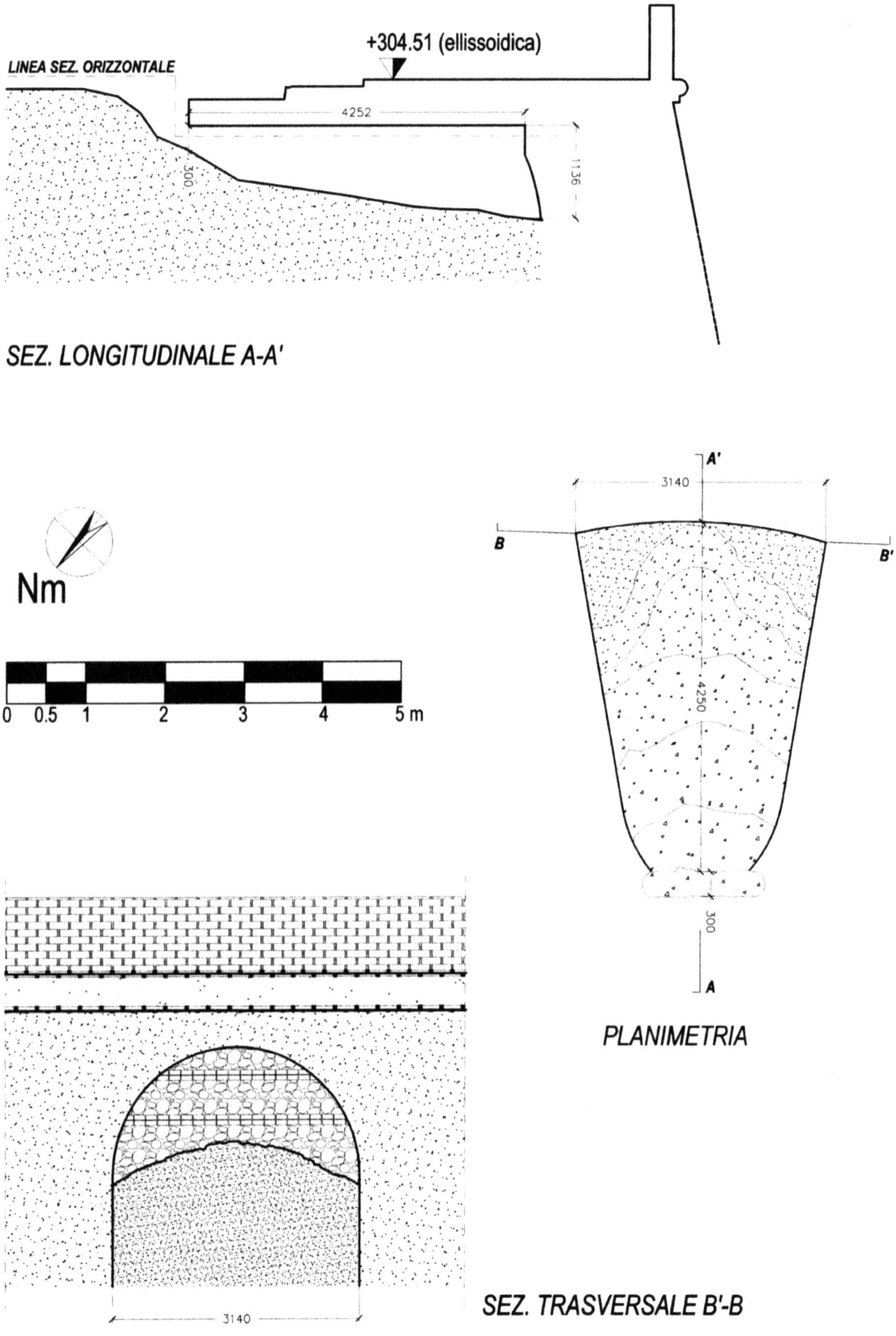

Fig. III.35. *Fourth Cannon Room (CA 00006 PI TO) (Ass.ne S.C.A.M. Archive).*

Description: the structure lies below the semi-circular parapet of the Donjon and is on the radial axis. The chamber is composite in shape and has a barrel vault, formed by a trapeze, the base of which supports a section of circle, the curved side of which follows the internal profile of the semi-donjon while the lower base faces the centre of the structure. The entrance opens under the edge of the chemin-de-ronde and was found blocked by a stone slab. It lateral walls, a few centimetres from the entrance, follow a narrowing course; the arch crown is damaged. At least one part of the buried section and the closure mechanism were affected when the top of the Donjon was converted for rose cultivation purposes. The room is 4.25 m long and 3.14 m wide at its base; a small 30 cm wide trench was made to allow access. Buried to just above the piers, it is thought that the room may in fact be slightly larger. The piers and the roof are made of brick, while the infill wall is made of regular sections of pebbles, broken rocks and stones and occasional stone fragments, alternating with regular sections of bricks.
Interpretation: room under the parapet, which may have been used to shelter a piece of artillery and/or to keep munitions in the event of an attack.
Dating: built at the same time as the Donjon, it dates to between the third and fourth quarters of the XVII century.
Notes: the probe descends into the buried section by more than one metre and suggests that the height of the room is in excess of 2 m; it is similar to the other Cannon Room. Except for a modest collapse at the entrance, the structure presents no damage. There are several small formations on the roof. In order to conduct a complete investigation, the structure and the entire upper section of the Donjon should be cleared from within.
Bibliography: Padovan D., Padovan G., Bordignon L., Ottino M. 1997, *La Fortezza di Verrua Savoia*, in Club Alpinistico Triestino - Gruppo Grotte Sezione Ricerche e Studi su Cavità Artificiali, *Atti del IV Convegno Nazionale sulle Cavità Artificiali – Osoppo 30/31 Mggio - 1 Giugno 1997*, Trieste 1997, pgs. 187-208.
Padovan G., *Verrua Savoia: indagini sotterranee*, in Associazione Culturale "Amici della Biblioteca" di Crescentino, *Atti del Convegno Storico "Terre sul Po dal Medioevo alla Resistenza" – Crescentino 2-3 Ottobre 1988*, Crescentino 2002, pgs. 213-243.
Data ownership: Associazione Speleologia Cavità Artificiali Milano (S.C.A.M.).
Compiled by: Gianluca Padovan (Ass.ne S.C.A.M.).

III.3.14 - *CA 00007 PI TO; Draw-Bridge Cistern*

Cadastral number: CA 00007 PI TO
Denomination: Draw-Bridge Cistern (figs. III.36 and III.37)
Region-country: Piedmont, Italy
Province: Turin
Municipality: Verrua Savoia
Locality: Verrua Hill
Location: Terrace Bastion
Ownership: private
Cartography: CTR 1:10.000
Geological unit: marls, sandstones and Pliocene limestone resting on an Eocene Clay Shale Formation
Altitude: //
Position: //
Context: within the Terrace Bastion
Operations conducted: exploration, planimetric survey, photographic service
Work carried out by: Associazione Speleologia Cavità Artificiali Milano (S.C.A.M.); 2002, 2003
Warnings: care should be taken with the upper manhole
Typology: 2 c - icehouse converted into a water-storage cistern
Description: the entrance consists of a short corridor, branching off from the cobbled road which leads from the Castle Gate to the square behind the "bomb-proof" Barracks. Dimensions: the internal surface is of 6.84 m and is 6.55 m high at the vault and 7.05 m high at the top of the shaft. Hydraulic mortar lines the exposed brick structure to almost the top the conch vault springer. The base, also lined with hydraulic mortar, appears to be perfectly even and is almost entirely covered by soil and detritus presenting water to no more than 0.4 m. In a western, internally facing direction is a room with a pointed vault, which is blocked off by bricks. A second, south-facing room, which has been rendered impermeable and is also blocked off, opens out at ground level. At the bottom of the room is a metal filter. Immediately above this, is a third room, again blocked off by bricks. At the centre of the vault is a quadrangular well, closed at the summit by a stone cover. Footholds: none found. Piping: there is a large drainpipe, consisting of two overlapping tiles at the blockage of the first room.
Interpretation: although similar in appearance to an icehouse, this was used as a cistern in the XX century, possibly for the storage of meteoric water and would supply the Monferrato aqueduct from the 1920s. In any case, the building would suggest the existence of an air gap for thermal insulation purposes. The three blocked rooms indicate

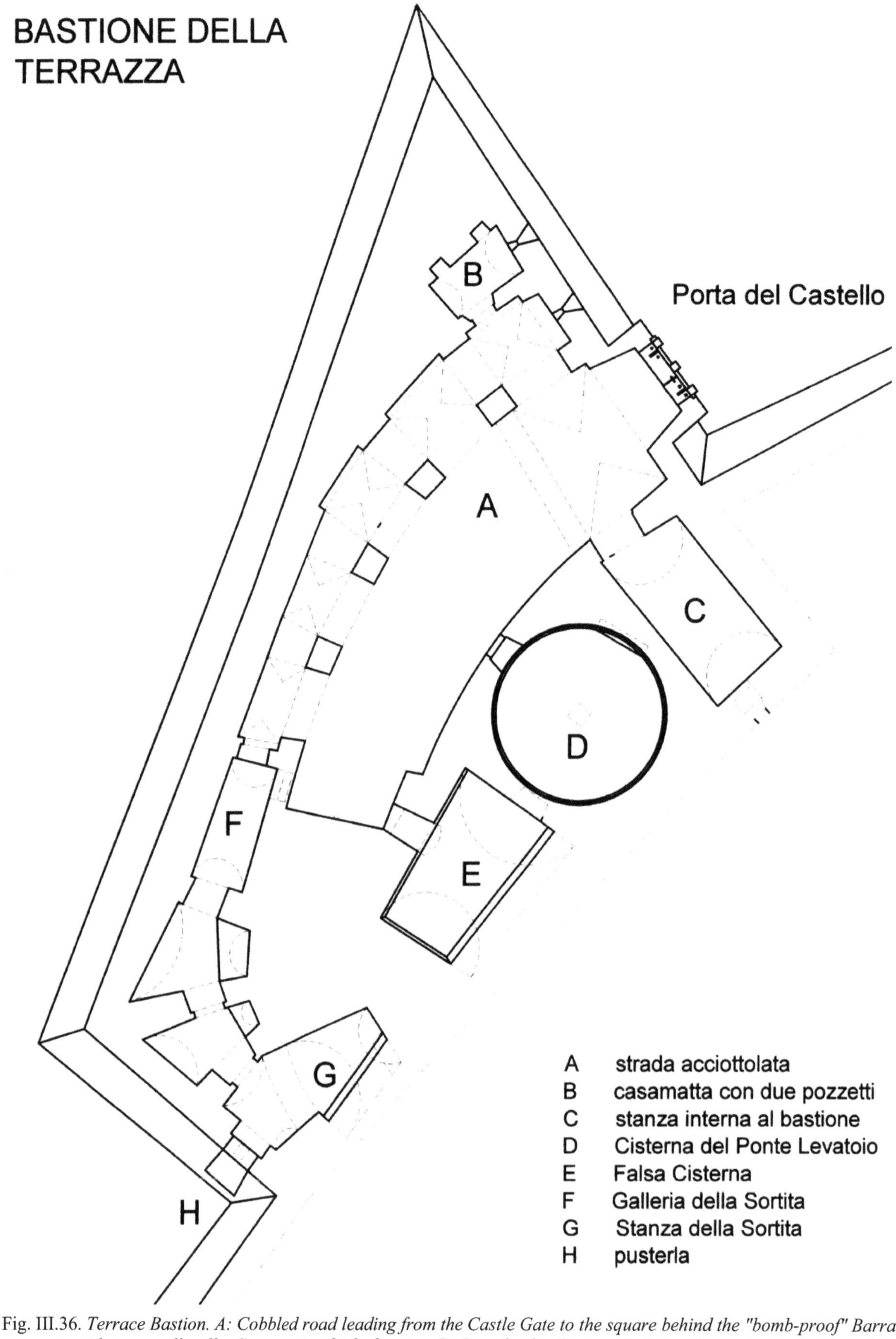

Fig. III.36. *Terrace Bastion. A: Cobbled road leading from the Castle Gate to the square behind the "bomb-proof" Barracks. B: casemate with two small wells. C: room inside the bastion. D: Drawbridge Cistern. E: False Cistern F: Sortie Tunnel. G: Sortie Room. H: postern (Ottino 1996. Bordignon 2000).*

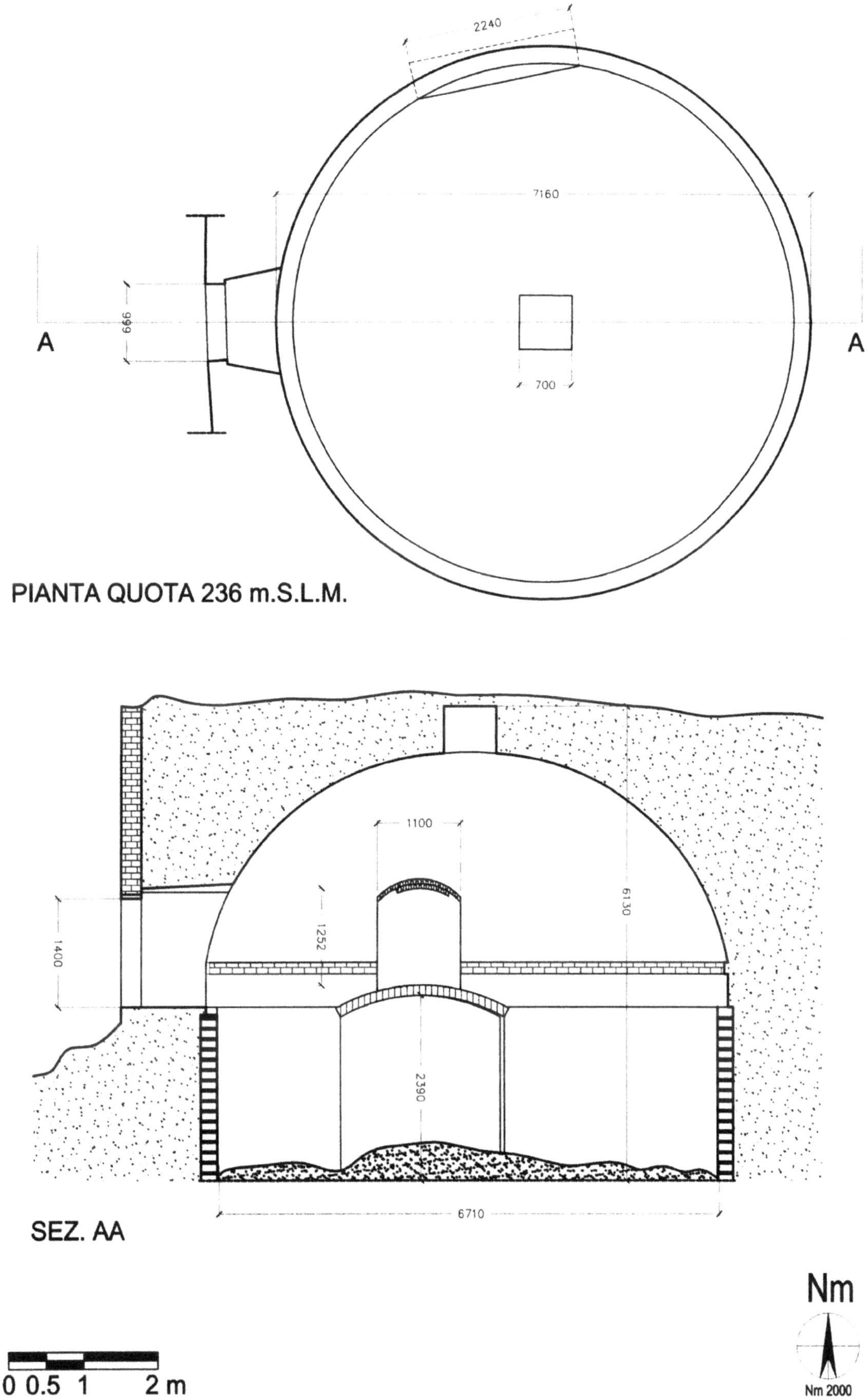

Fig. III.37. *Drawbridge Cistern (CA 00007 PI TO) (Ass.ne S.C.A.M. Archive).*

communication with other environments; the lower room with the filter in particular, would suggest communication with a further water storage chamber.
Dating: it would appear that the structure was built at the same time as the Terrace Bastion (latter part of the XVII century), however, it cannot be excluded that the structure may slightly predate the Bastion. In the case of the latter, the restructure of the Fortress' central area would have maintained the structure of the cistern and been built around it. Its re-utilisation as a cistern appears to date back to the first part of the XX century.
Notes: cleaning of the cistern and removal of infiltrations would guarantee the recovery of a vast environment of particular interest.
Bibliography: Padovan D., Padovan G., Bordignon L., Ottino M. 1997, *La Fortezza di Verrua Savoia*, in Club Alpinistico Triestino - Gruppo Grotte Sezione Ricerche e Studi su Cavità Artificiali, *Atti del IV Convegno Nazionale sulle Cavità Artificiali – Osoppo 30/31 Mggio - 1 Giugno 1997*, Trieste 1997, pgs. 187-208.
Padovan G., *Verrua Savoia: indagini sotterranee*, in Associazione Culturale "Amici della Biblioteca" di Crescentino, *Atti del Convegno Storico "Terre sul Po dal Medioevo alla Resistenza" – Crescentino 2-3 Ottobre 1988*, Crescentino 2002, pgs. 213-243.
Data ownership: Associazione Speleologia Cavità Artificiali Milano (S.C.A.M.).
Compiled by: Gianluca Padovan (Ass.ne S.C.A.M.).

III.3.15 - *CA 00008 PI TO; Fifth Cannon Room (Merlon Room)*

Cadastral number: CA 00008 PI TO
Denomination: Fifth Cannon Room (Merlon Room) (fig. III.38)
Region-country: Piedmont, Italy
Province: Turin
Municipality: Verrua Savoia
Locality: Verrua Hill
Location: "Bomb-proof" Barracks
Ownership: private
Cartography: CTR 1:10.000
Geological unit: marls, sandstones and Pliocene limestone resting on an Eocene Clay Shale Formation
Altitude: (ellipsoid) 304.01 m
Position: U.T.M. 429361.0504 E; 5002680.3617 N
Context: within the Donjon structure, upper area
Operations conducted: exploration, planimetric survey, photographic service
Work carried out by: Associazione Speleologia Cavità Artificiali Milano (S.C.A.M.); 2002, 2003
Warnings: structural collapse at the entrance
Typology: 6 - shelter
Description: the structure is situated below the building used as a lemon-house and under the last surviving merlon of the Donjon defensive structure. Its base has been significantly raised by soil and debris. Access is gained to the structure by means of a basic excavation in the detritus and debris; it is 0.47 m wide and the lateral walls are of 1.71 m and 1.8 m while the base is of 3 m. Now filled with soil, it is thought that the room may be significantly larger and longer than it currently appears. The piers and the vaulted roof are made of brick, while the infill wall is made of regular sections of pebbles, broken rocks and stones and occasional stone fragments, alternating with regular sections of bricks.
Interpretation: room, originally situated under the parapet, which may have been used to shelter a piece of artillery and/or to keep munitions in the event of an attack.
Dating: built at the same time as the Donjon, it appears to date back to between the third and fourth quarters of the XVII century.
Notes: it is thought that the room is more than 2 m high; apart from the collapse at the entrance, the structure presents no damage. In order to conduct a complete investigation, the structure and the entire upper section of the Donjon should be cleared from within.
Bibliography: none.
Data ownership: Associazione Speleologia Cavità Artificiali Milano (S.C.A.M.).
Compiled by: Gianluca Padovan (Ass.ne S.C.A.M.).

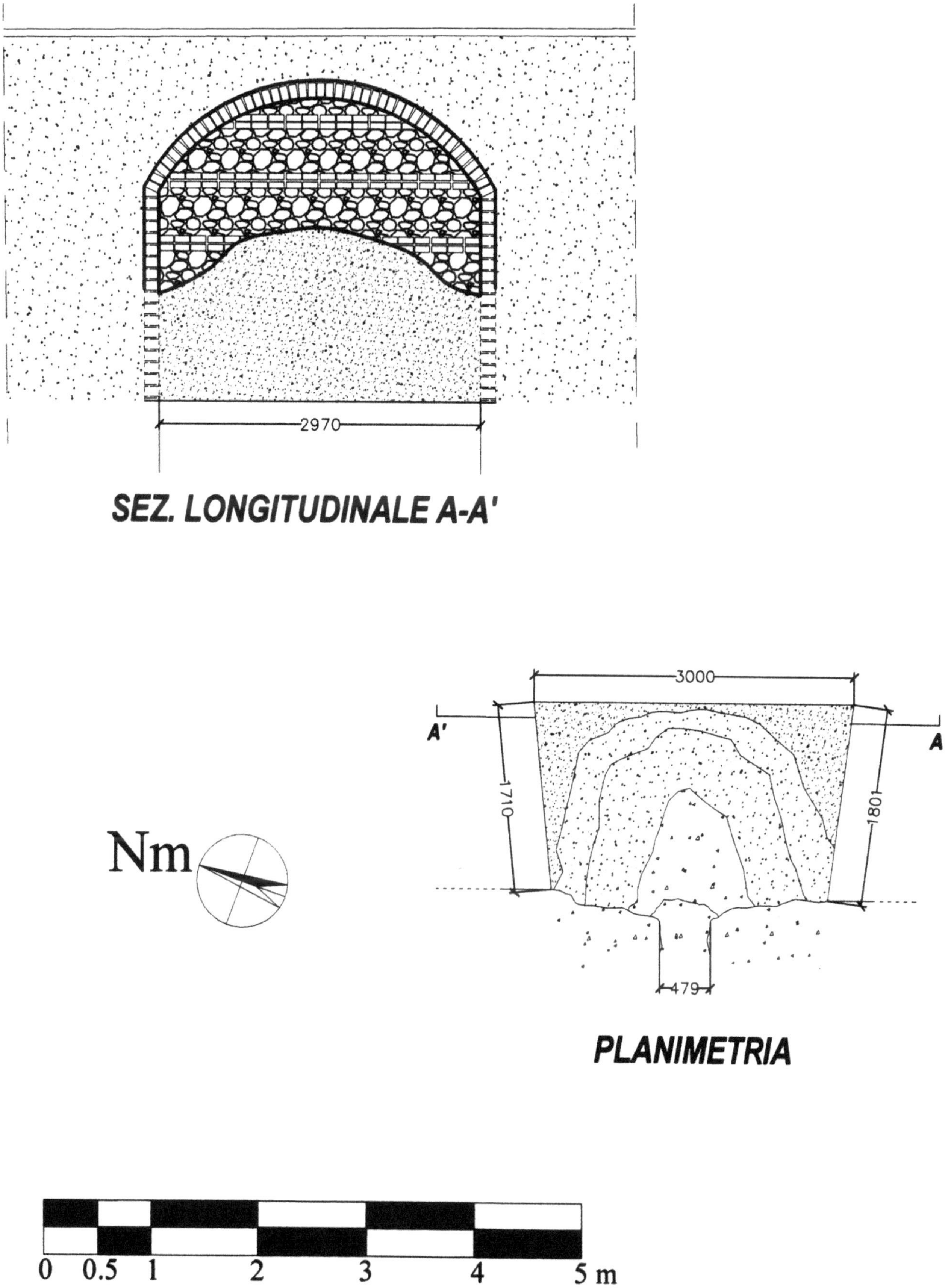

Fig. III.38. *Fifth Cannon Room. Also known as the Merlon Room as it lies below the Donjon's sole surviving Merlon (CA 00008 PI TO) (Ass.ne S.C.A.M. Archive).*

III.3.16 - *CA 00009 PI TO; False Cistern*

Cadastral number: CA 00009 PI TO
Denomination: False Cistern (figs. III.39 and III.40)
Region-country: Piedmont, Italy
Province: Turin
Municipality: Verrua Savoia
Locality: Verrua Hill
Location: within the Terrace Bastion
Ownership: private
Cartography: CTR 1:10.000
Geological unit: marls, sandstones and Pliocene limestone resting on an Eocene Clay Shale Formation
Altitude: //
Position: //
Context: within the Terrace Bastion
Operations conducted: exploration, planimetric survey, photographic service
Work carried out by: Associazione Speleologia Cavità Artificiali Milano (S.C.A.M.); 2003
Warnings: care should be taken in the entrance area
Typology: 6 - military soutterain, converted to typology 2 c - cistern
Description: the structure is located within the Bastion and consists of a deep opening with a sloping floor, giving access to a trapezoidal chamber, part of which is blocked up. The opening is accessed from the cobbled road that leads from the Castle Gate to the square behind the "bomb-proof" Barracks. Its dimensions are as follows: 4.3 x 4.58 x 3 x 3.71 m; 3.41 m at the springer and 7.23 m at the top of the vaulted roof. The walls are in exposed brick to almost as far as the top of the springer of the barrel vault, which is lined in hydraulic mortar. The base, also lined in hydraulic mortar, appears to be level. It almost entirely covered by soil and detritus, and lightly slopes in a southern direction. There is no water inside the cistern. To the south, towards the Drawbridge Cistern, is a room with a blocked up drop vault, which does not communicated with the adjacent cistern. Footholds: none found. Piping: none visible.
Interpretation: the blocked wall facings would suggest that that the original structure of the room was different and that the rooms, particularly the Sortie Room, communicated with other environments. The internal wall is scarped and this suggests that it may originally have been the curtain wall of the previous fortification, reutilised in later conversions.
Dating: in all probability, the original structure was built at the same time as the Terrace Bastion (latter part of the XVII century), while its conversion into a storage tank for meteoric water is undoubtedly recent and dates back to the period the area was used for greenhouse purposes in the second part of the XX century.
Notes: in order to check for the existence of possible communication with adjacent environments, the detritus should be cleared and a survey of the fill materials carried out. Graffiti from the Crescentino fortification survived, almost in its entirety, on the wall to the left of the current entrance until 1999. Unfortunately a few parts are missing due to sulphation and a few parts in the centre have been damaged due the moss which covers large portions of the wall. Plotted by survey, it reveals quadrilateral system valley fornication with small corner bastions and a fifth bastion half-way along the upper curtain. Defence was completed by two rectangular central cores on the right and left curtains, almost on the street axis. These may have been drawbridge pillars, used to protect fortification entrances. The inner section is enclosed by a quadrangular perimeter and is divided into two distinct chessboard-divided "areas". The upper area is the smaller of the two and does not touch the perimeter, from which it is separated by a series of inverted V marks, the significance of which is unknown. The other is larger and is as wide as the quadrilateral that encloses it and its lower side coincides with the quadrilateral. There are two circles in the drawbridge pillar area, which may represent the number of towers or lookout positions. The external area presents numerous "signs", which are not always easily interpreted. In relation to the triangular signs, these undoubtedly indicate structures, probably terrepleins, namely ravelins or lunettes, while the circular signs could indicate "fortlets". Outwith the bastioned quadrilateral and in its defence are the following works, listed clockwise from left to right: a large ravelin, the lower side of which protects the inner curtain; two coupled ravelins (with graffiti under the second ravelin depicting a mortar ball, its fuse alight); a second large ravelin with rear-lying central core and a further lateral ravelin; a third large ravelin with a well-defined lateral circular fortlet. It appears that the works were surrounded by a palisade or more probably a terreplein with rear-lying ditch. It is thought that the graffiti represents the Crescentino fortification as seen from above from Verrua Fortress. The fortified Crescentino system is thought to date to the XVI century: this is clearly conflicts with the previously assumed mid XVII century construction of the False Cistern. An accurate investigation of the wall structures of the Terrace Bastion is required and brick samples should undergo thermoluminence testing.

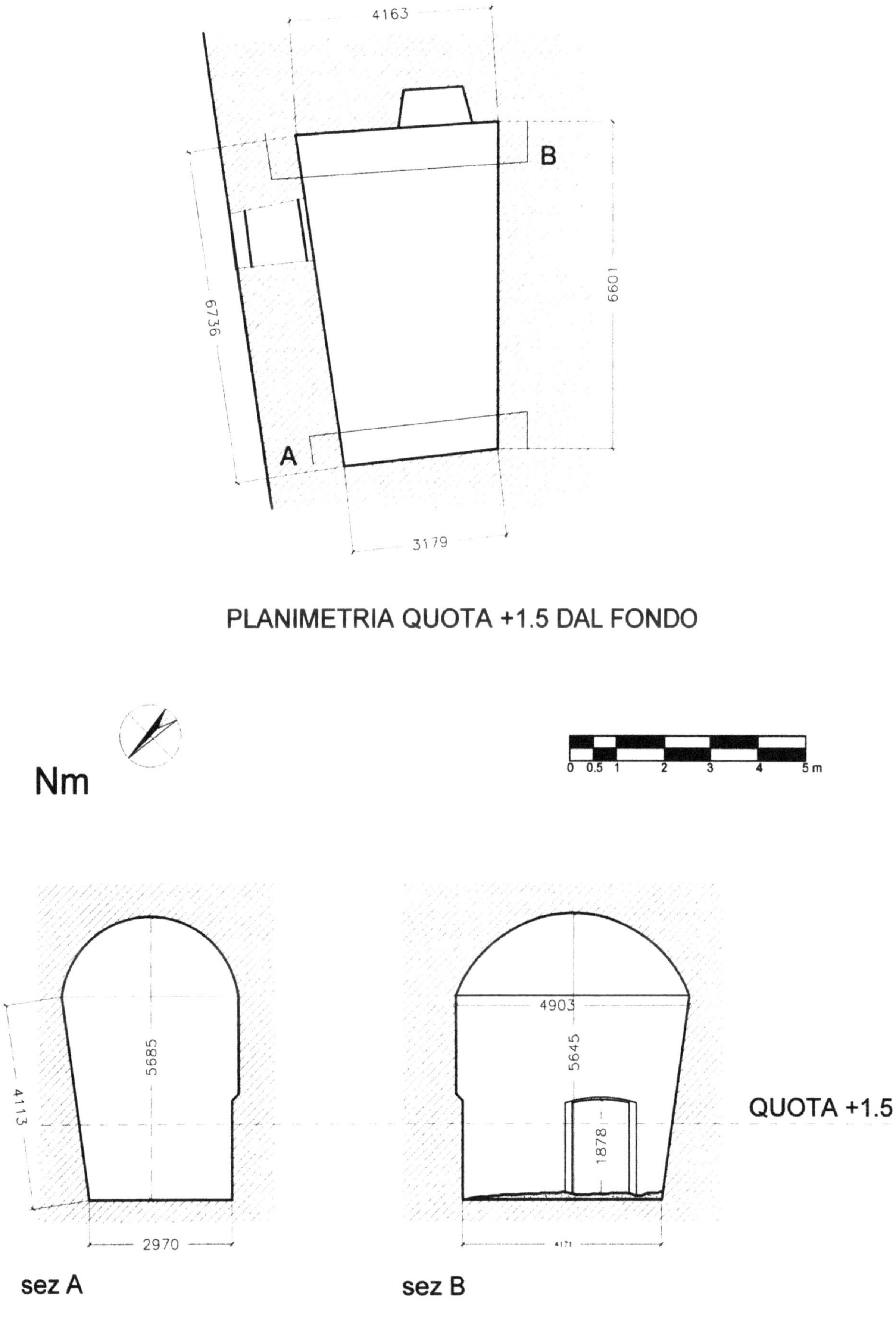

Fig. III.39. *False Cistern (CA 00009 PI TO) (Ass.ne S.C.A.M. Archive).*

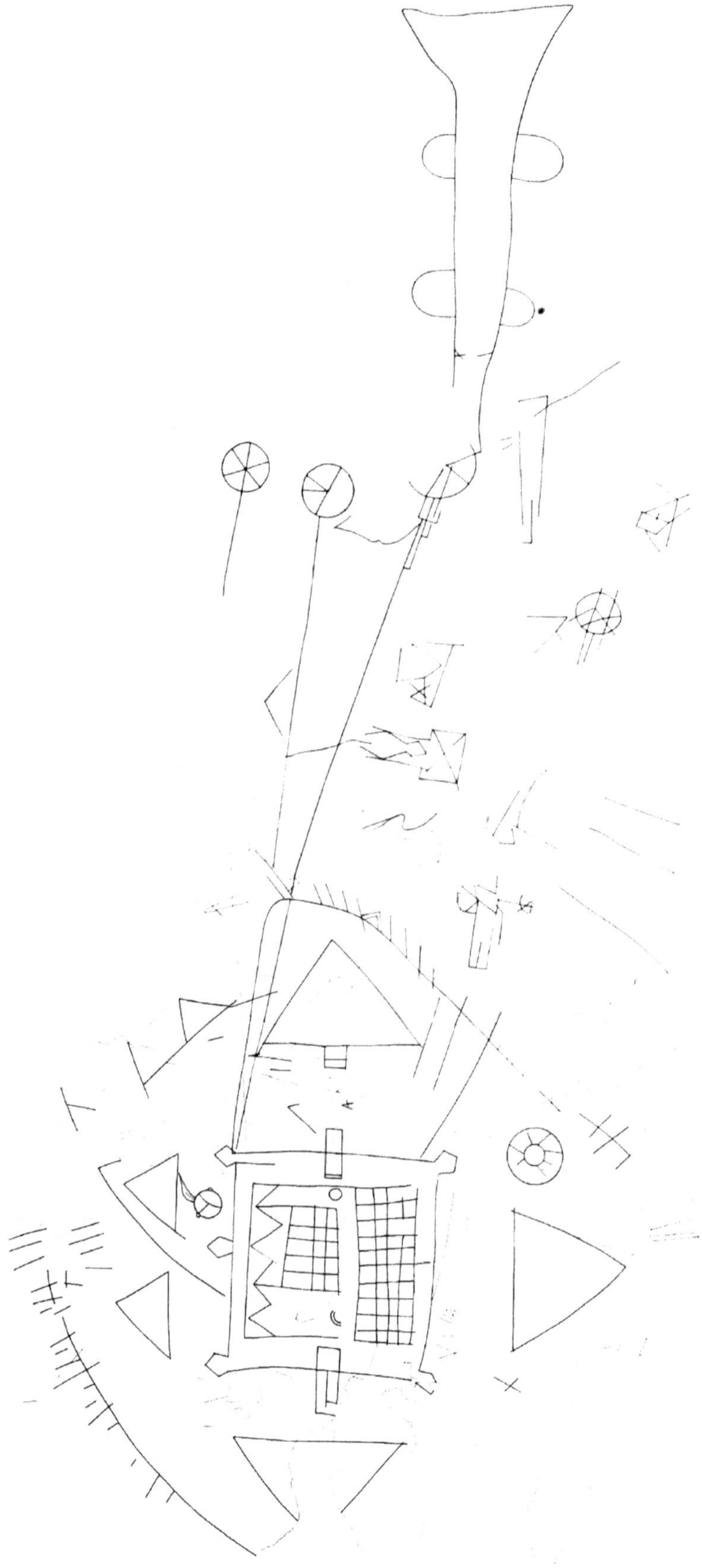

Fig. III.40. *Graphic reconstruction of the graffiti found to the left of the False Cistern's entrance, which shows the township of Crescentino in the XVI century (CA 00007 PI TO) (Ass.ne S.C.A.M. Archive).*

Bibliography: none.
Data ownership: Associazione Speleologia Cavità Artificiali Milano (S.C.A.M.).
Compiled by: Gianluca Padovan (Ass.ne S.C.A.M.).

III.3.17 - *CA 00010 PI TO; Sortie Tunnel*

Cadastral number: CA 00010 PI TO
Denomination: Sortie Tunnel (fig. III.41)
Region-country: Piedmont, Italy
Province: Turin
Municipality: Verrua Savoia
Locality: Verrua Hill
Location: within the Terrace Bastion
Ownership: private
Cartography: CTR 1:10.000
Geological unit: marls, sandstones and Pliocene limestone resting on an Eocene Clay Shale Formation
Altitude: //
Position: //
Context: part of the internal structure of the Terrace Bastion
Operations conducted: exploration, planimetric survey, photographic service
Work carried out by: Associazione Speleologia Cavità Artificiali Milano (S.C.A.M.); 2003
Warnings: the large bat colony should not be disturbed
Typology: 6 - sortie tunnel
Description: this consists of four communicating environments. The secondary Stronghold gate is in the first environment, known as the Sortie Tunnel. The entrance is situated on the right side of the road leading up from the Gate to the "bomb-proof" Barracks square. The tunnel extends for 18 m between the entrance and the Gate; its total length is 24 m. There is a blocked up room to the upper left of the room. The second room, and to a lesser extent the third room, now house a large colony of bats. The fourth is well structured and was evidently built over various construction phases, during the course of which and at a time yet to be determined communication with the rear-lying room, known as the False Cistern, was blocked. A gate, once equipped with drawbridge, can be seen on one of the walls.
Interpretation: communication tunnel situated within the bastion.
Dating: the structure was built at the same time as Terrace Bastion (latter part of the XVII century).
Notes: the 5-6 m high ceilings indicated that the structure was had an intermediate floor, as also suggested by the blocked entrance to the first room. A colony of at least one thousand *Myotis blythii* (Vespertilio di Monticelli) bats resides within the structure.
Bibliography: none.
Data ownership: Associazione Speleologia Cavità Artificiali Milano (S.C.A.M.).
Compiled by: Gianluca Padovan (Ass.ne S.C.A.M.).

III.3.18 - *CA 00011 PI TO; Celestino Tunnel*

Cadastral number: CA 00011 PI TO
Denomination: Celestino Tunnel (figs. III.42, III.43, III.43.a, b, c and III.44)
Region-country: Piedmont, Italy
Province: Turin
Municipality: Verrua Savoia
Locality: Verrua Hill
Location: situated along the edge of the current quarry, by which it is intercepted
Ownership: private
Cartography: CTR 1:10.000
Geological unit: marls, sandstones and Pliocene limestone resting on an Eocene Clay Shale Formation
Altitude: 227 m a.s.l.; ellipsoid vault altitude 275, 8009 m, ellipsoid cavity altitude 274, 8229 m
Position: (topographic GPS) vault 45°10'24.14008" N, 8°06'08.43461" E
Context: within the stronghold, in the Borgo Alto area; it is situated in the depression of Verrua Hill (no longer readable due to quarry works) and lies behind the Bastion of San Francesco in Paola
Operations conducted: exploration, planimetric survey, photographic service
Work carried out by: Associazione Speleologia Cavità Artificiali Milano (S.C.A.M.); 2003

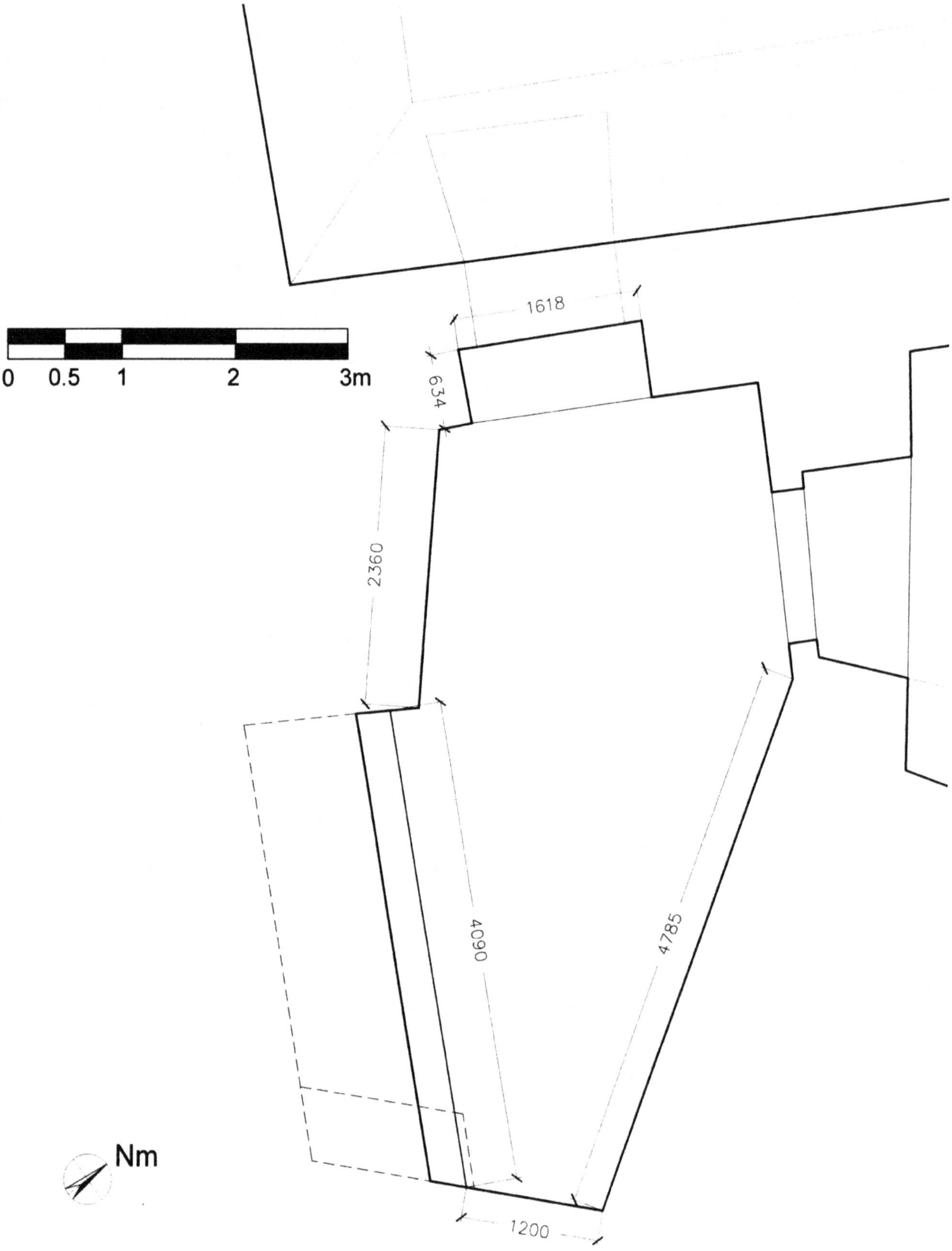

Fig. III.41. *Sortie Room with the Castle's False Gate, which was once equipped with a drawbridge (CA 00010 PI TO) (Ass.ne S.C.A.M. Archive).*

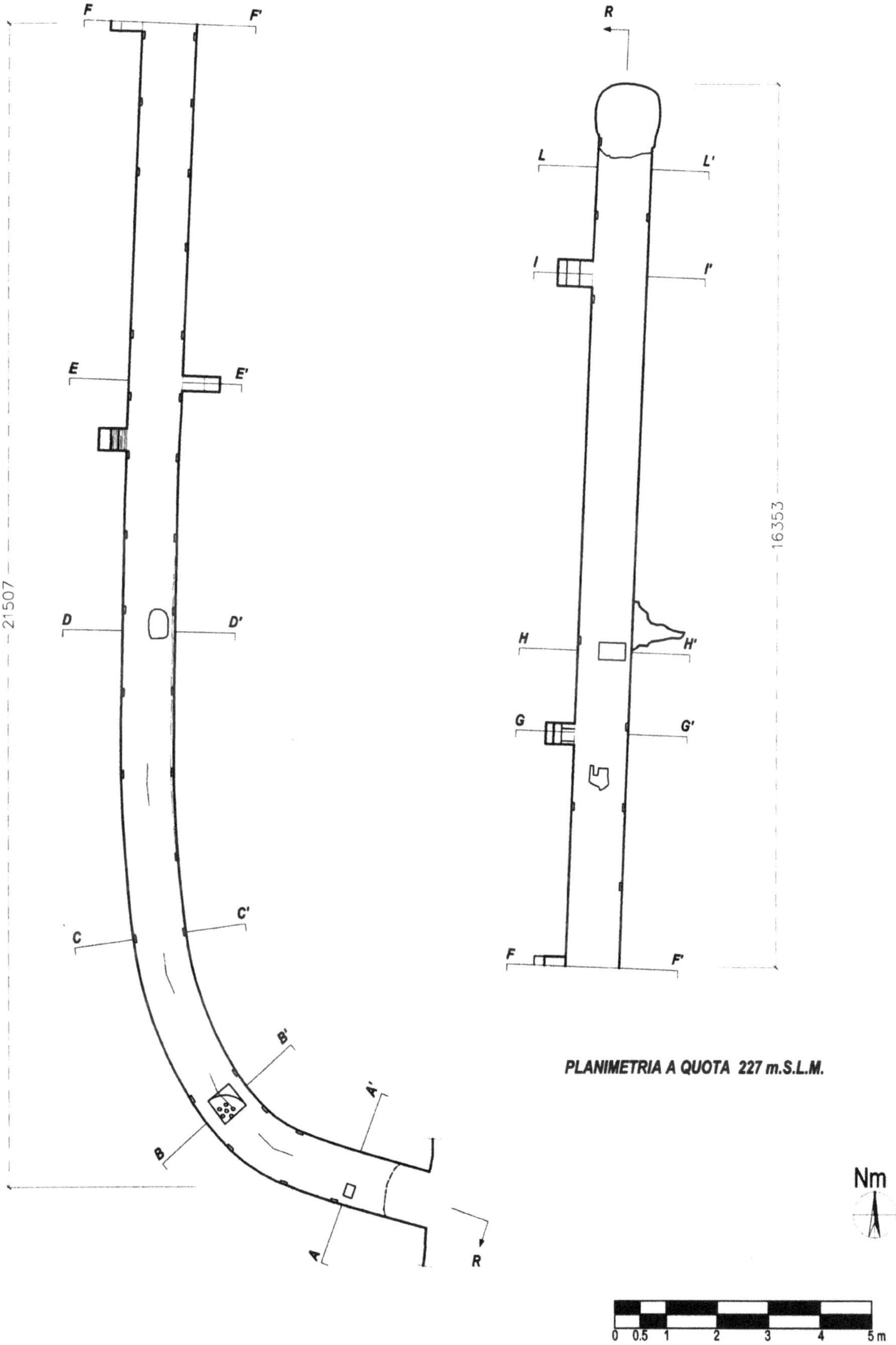

Fig. III.42. *Celestino Tunnel (CA 00011 PI TO) (Ass.ne S.C.A.M. Archive).*

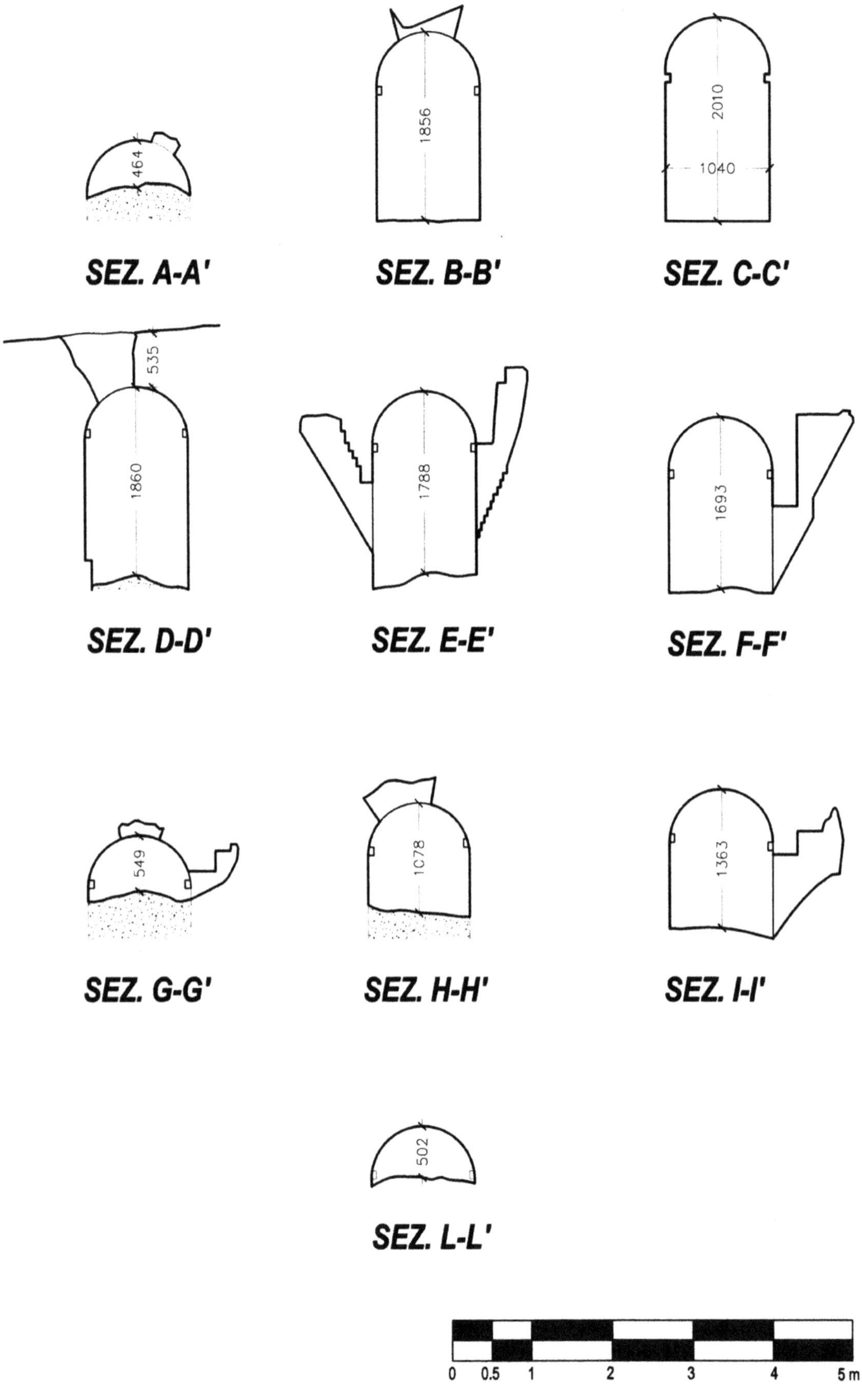

Fig. III.43. *Celestino Tunnel; sections (CA 00011 PI TO) (Ass.ne S.C.A.M. Archive).*

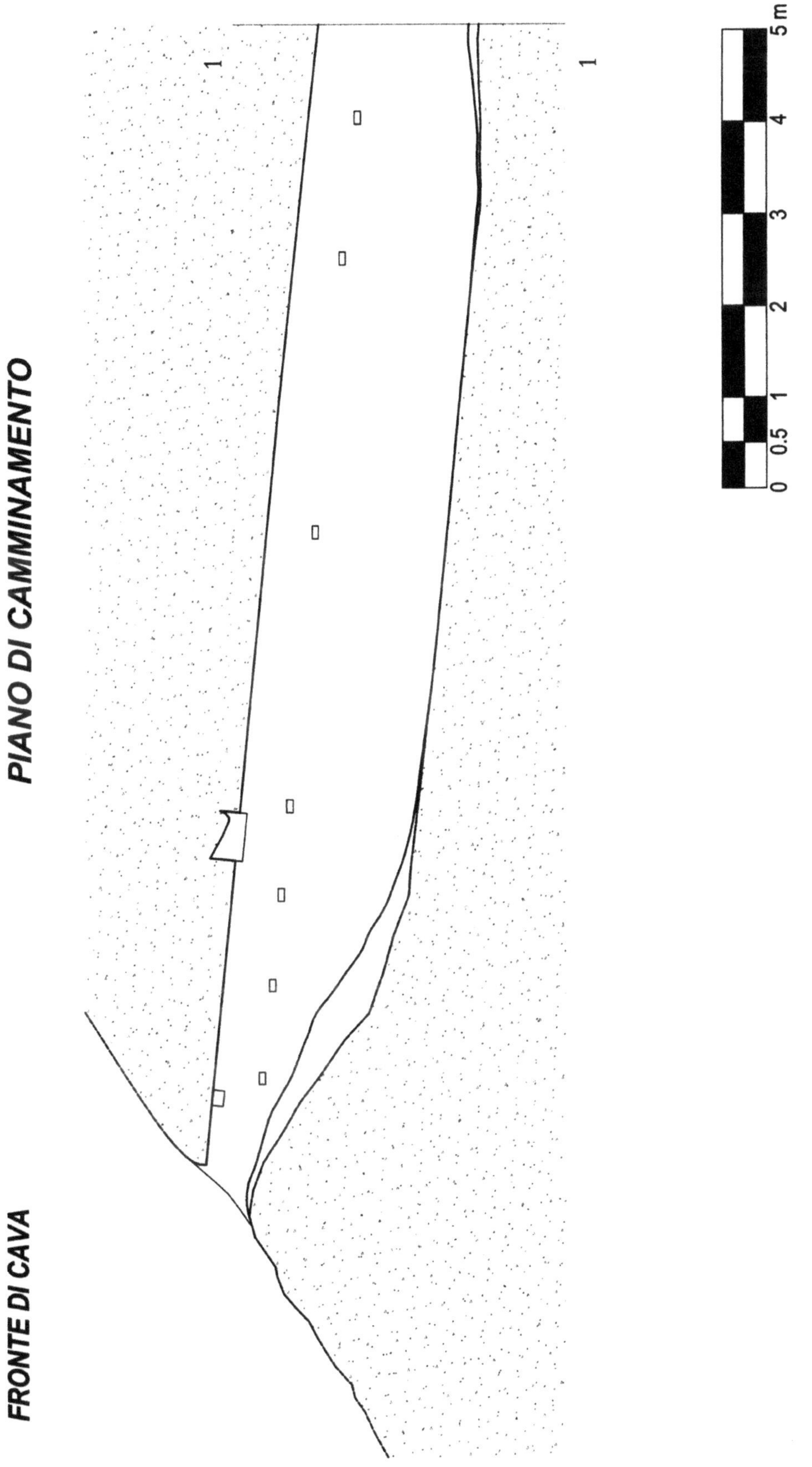

Fig. III.43.a. *Celestino Tunnel; section (CA 00011 PI TO) (Ass.ne S.C.A.M. Archive).*

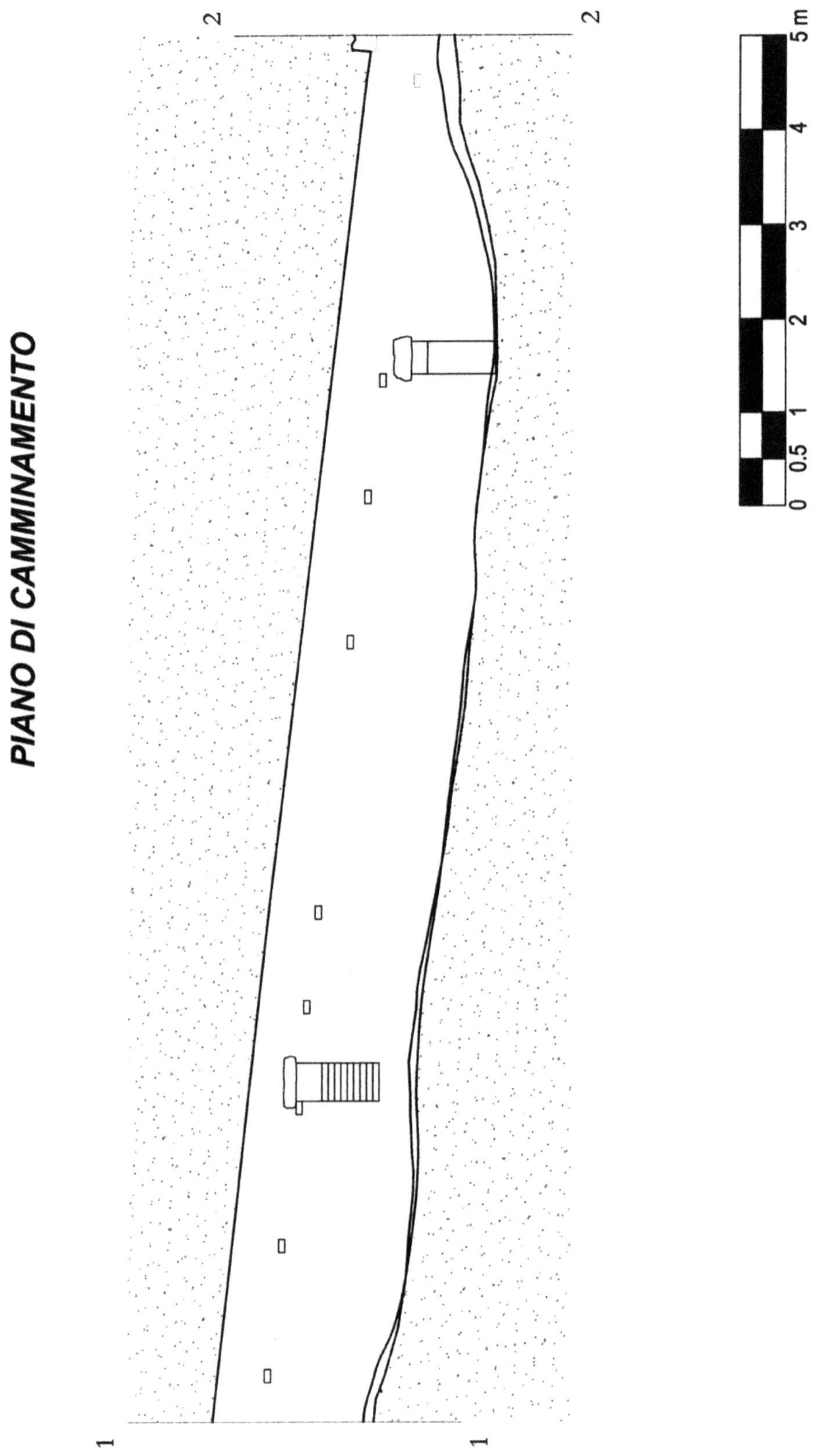

Fig. III.43.b. *Celestino Tunnel; section (CA 00011 PI TO) (Ass.ne S.C.A.M. Archive).*

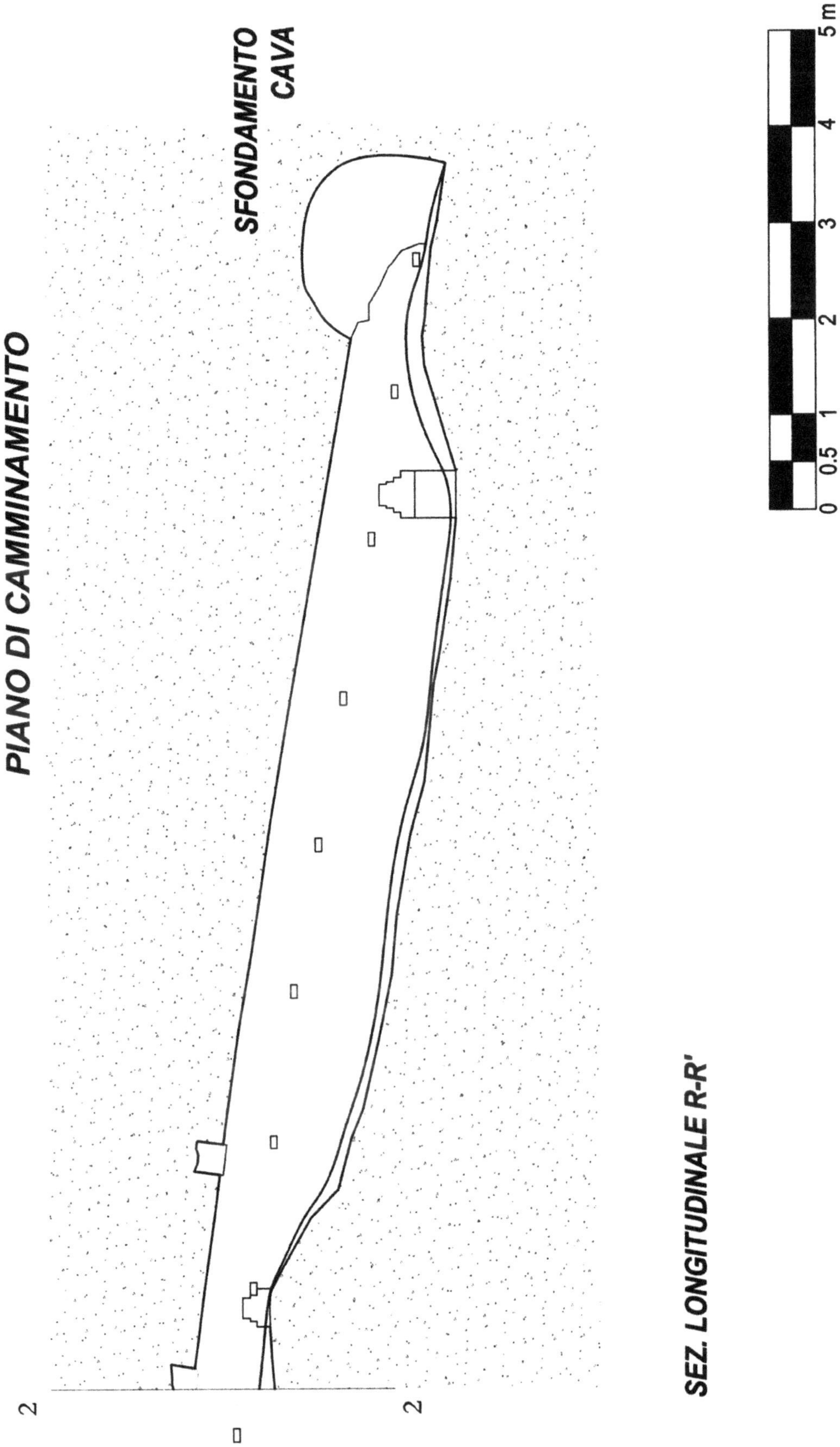

Fig. III.43.c. *Celestino Tunnel; section (CA 00011 PI TO) (Ass.ne S.C.A.M. Archive).*

Fig. III.44. *First section of the Celestino Tunnel (CA 00011 PI TO). The structure and the alternating bricks used to provide roof support (during the construction phase) are reminiscent of the countermine system in Turin's Citadel (photo G. Padovan).*

Warnings: possible internal collapse of the current tunnel bottom
Typology: 6 - communication tunnel
Description: consists of a brick tunnel with ventilation shafts. Today the tunnel opens onto the quarry face, and is unequivocally divided by the progression of the face; a second entrance has been uncovered by breaking a wall in the arch crown. The overall length of the environment is 41 m; its average width is 1.04 m and it is more than 2 m high. The masonry structure was built within a trench or within a hillside terrace and then covered, making it underground and safe from shellfire.
Interpretation: this presumably linked the Supply Barracks and the gate by the same name with the Battery of the Old Church.
Dating: the structure should date back to the fortification's restoration following the 1625 siege. More precisely, given its position, it may date back to the third quarter of the XVII century and to the construction of the triple bastion stronghold.
Notes: this section of tunnel, as also suggested by the other sections uncovered along the quarry face (00012 PI TO, 00013 PI TO, 00014 PI TO in-phase with 00027 PI TO), appears to be part of a structured system for the movement of troops and supplies between the Supply Gate and the bastioned fortifications defending the southeast, south and southwest flanks of the stronghold, or those likely to be assailed in the event of siege, as took place during the XVII and XVIII centuries. In 2003, the speleologist Enrico Lana carried out several investigations of the fauna within the Tunnel.
Bibliography: none.
Data ownership: Associazione Speleologia Cavità Artificiali Milano (S.C.A.M.).
Compiled by: Gianluca Padovan (Ass.ne S.C.A.M.).

III.3.19 - *CA 00012 PI TO; Room of the Opening*

Cadastral number: CA 00012 PI TO
Denomination: Room of the Opening (figs. III.45, III.46 and III.46.a)
Region-country: Piedmont, Italy
Province: Turin
Municipality: Verrua Savoia
Locality: Verrua Hill
Location: situated along the edge of the current quarry, by which it has been intercepted
Ownership: private
Cartography: CTR 1:10.000
Geological unit: marls, sandstones and Pliocene limestone resting on an Eocene Clay Shale Formation
Altitude: 197 m a.s.l.
Position: (GPS investigation) east 1429481, west 5002835
Context: within the stronghold, situated in the Barracks area of the Supply Gate, northwest of the Gate and the enceinte walls
Operations conducted: exploration, planimetric survey, photographic service
Work carried out by: Associazione Speleologia Cavità Artificiali Milano (S.C.A.M.); 2003
Warnings: possible collapse of the remains of the tunnel
Typology: 6- communication tunnel with lateral chamber
Description: consists of a brick tunnel section, now almost completely buried, of which thanks to the demolition work of the quarry, only the internal piers and part of the barrel vault remain. There is a 0.81 m corridor in the wall, which gives access to a trapezoidal room, within the lower fill of which, is a 0.24x0.18 m opening which is 0.38 m deep and communicates with the TE.S.E.S. Tunnel (CA 00027 PI TO). The room measures 2.28 x 5.14 x 3.12 x 5.09 m. Originally, it would have been more than 4 m high, but quarry mud has raised its base. The walls and the barrel vault are made of brick while the fill-walls consist of fairly regular rows of stones, pebbles and stone fragments, alternated by regular rows of bricks.
Interpretation: every aspect of the tunnel is the same as in the Celestino Tunnel (CA 00011 PI TO), suggesting that they belong to the same complex.
Dating: the structure should date back to the fortification's restoration following the 1625 siege. More precisely, given its position, it may date back to the third quarter of the XVII century and the construction of the triple bastion stronghold.
Notes: the opening, from which the name Room of the Opening arises, would have been used for the purposes of air exchange as the environment presented no ventilation shafts and its only air inlet was a corridor, which communicated with the tunnel. Clearing the quarry face would uncover the tunnel section in question and its cleaning and consolidation would permit its full investigation and confirm the existence of any communicating environments.

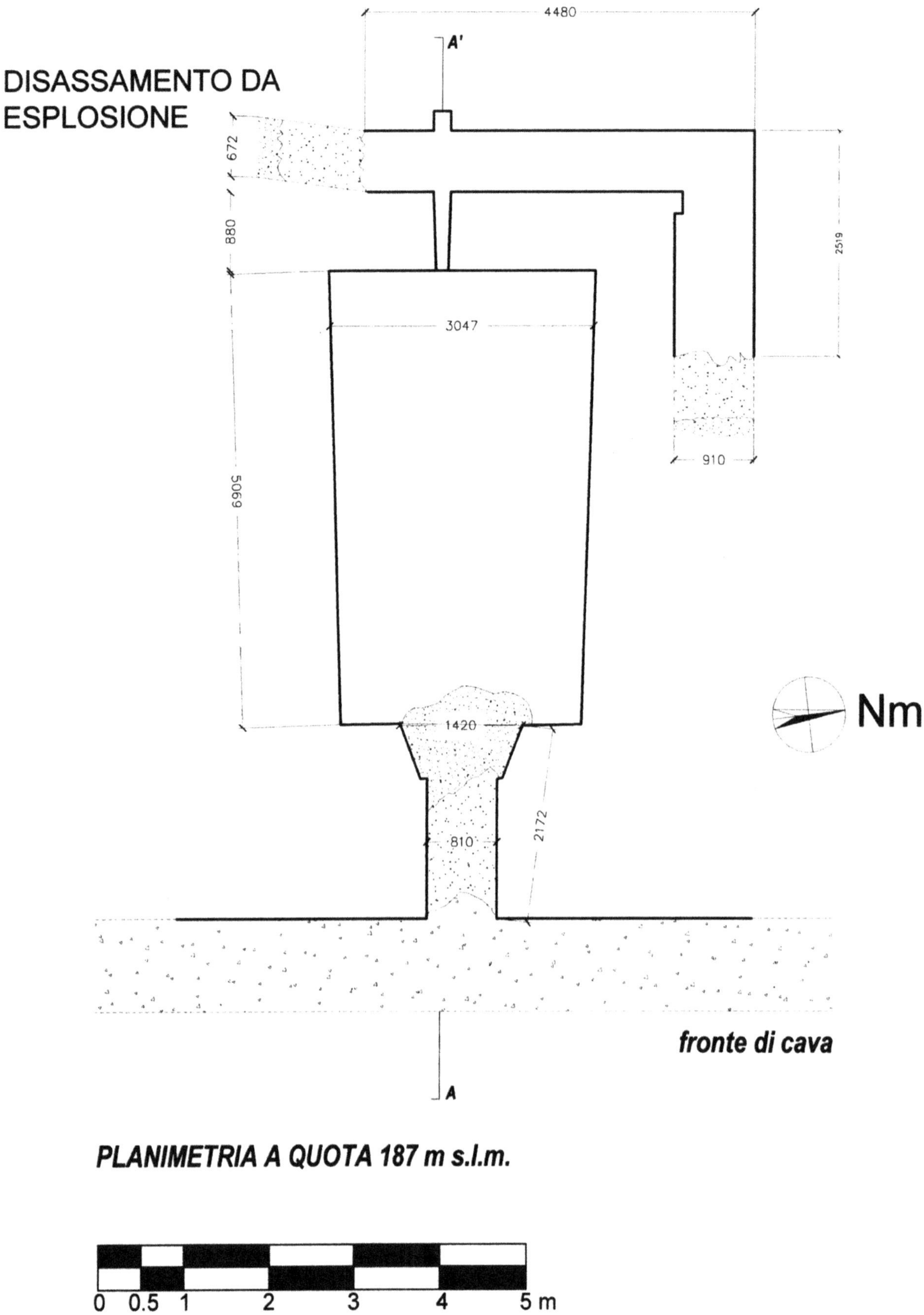

Fig. III.45. *Stanza del Foro (CA 00012 PI TO) and TE.S.E.S. Tunnel (CA 00027 PI TO); plan (Ass.ne S.C.A.M. Archive).*

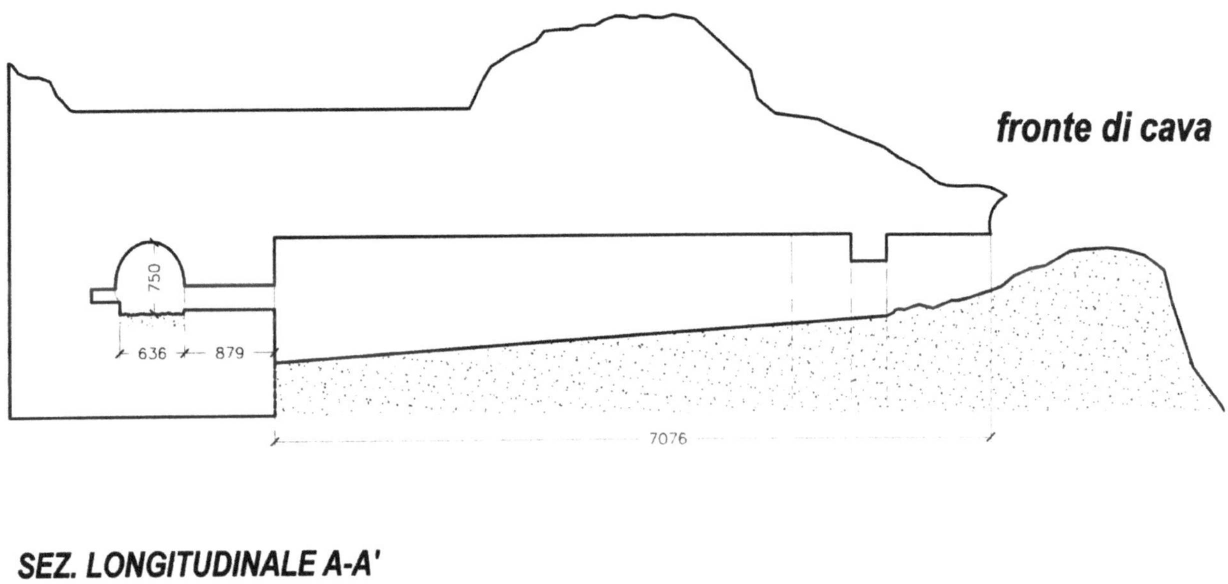

Fig. III.46. *Stanza del Foro (CA 00012 PI TO) and TE.S.E.S. Tunnel (CA 00027 PI TO); section (Ass.ne S.C.A.M. Archive).*

Fig. III.46.a. *Stanza del Foro (CA 00012 PI TO) (photo G. Padovan).*

Removing quarry mud from the room would shed further light on the surviving structure and reveal any communicating underground environments.
Bibliography: none.
Data ownership: Associazione Speleologia Cavità Artificiali Milano (S.C.A.M.).
Compiled by: Gianluca Padovan (Ass.ne S.C.A.M.).

III.3.20 - *CA 00013 PI TO; First Tunnel Section*

Cadastral number: CA 00013 PI TO
Denomination: First Tunnel Section (fig. III.47)
Region-country: Piedmont, Italy
Province: Turin
Municipality: Verrua Savoia
Locality: Verrua Hill
Location: situated along the edge of the current quarry, by which it has been intercepted
Ownership: private
Cartography: CTR 1:10.000
Geological unit: marls, sandstones and Pliocene limestone resting on an Eocene Clay Shale Formation
Altitude: //
Position: //
Context: within the stronghold, in the Borgo Alto area; situated in a hill depression, the tunnel section and the room to which it connects are on the hypothetical, yet plausible, original site of the Celestino Tunnel (CA 00011 PI TO)
Operations conducted: exploration, planimetric survey, photographic service
Work carried out by: Associazione Speleologia Cavità Artificiali Milano (S.C.A.M.); 2003
Warnings: possible roof collapse
Typology: 6- communication tunnel with lateral chamber
Description: the structure consists of a room, which has undergone at least one modification and which communicates with what remains of a tunnel of approximately 2 m in height. This was part of the same underground system as the Celestino Tunnel (CA 00011 PI TO). The entrance is on the quarry face and the walls were evidently destroyed by quarry expansion. The tunnel is 4.84 m wide, 5.6 m long and was originally more than 2 m high. The masonry structure was built within a trench or hillside terrace and then covered, making it underground and safe from shellfire. The tunnel is almost entirely full of soil, debris and quarry mud. It is thought that a short corridor, once led from the tunnel walkway to the room, which would have had at least one air shaft and a further continuation, which was subsequently blocked. An archway can be seen emerging from the debris on the left hand side, however it has not yet been established whether this is an arch discharge or the entrance archway to a further room.
Interpretation: every aspect of the tunnel is the same as in the Celestino Tunnel, suggesting that they belong to the same complex.
Dating: the structure should date back to the fortification's restoration following the 1625 siege. More precisely, given its position, it may date back to the third quarter of the XVII century and the construction of the triple bastion stronghold.
Notes: the room probably communicated with another environment or even a stairwell or incline providing access to the surface. Clearing the quarry face would uncover the tunnel section which conceals the room and would shed light on its function within the underground complex. Clearing the easily restored environment would confirm the existence of adjacent tunnels or underground rooms.
Bibliography: none.
Data ownership: Associazione Speleologia Cavità Artificiali Milano (S.C.A.M.).
Compiled by: Gianluca Padovan (Ass.ne S.C.A.M.).

III.3.21 - *CA 00014 PI TO; Second Tunnel Section*

Cadastral number: CA 00014 PI TO
Denomination: Second Tunnel Section (fig. III.48)
Region-country: Piedmont, Italy
Province: Turin
Municipality: Verrua Savoia
Locality: Verrua Hill
Location: situated along the edge of the current quarry, by which it is intercepted
Ownership: private

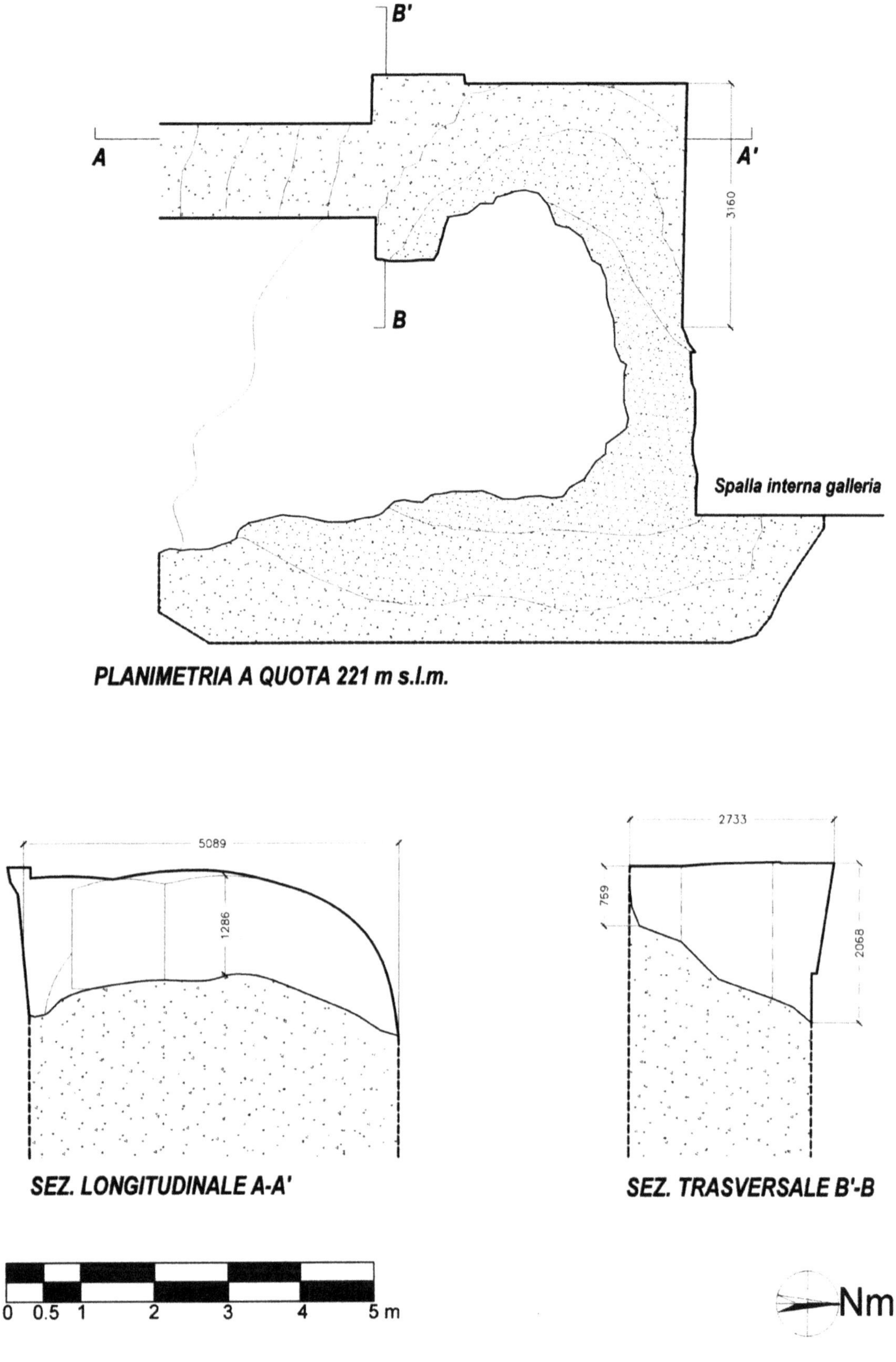

Fig. III.47. *First Tunnel Section (CA 00013 PI TO) (Ass.ne S.C.A.M. Archive).*

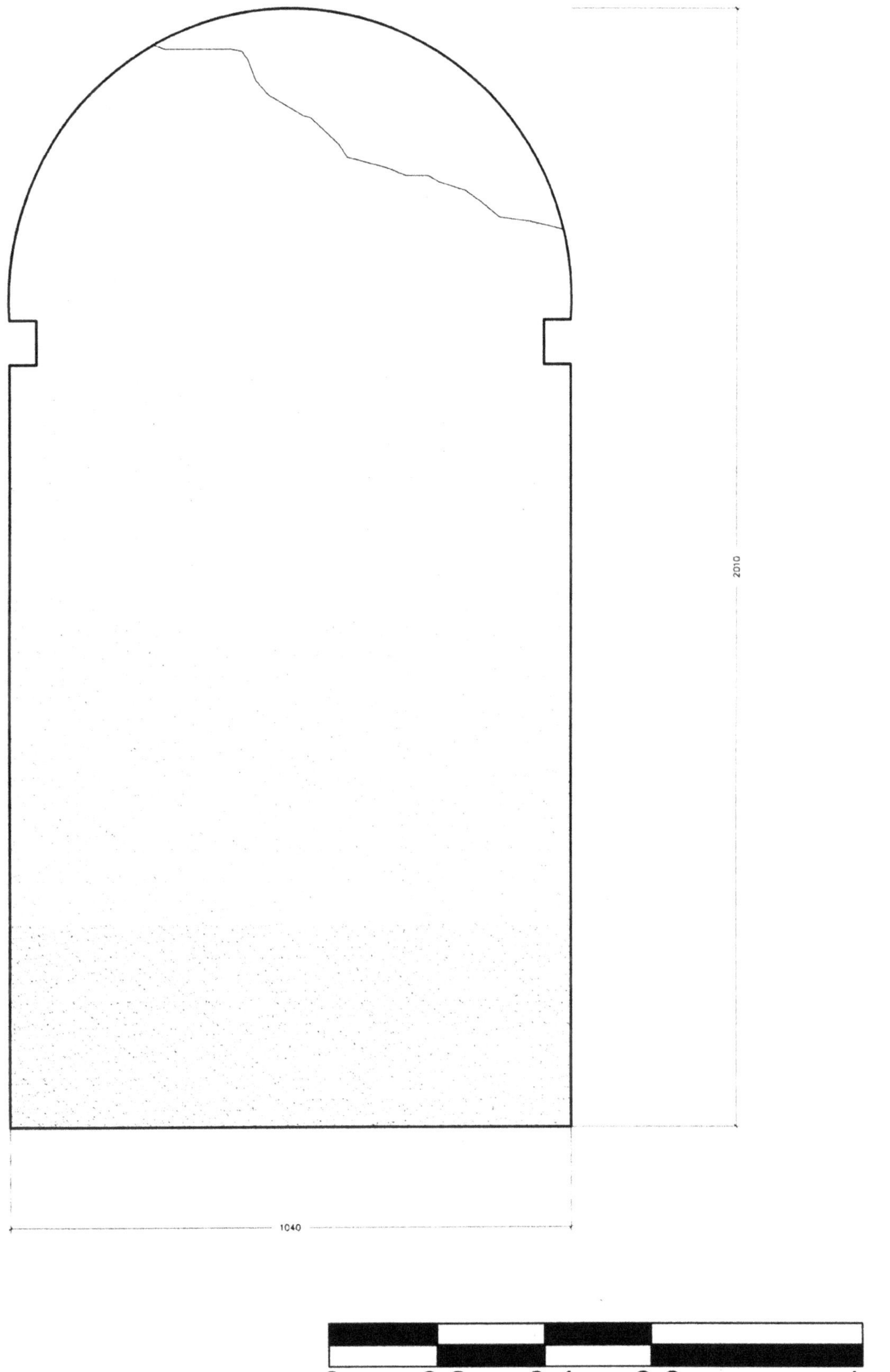

Fig. III.48. *Second Tunnel Section (CA 00014 PI TO) (Ass.ne S.C.A.M. Archive).*

Cartography: CTR 1:10.000
Geological unit: marls, sandstones and Pliocene limestone resting on an Eocene Clay Shale Formation
Altitude: //
Position: //
Context: located the structure is within the stronghold, in the Borgo Alto area; situated in a hill depression, the tunnel section is on the hypothetical, yet plausible, original site of the Celestino Tunnel (CA 00011 PI TO)
Operations conducted: exploration, planimetric survey, photographic service
Work carried out by: Associazione Speleologia Cavità Artificiali Milano (S.C.A.M.); 2003
Warnings: possible roof collapse
Typology: 6 - communication tunnel
Description: the entrance is on the quarry face and the tunnel has been interrupted by quarry expansion. The structure is at least ten metres long, is just over a metre wide and is at least 2 m high; the masonry structure was built within a trench or a hillside terrace and then covered to make it underground and safe from shellfire. It appears to be exactly the same as the Celestino Tunnel, even in its dimensions and was probably a continuation of said tunnel.
Interpretation: this section of tunnel was uncovered during quarry works and was almost completely buried. It is probably part of the same underground system as the Celestino Tunnel.
Dating: the structure appears date back to the restoration of fortification following the 1625 siege and more precisely, given its position, to the third quarter of the XVII century and the construction of the triple bastion stronghold.
Note: clearing the quarry face would reveal the tunnel section in question and its clearance would permit its full investigation and confirm the existence of any communicating environments.
Bibliography: none.
Data ownership: Associazione Speleologia Cavità Artificiali Milano (S.C.A.M.).
Compiled by: Gianluca Padovan (Ass.ne S.C.A.M.).

III.3.22 - *CA 00015 PI TO; Chlorine Cistern*

Cadastral number: CA 00015 PI TO
Denomination: Chlorine Cistern (figs. III.49, III.50 and III.51)
Region-country: Piedmont, Italy
Province: Turin
Municipality: Verrua Savoia
Locality: Verrua Hill
Location: situated within Borgo Alto
Ownership: private
Cartography: CTR 1:10.000
Geological unit: marls, sandstones and Pliocene limestone resting on an Eocene Clay Shale Formation
Altitude: (ellipsoid) 275.7239 m a.s.l.
Position: (topographic GPS) 45°10'24.24189" N, 8°06'07.76795" E
Context: situated on the southeast confines of Borgo Alto, the cistern partially englobes the Filled Well (CA 00016 PI TO); it was originally located within a building, which was destroyed by quarry works
Operations conducted: exploration, planimetric survey, photographic service
Work carried out by: Associazione Speleologia Cavità Artificiali Milano (S.C.A.M.) and Gruppo Grotte C.A.I. Novara; 2003
Warnings: poor air circulation and there are carcasses of small animals in the water
Typology: 2c - cistern
Description: consisting of a brick shaft, surrounded by the masonry remains of the building, the cistern gives access to a circular chamber which in turn contains a well shaft. The current entrance consists of a quadrangular breach in the wall, which perforates the shaft and allows access to the storage chamber. The breach was recently closed by a welded metal door. The shaft is an overall 3.31 m with internal diameter of 0.87 m, while the storage chamber is currently 6.35 m deep (in respect of the surface) with 4.21 m diameter. The storage chamber is brick-lined and has four protruding water inlet conduits.
Interpretation: the cistern was probably used to collect meteoric water; it would have exclusively served the building within which it was contained.
Dating: predates the Filled Well encompassed within the cistern.
Notes: given the vastness of the overall site, there would have been numerous water supply structures. Early XX century photographs attest that there were once three buildings in this area, buildings which were demolished by the quarry.
Bibliography: none.

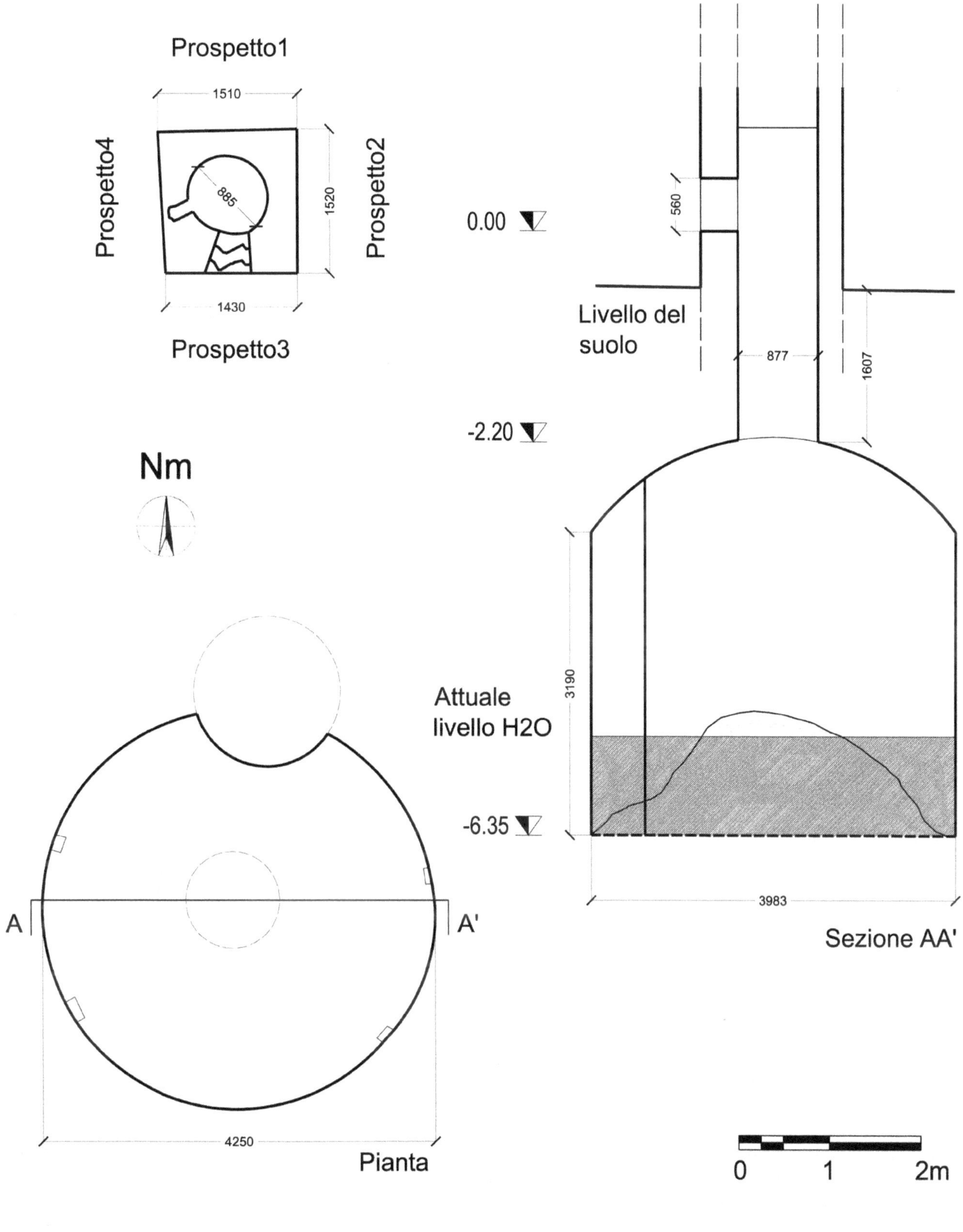

Fig. III.49. *Chlorine Cistern (CA 000015 PI TO) (Ass.ne S.C.A.M. Archive).*

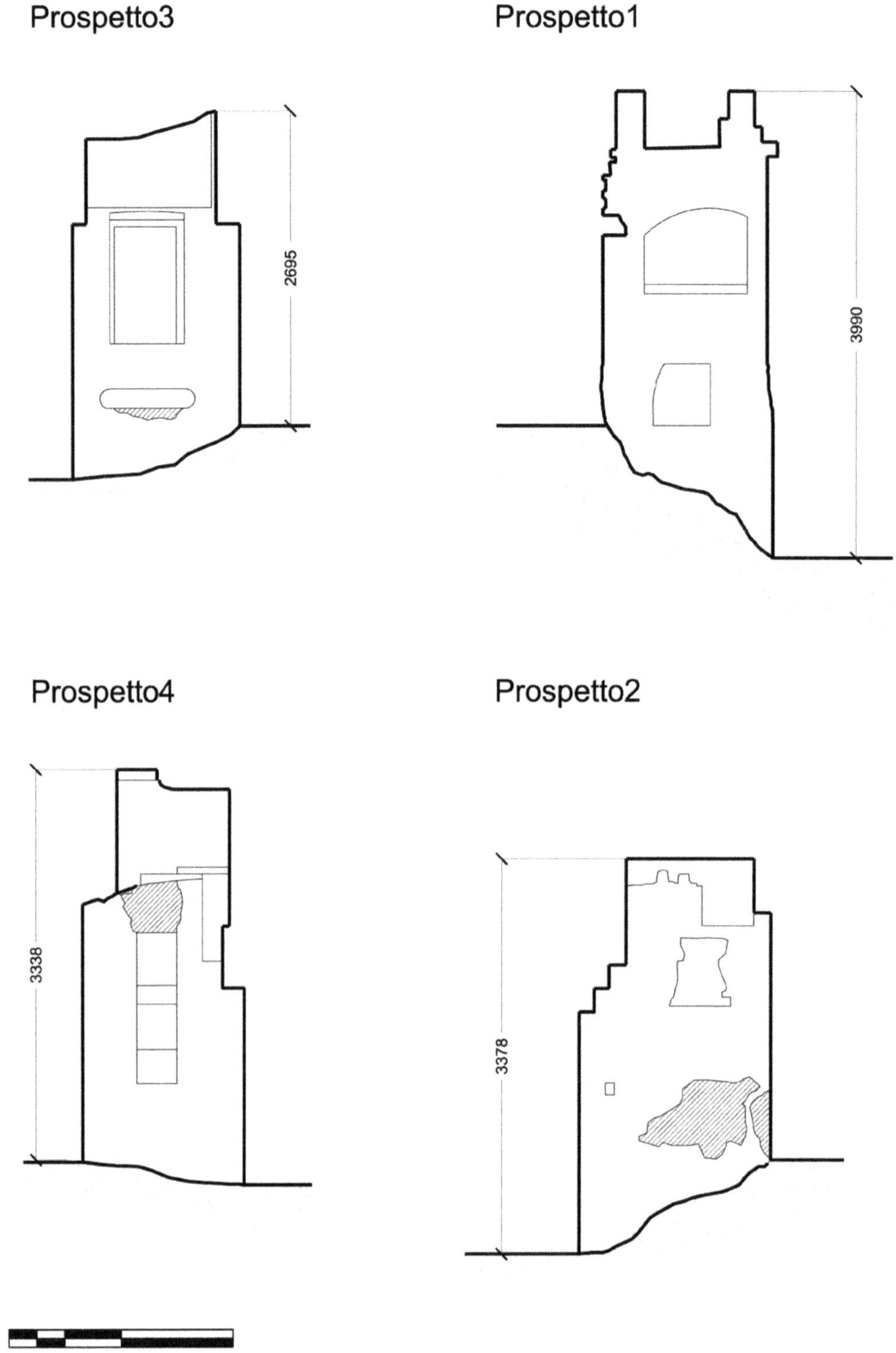

Fig. III.50. *Chlorine Cistern (CA 000015 PI TO) (Ass.ne S.C.A.M. Archive).*

Fig. III.51. *Entrance to the Chlorine Cistern (CA 00015 PI TO) (photo G. Padovan).*

Data ownership: Associazione Speleologia Cavità Artificiali Milano (S.C.A.M.).
Compiled by: Gianluca Padovan (Ass.ne S.C.A.M.).

III.3.23 - *CA 00016 PI TO; Filled Well*

Cadastral number: CA 00016 PI TO
Denomination: Filled Well (fig. III.52)
Region-country: Piedmont, Italy
Province: Turin
Municipality: Verrua Savoia
Locality: Verrua Hill
Location: situated within the Borgo Alto area
Ownership: private
Cartography: CTR 1:10.000
Geological unit: marls, sandstones and Pliocene limestone resting on an Eocene Clay Shale Formation
Altitude: (ellipsoid) 275.7239 m a.s.l.
Position: (topographic GPS) 45°10'24.24189" N, 8°06'07.76795" E
Context: situated at the southwest confines of Borgo Alto, the structure is partially contained within the Chlorine Cistern (CA 00015 PI TO)
Operations conducted: photographic service
Work carried out by: Associazione Speleologia Cavità Artificiali Milano (S.C.A.M.); 2003
Warnings: possible structural collapse
Typology: 7
Description: vertical perforation in the ground. The structure has a circular, brick opening and an external section of the shaft is visible from the Chlorine Cistern storage room (CA 00015 PI TO). The presence of soil prevents its full exploration.
Interpretation: it can be assumed that the structure is an ordinary, albeit cylindrical well without excluding the possibility of there being lateral branches for improved drainage. It could also be the entrance shaft to an underlying storage chamber or the ventilation shaft of an underground transit or storage system for war materials.
Dating: predates the Chlorine Cistern.
Notes: the stratigraphic excavation of this area of Borgo Alto, spared by quarry extraction, is recommended. Given the vastness of the site, there would once have been numerous water supply structures. Although not recorded in the old maps and consequently unknown, it has now been established that Verrua Fortress had underground transit systems.
Bibliography: none.
Data ownership: Associazione Speleologia Cavità Artificiali Milano (S.C.A.M.).
Compiled by: Gianluca Padovan (Ass.ne S.C.A.M.).

III.3.24 - *CA 00017 PI TO; First Countermine Tunnel*

Cadastral number: CA 00017 PI TO
Denomination: First Countermine Tunnel (figs. III.53.a, III.53.b, III.54, III.54.a, III.54.b, III.55, III.55.a and III.55.b)
Region-country: Piedmont, Italy
Province: Turin
Municipality: Verrua Savoia
Locality: Verrua Hill
Location: situated outwith the west re-entrant flank of the Fortress
Ownership: private
Cartography: CTR 1:10.000
Geological unit: marls, sandstones and Pliocene limestone resting on an Eocene Clay Shale Formation
Altitude: 214 m a.s.l.
Position: (GPS investigation) east 1429488, west 5002496
Context: currently extends between the Major's Garden Bastion and the area where the Third Village once stood
Operations conducted: exploration, planimetric survey, photographic service
Work carried out by: Associazione Speleologia Cavità Artificiali Milano (S.C.A.M.); 2003
Warnings: extremely poor air circulation and possible structural collapse
Typology: 6 - countermine tunnel

Fig. III.52. *Filled Well. The entrance was uncovered by clandestine digs (CA 00016 PI TO) (photo G. Padovan).*

Fig. III.53.a. *Internal section of the First Countermine Tunnel (CA 00017 PI TO) (photo G. Padovan).*

Fig. III.53.b. *Section of the internal room of the First Countermine Tunnel (CA 00017 PI TO) (photo G. Padovan).*

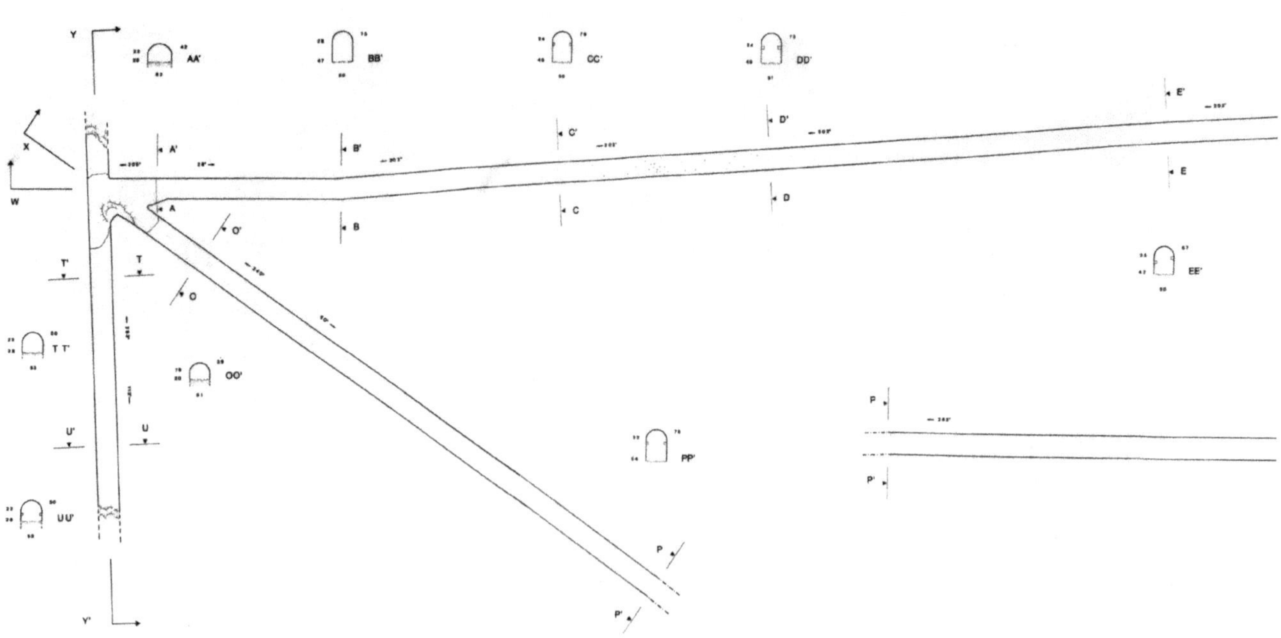

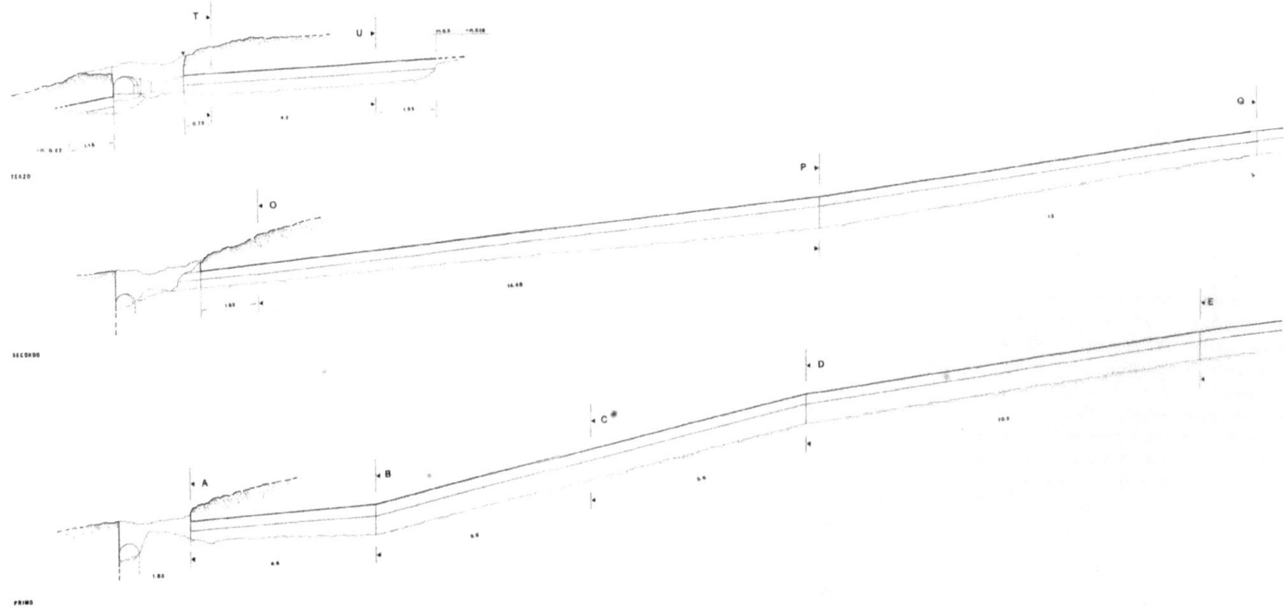

Fig. III.54. *Countermine Tunnel System plans (CA 00017 PI TO; CA 00018 PI TO; CA 00019 PI TO; CA 00035 PI TO; CA 00036 PI TO) (Ass.ne S.C.A.M. Archive).*

Fig. III.55. *Countermine Tunnel System section (CA 00017 PI TO; CA 00018 PI TO; CA 00019 PI TO; CA 00035 PI TO; CA 00036 PI TO) (Ass.ne S.C.A.M. Archive).*

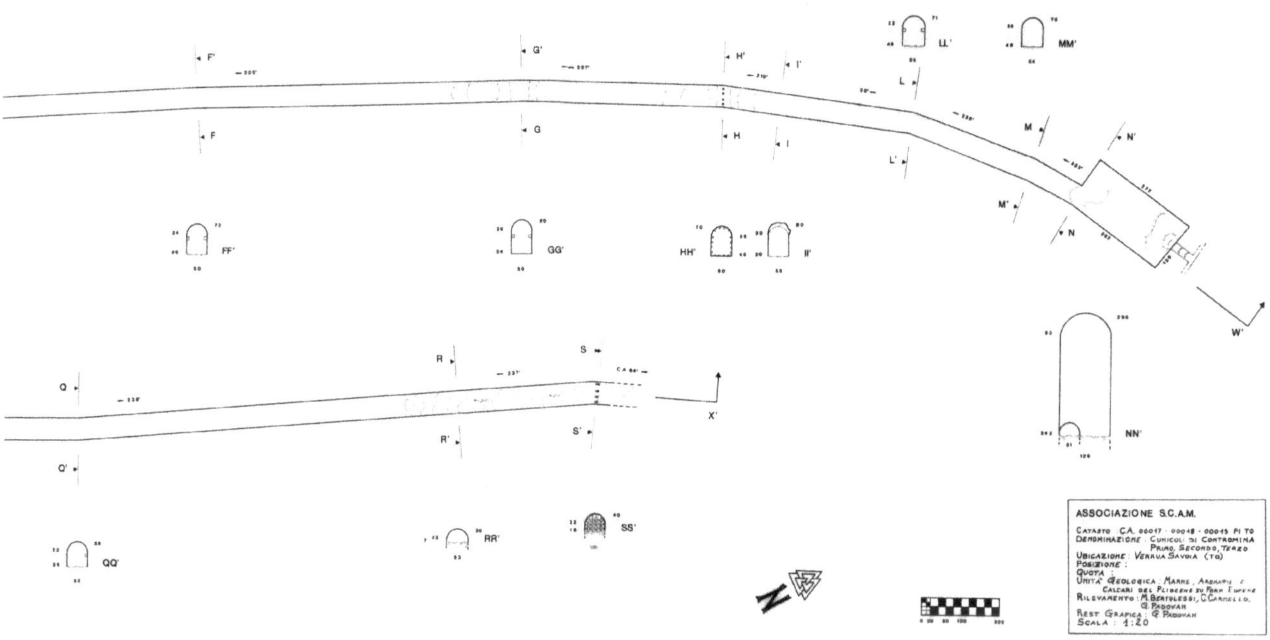
FF'
GG'
HH'
II'
LL'
MM'
NN'
QQ'
RR'
SS'
W'
X'
ASSOCIAZIONE S.C.A.M.
CATASTO : C.A. 00017 - 00048 - 00045 PI TO
DENOMINAZIONE : CUNICOLI DI CONTROMINA PRIMO, SECONDO, TERZO
UBICAZIONE : VERRUA SAVOIA (TO)
POSIZIONE :
QUOTA :
RILEVAMENTO : M. BERTOLESSI, C. CARMELLO, G. PADOVAN
REST. GRAFICA : G. PADOVAN
SCALA : 1:20

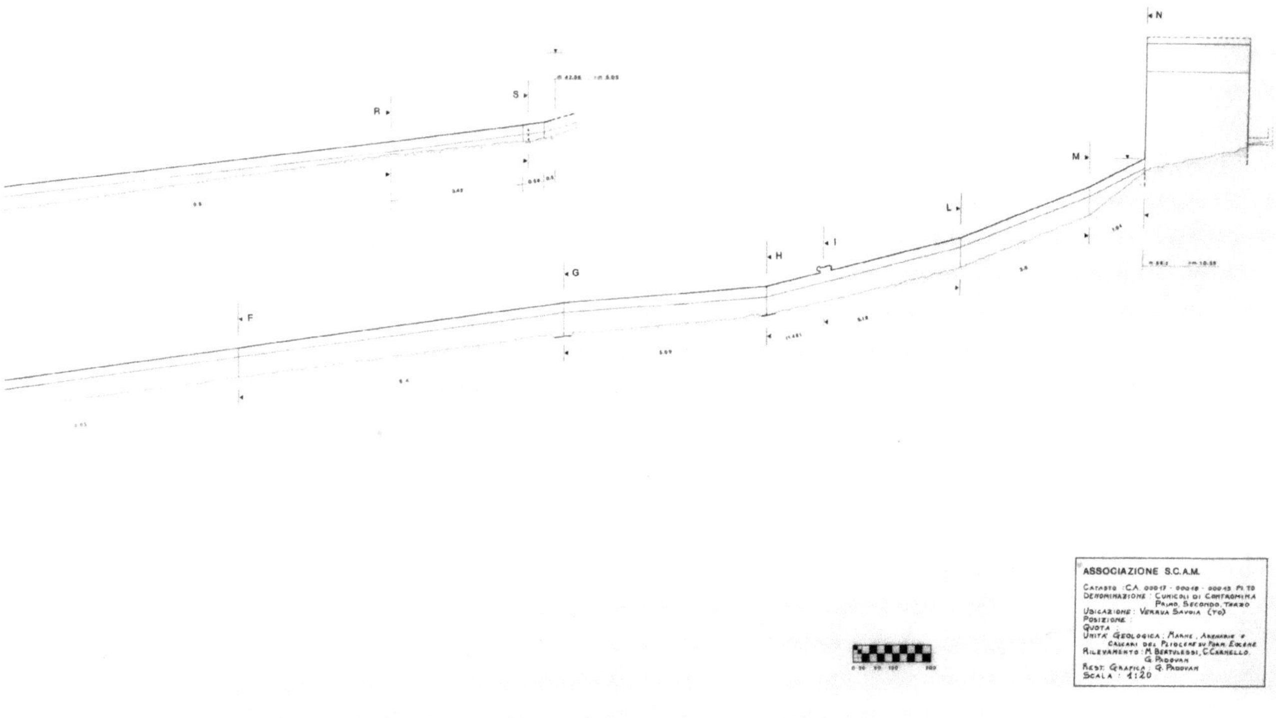
ASSOCIAZIONE S.C.A.M.
DENOMINAZIONE : CUNICOLI DI CONTROMINA PRIMO, SECONDO, TERZO
UBICAZIONE : VERRUA SAVOIA (TO)
POSIZIONE :
QUOTA
RILEVAMENTO : M. BERTOLESSI, C. CARMELLO, G. PADOVAN
REST. GRAFICA : G. PADOVAN
SCALA : 1:20

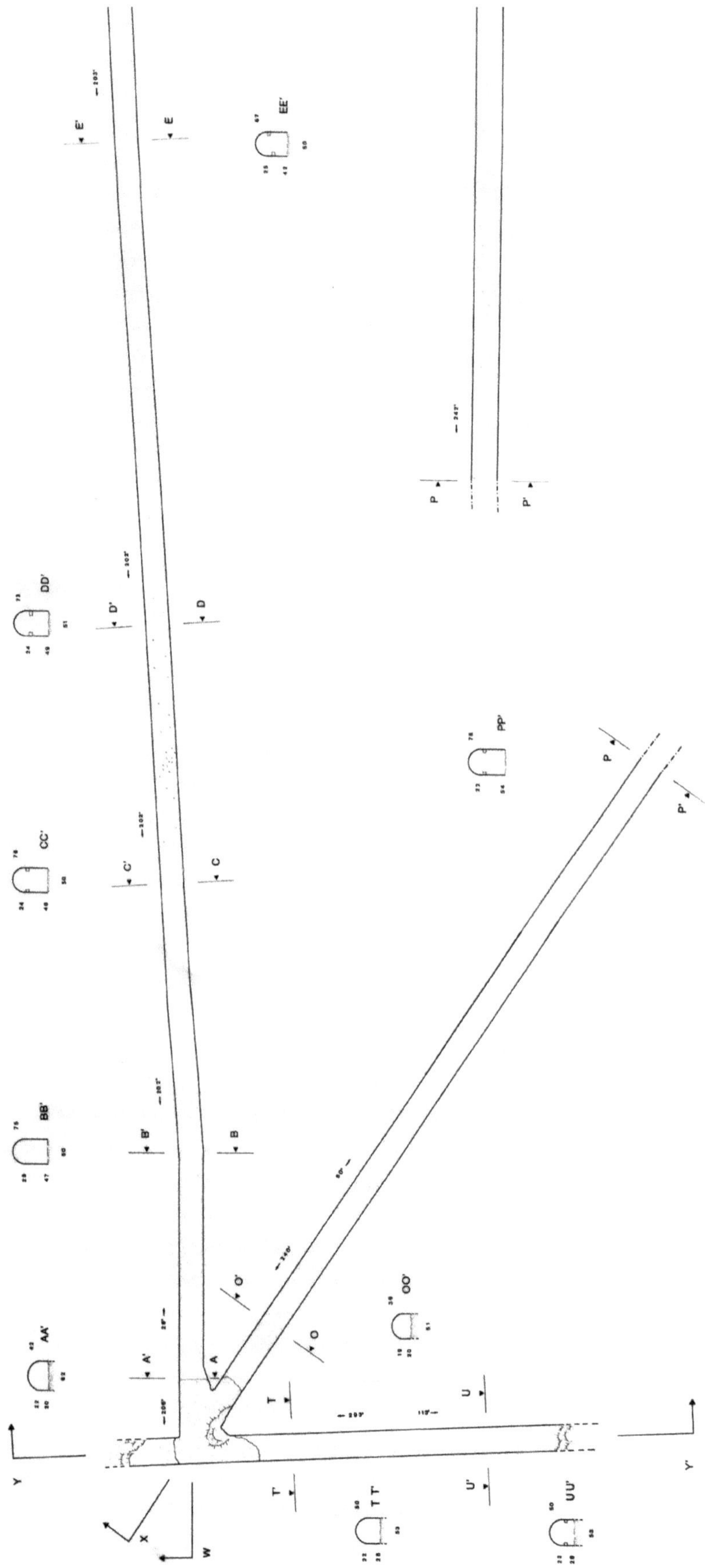

Fig. III.54.a. *For a better view: Countermine Tunnel System plans, southern part.*

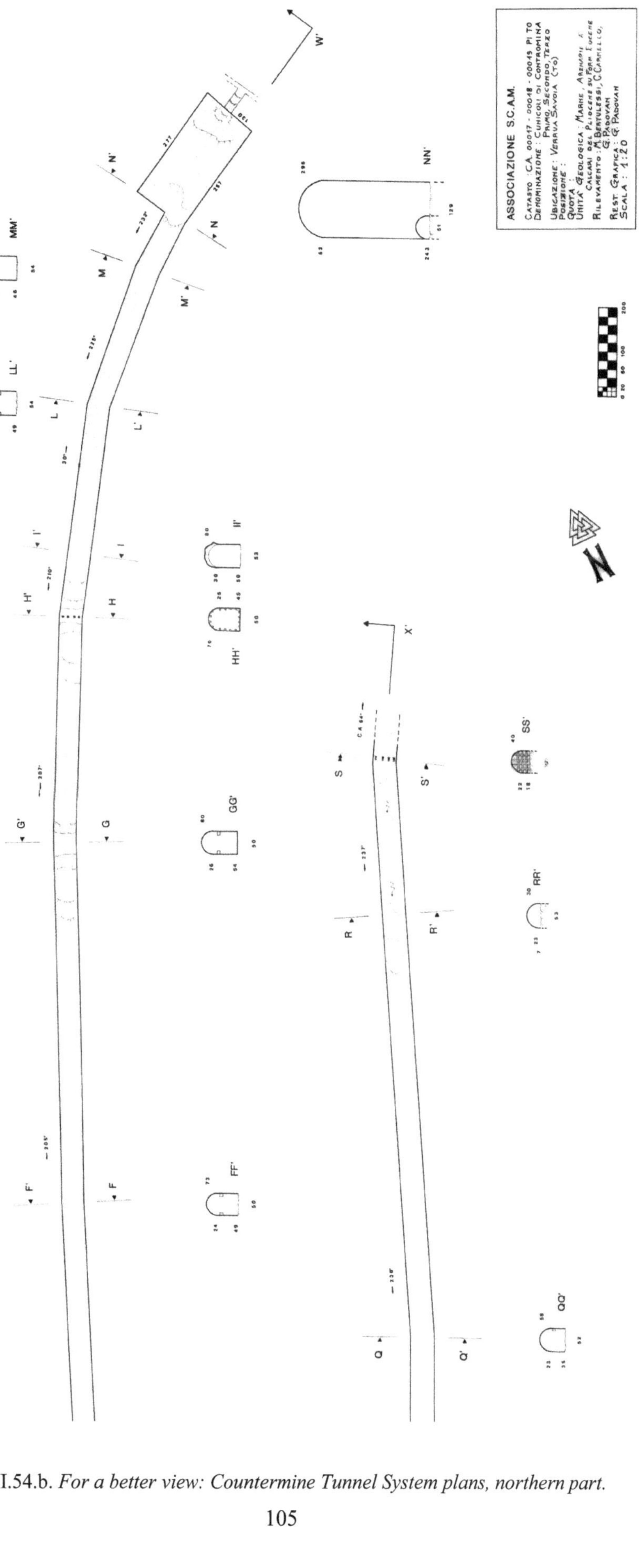

Fig. III.54.b. *For a better view: Countermine Tunnel System plans, northern part.*

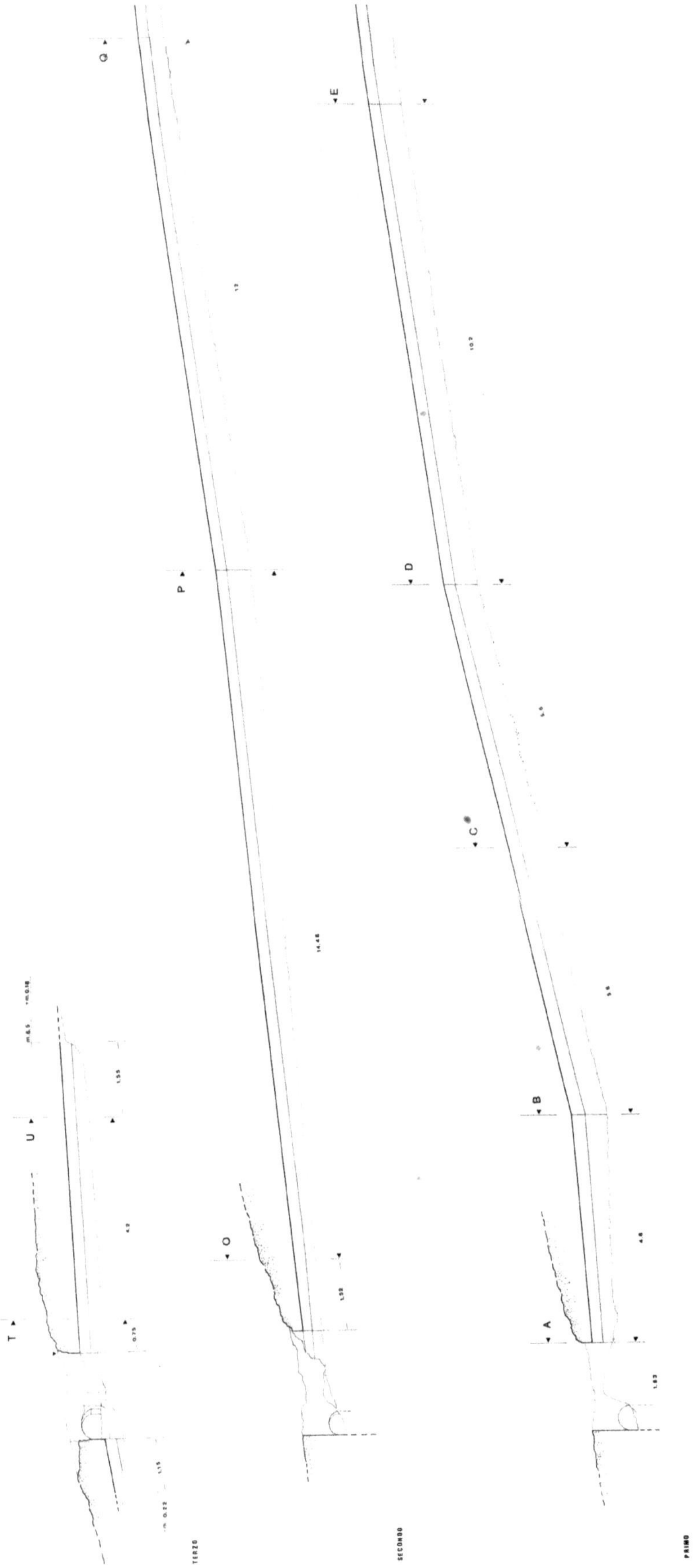

Fig. III.55.a. *For a better view: Countermine Tunnel System section, southern part.*

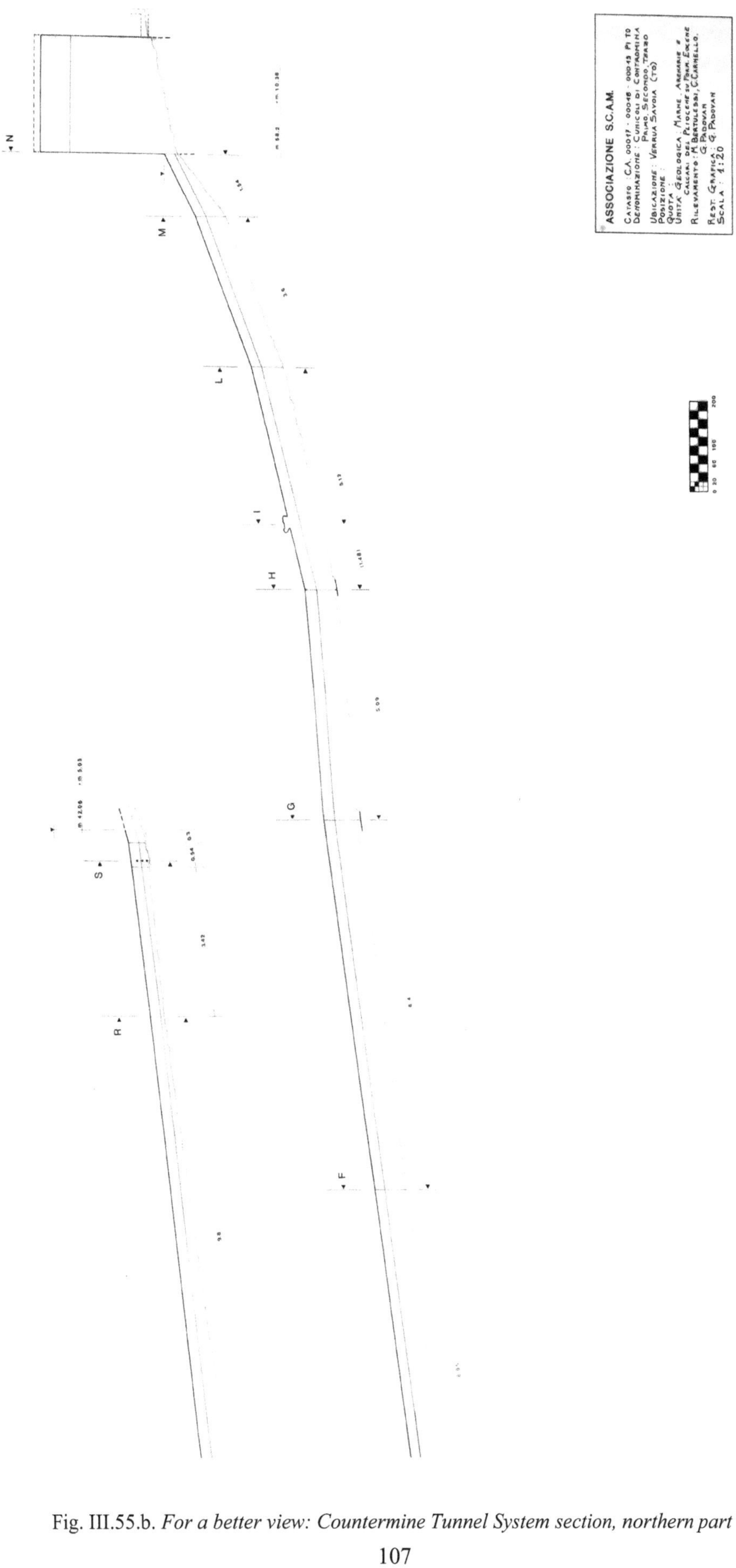

Fig. III.55.b. *For a better view: Countermine Tunnel System section, northern part*

Description: the entrance is situated in the Small Quadrivium Room (CA 00035 PI TO), next the south-eastern edge of the plain directly below the west re-entrant flank of the Fortress. The original entrance has not been located. It consists of a 0.62 m wide tunnel (a few tens of centimetres further on it is 0.5 m wide and 0.42 m high at the soil level) and a trapezoidal room. The overall length of the tunnel is 58.2 m and the distance between the lowest and highest point of the tunnel is +10.38 m, although it should be taken into account that in the absence of the soil filling, the distance would be greater. Excluding a short initial section, one side of which vaguely curves towards the Small Quadrivium Room, its width varies between 0.5 m and 0.54 m, while its height in the soil-free section free is 0.8 m. The structure is subterranean with exposed brick roof and walls and its base is covered by sediment and detritus: in one section, it is evident that the base is flat, made of bricks and covered by a 1 cm layer of whitish hydraulic mortar. Both the tunnel and the terminal room have barrel vaults. The bricks used to provide roof support protrude at regular intervals along the tunnel walls, just under the springer. There are no such bricks within the room. Positioned in a 26° direction, the tunnel proceeds in almost rectilinear fashion almost to its termination and then curves in a 52° direction. At 47.94 m from the entrance, the structure was once closed by an iron grille (now missing), which was inserted after the tunnel was built, through a breach in the wall and roof. The roof shows evident signs of a rough brick and pebble overlay covered with fairly strong grey hydraulic mortar. The vaguely trapezoidal room has a high ceiling and its lower fill presents a hole made at the same time as the structure itself and which appears to lead to an air space; the room measures 1.29 x 2.77 x 1.3 x 2.57 m and at its highest point and at any rate at the fill, it reaches a height of 2.96 m. The wall facing appears to have been rebuilt and there are signs of at least two subsequent building phases, indicating that the original roof or part of the northwest wall may have caved in. In all probability, an area leading to the original entrance is situated under the soil fill. Dripping water containing silt emerges from the opening; small formations on the roof and floor merge along the tunnel section nearest the room.
Interpretation: this is a perfect example of a countermine tunnel, which is probably part of the system used during the 1625 siege.
Dating: this structure predates the building phase of the second and third quarters of the XVII century. It could date back to the period of time between the end of XVI century and the first quarter of the XV century.
Notes: The later installation of the grille could indicate that the final Fortress design did not envisage the use of the previous system, but that the system was in any case retained to cover all eventualities. In any case, its continuation is to be sought within the room. In all probability, the tunnel was built within a trench and then covered in the same way as several other countermine tunnels in Turin. It would seem that the Small Quadrivium Room was uncovered in the mid 1990s when excavators eradicated the remaining tree stumps from the coppice, to be used as a vineyard. Shortly afterwards the grille was sawn and removed by hands unknown.
Bibliography: Padovan D., Padovan G., Bordignon L., Ottino M. 1997, *La Fortezza di Verrua Savoia*, in Club Alpinistico Triestino - Gruppo Grotte Sezione Ricerche e Studi su Cavità Artificiali, *Atti del IV Convegno Nazionale sulle Cavità Artificiali – Osoppo 30/31 Mggio - 1 Giugno 1997*, Trieste 1997, pgs. 187-208.
Padovan G., *Verrua Savoia: indagini sotterranee*, in Associazione Culturale "Amici della Biblioteca" di Crescentino, *Atti del Convegno Storico "Terre sul Po dal Medioevo alla Resistenza" – Crescentino 2-3 Ottobre 1988*, Crescentino 2002, pgs. 213-243.
Data ownership: Associazione Speleologia Cavità Artificiali Milano (S.C.A.M.).
Compiled by: Gianluca Padovan (Ass.ne S.C.A.M.).

III.3.25 - *CA 00018 PI TO; Second Countermine Tunnel*

Cadastral number: CA 00018 PI TO
Denomination: Second Countermine Tunnel (figs. III.54, III.54.a, III.54.b, III.55, III.55.a and III.55.b)
Region-country: Piedmont, Italy
Province: Turin
Municipality: Verrua Savoia
Locality: Verrua Hill
Location: situated outwith the west re-entrant flank of the Fortress
Ownership: private
Cartography: CTR 1:10.000
Geological unit: marls, sandstones and Pliocene limestone resting on an Eocene Clay Shale Formation
Altitude: 214 m a.s.l.
Position: (GPS investigation) east 1429488, west 5002496
Context: currently extends between the fausse-braye, opposite the curtain located between the Major's Garden Bastion and the Santa Maria Bastion and the area the Third Village once was
Operations conducted: exploration, planimetric survey, photographic service
Work carried out by: Associazione Speleologia Cavità Artificiali Milano (S.C.A.M.); 2003

Warnings: extremely poor air circulation and possible structural collapse
Typology: 6 - countermine tunnel
Description: the entrance is situated in the Small Quadrivium Room (CA 00035 PI TO), next the south-eastern edge of the plain below the west re-entrant flank of the Fortress; the location of the original entrance is unknown. It consists of a tunnel, 41.22 m of which is accessible. Thereafter, significant limestone formations envelop a grille, after which the tunnel continues for a further few metres; the distance between the lowest and highest point of the tunnel is +5 m. The width varies between 0.51 m and 0.54 m, while the height, taking the soil fill into account, does not exceed 0.76 m. At the point where the grille is positioned, the tunnel dimensions are 0.52 x 0.4 m. This is a subterranean structure with exposed brick roof and walls and a base covered by sediment and detritus; just over 4 m from the grille, the base is covered by a thick, whiter than white calcite formations. The tunnel has a barrel vault. The bricks used to provide roof support protrude at regular intervals just under the springer. The first tunnel section points in a 60° direction. It then turns a few degrees north and after the grille, curves again to 64°. Further along it turns sharply east. An almost constant rivulet of water runs through the tunnel.
Interpretation: this is a perfect example of a countermine tunnel, which is probably part of the system used during the 1625 siege.
Dating: this structure predates the building phase of the second and third quarters of the XVII century. It could date back to the period of time between the end of XVI century and the first quarter of the XV century.
Notes: in all probability, the tunnel was built within a trench and then covered in the same way as several other countermine tunnels in Turin. It would seem that the Small Quadrivium Room was uncovered in the mid 1990s when excavators eradicated the remaining tree stumps from the coppice, to be used as a vineyard. The excavators again intervened in 1996 and eliminated a small cement tank containing remnants of verdigris as well as removing wall remains and intercepting the Third Countermine Tunnel (CA 00019 PI TO), approximately 20 m of which was practicable at the time.
Bibliography: Padovan D., Padovan G., Bordignon L., Ottino M. 1997, *La Fortezza di Verrua Savoia*, in Club Alpinistico Triestino - Gruppo Grotte Sezione Ricerche e Studi su Cavità Artificiali, *Atti del IV Convegno Nazionale sulle Cavità Artificiali – Osoppo 30/31 Mggio - 1 Giugno 1997*, Trieste 1997, pgs. 187-208.
Padovan G., *Verrua Savoia: indagini sotterranee*, in Associazione Culturale "Amici della Biblioteca" di Crescentino, *Atti del Convegno Storico "Terre sul Po dal Medioevo alla Resistenza" – Crescentino 2-3 Ottobre 1988*, Crescentino 2002, pgs. 213-243.
Data ownership: Associazione Speleologia Cavità Artificiali Milano (S.C.A.M.).
Compiled by: Gianluca Padovan (Ass.ne S.C.A.M.).

III.3.26 - *CA 00019 PI TO; Third Countermine Tunnel*

Cadastral number: CA 00019 PI TO
Denomination: Third Countermine Tunnel (figs. III.54, III.54.a, III.54.b, III.55, III.55.a, III.55.b and III.56)
Region-country: Piedmont, Italy
Province: Turin
Municipality: Verrua Savoia
Locality: Verrua Hill
Location: situated outwith the west re-entrant flank of the Fortress
Ownership: private
Cartography: CTR 1:10.000
Geological unit: marls, sandstones and Pliocene limestone resting on an Eocene Clay Shale Formation
Altitude: 214 m a.s.l.
Position: (GPS investigation) east 1429488, west 5002496
Context: currently extends between the Santa Maria Bastion and the area where the Third Village once was
Operations conducted: exploration, planimetric survey, photographic service
Work carried out by: Associazione Speleologia Cavità Artificiali Milano (S.C.A.M.); 2003
Warnings: poor air circulation and current structural collapse
Typology: 6 - countermine tunnel
Description: the entrance is situated in the Small Quadrivium Room (CA 00035 PI TO), next the south-eastern edge of the plain below the west re-entrant flank of the Fortress; the location of the original entrance is unknown. The entrance is 0.53 m wide and 0.5 m high at the level of the soil build-up; 6.5 m of the tunnel are accessible and it maintains the aforementioned measurements throughout. 25 m of the tunnel were accessible until just a few months before the survey was carried out. The roof and walls of this underground structure are of exposed brick and its base is covered by sediment and detritus; it has a barrel vault. The bricks used to provide roof support protrude at regular intervals just under the springer. The tunnel follows a 113° course.

Fig. III.56. *Internal section of the Third Countermine Tunnel (CA 00019 PI TO) (photo G. Padovan).*

Dating: this structure predates the building phase of the second and third quarters of the XVII century. It could date back to the period of time between the end of XVI century and the first quarter of the XV century.
Notes: the structure would appear to be the same as the First Countermine Tunnel. If this were the case, this structure would also lead to a room, connected to a "capital tunnel". In all probability, the tunnel was built within a trench and then covered in the same way as several other countermine tunnels in Turin. It would seem that the Small Quadrivium Room was uncovered in the mid 1990s when excavators eradicated the remaining tree stumps from the coppice, to be used as a vineyard. The excavators again intervened in 1996 and eliminated a small cement tank containing remnants of verdigris as well as removing wall remains and intercepting this tunnel, approximately 20 m of which was practicable at the time.
Bibliography: Padovan D., Padovan G., Bordignon L., Ottino M. 1997, *La Fortezza di Verrua Savoia*, in Club Alpinistico Triestino - Gruppo Grotte Sezione Ricerche e Studi su Cavità Artificiali, *Atti del IV Convegno Nazionale sulle Cavità Artificiali – Osoppo 30/31 Mggio - 1 Giugno 1997*, Trieste 1997, pgs. 187-208.
Padovan G., *Verrua Savoia: indagini sotterranee*, in Associazione Culturale "Amici della Biblioteca" di Crescentino, *Atti del Convegno Storico "Terre sul Po dal Medioevo alla Resistenza" – Crescentino 2-3 Ottobre 1988*, Crescentino 2002, pgs. 213-243.
Data ownership: Associazione Speleologia Cavità Artificiali Milano (S.C.A.M.).
Compiled by: Gianluca Padovan (Ass.ne S.C.A.M.).

III.3.27 - *CA 00020 PI TO; First Demolition Tunnel*

Cadastral number: CA 00020 PI TO
Denomination: First Demolition Tunnel (fig. III.57)
Region-country: Piedmont, Italy
Province: Turin
Municipality: Verrua Savoia
Locality: Verrua Hill

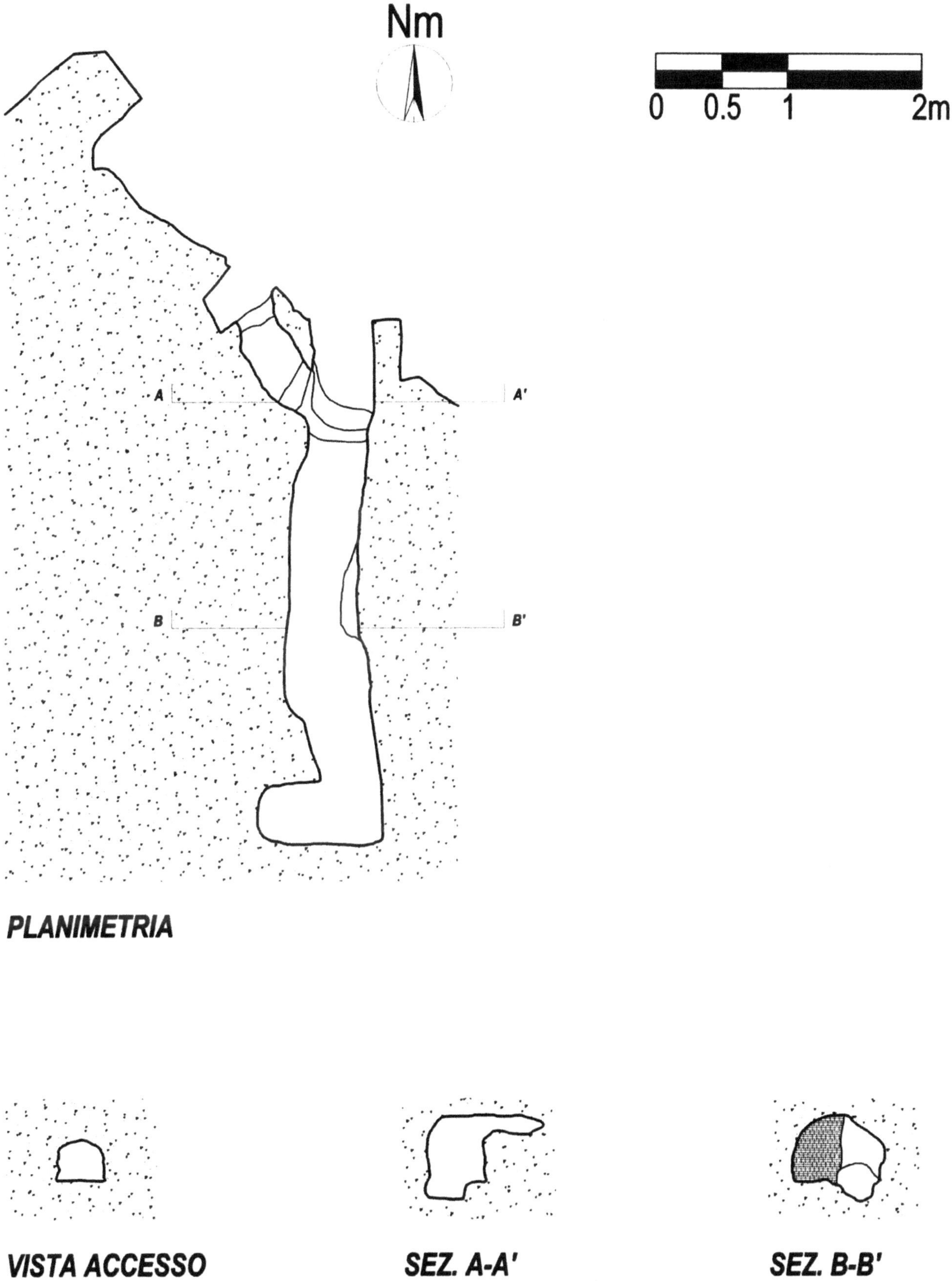

Fig. III.57. *First Demolition Tunnel (CA 00020 PI TO) (Ass.ne S.C.A.M. Archive).*

Location: Sergeants' Bastion
Ownership: private
Cartography: CTR 1:10.000
Geological unit: marls, sandstones and Pliocene limestone resting on an Eocene Clay Shale Formation
Altitude: //
Position: //
Context: built into the north corner of the external façade of the Sergeant's Bastion
Operations conducted: exploration, planimetric survey, photographic service
Work carried out by: Associazione Speleologia Cavità Artificiali Milano (S.C.A.M.); 2003
Warnings: wasps nests at the entrance with possible presence of animals inside
Typology: 6 - countermine tunnel
Description: this consists of a short lateral branch from the surface to the main tunnel and of an end demolition chamber, positioned almost in line with the tunnel's axis. It is situated near the corner of the bastion and was initially accessed through a breach in the wall and subsequently by carving the rock and following the inward-facing breach on the left. The chamber is carved directly in the rock and its base is slightly lower than that of the tunnel. Beyond the wall facing, the short tunnel section on the right is also carved into the rock and converges with the main branch. The bastion's external façade is positioned on a 113°-293° axis, while the tunnel proceeds in a 180° direction and the chamber in a 241° direction. The right entrance has been is no longer accessible while the main 0.54 x 0.44 m entrance can still be accessed. Both entrances show evidence of brick and cement fill, which have been almost completely removed. From the external façade of the bastion to the end of the demolition chamber, including the lateral branch, the tunnel extends for 8.45 m. At 2 m from the entrance, the tunnel is 0.72 x 0.66 m while towards its end it measures 0.68 x 0.72 m.
Interpretation: it is highly unlikely that the structure was excavated by Austrio-Piedmontese troops during the 1704-1705 sieges: providing the French besiegers with ready-made demolition chambers, ready to be charged, makes no sense. The tunnel may have been built by the French soldiers who partially demolished the Tunnel during its French occupation, between 1705 and 1706. However, it is more plausible that it was built later, at the same time as the other demolition tunnels (Second Demolition Tunnel CA 00021 PI TO and Third Demolition Tunnel CA 00022 PI TO). Following the War in the Alps which ended in 1796, with the Savoy being conquered by the French and the signing of the Peace Treaty in Paris in May of the same year, it was decided that several Savoia fortresses, such as Demonte Fort in Valle Stura and San Vittorio fort in Tortona, would be demolished. Another possibility is that the demolition chambers may have been created at this time and never utilised.
Dating: end of the XVIII century or start of the XIX.
Notes: within the structure, the breach in the external wall facing and signs left by the tools used to carve the rock are clearly visible. It should be understood that the bastion was not built but rather carved into the very sandstone rock and then lined. The brick and cement fill at the opposite corner of the Bastion could conceal the entrance to a further demolition tunnel (as yet uncovered).
Bibliography: none.
Data ownership: Associazione Speleologia Cavità Artificiali Milano (S.C.A.M.).
Compiled by: Gianluca Padovan (Ass.ne S.C.A.M.).

III.3.28 - *CA 00021 PI TO; Second Demolition Tunnel*

Cadastral number: CA 00021 PI TO
Denomination: Second Demolition Tunnel (fig. III.58)
Region-country: Piedmont, Italy
Province: Turin
Municipality: Verrua Savoia
Locality: Verrua Hill
Location: Terrace Bastion
Ownership: private
Cartography: CTR 1:10.000
Geological unit: marls, sandstones and Pliocene limestone resting on an Eocene Clay Shale Formation
Altitude: //
Position: //
Context: built into the north east corner of the external façade of the Terrace Bastion.
Operations conducted: exploration, planimetric survey, photographic service
Work carried out by: Associazione Speleologia Cavità Artificiali Milano (S.C.A.M.); 2003
Warnings: possible collapse particularly in the internal section

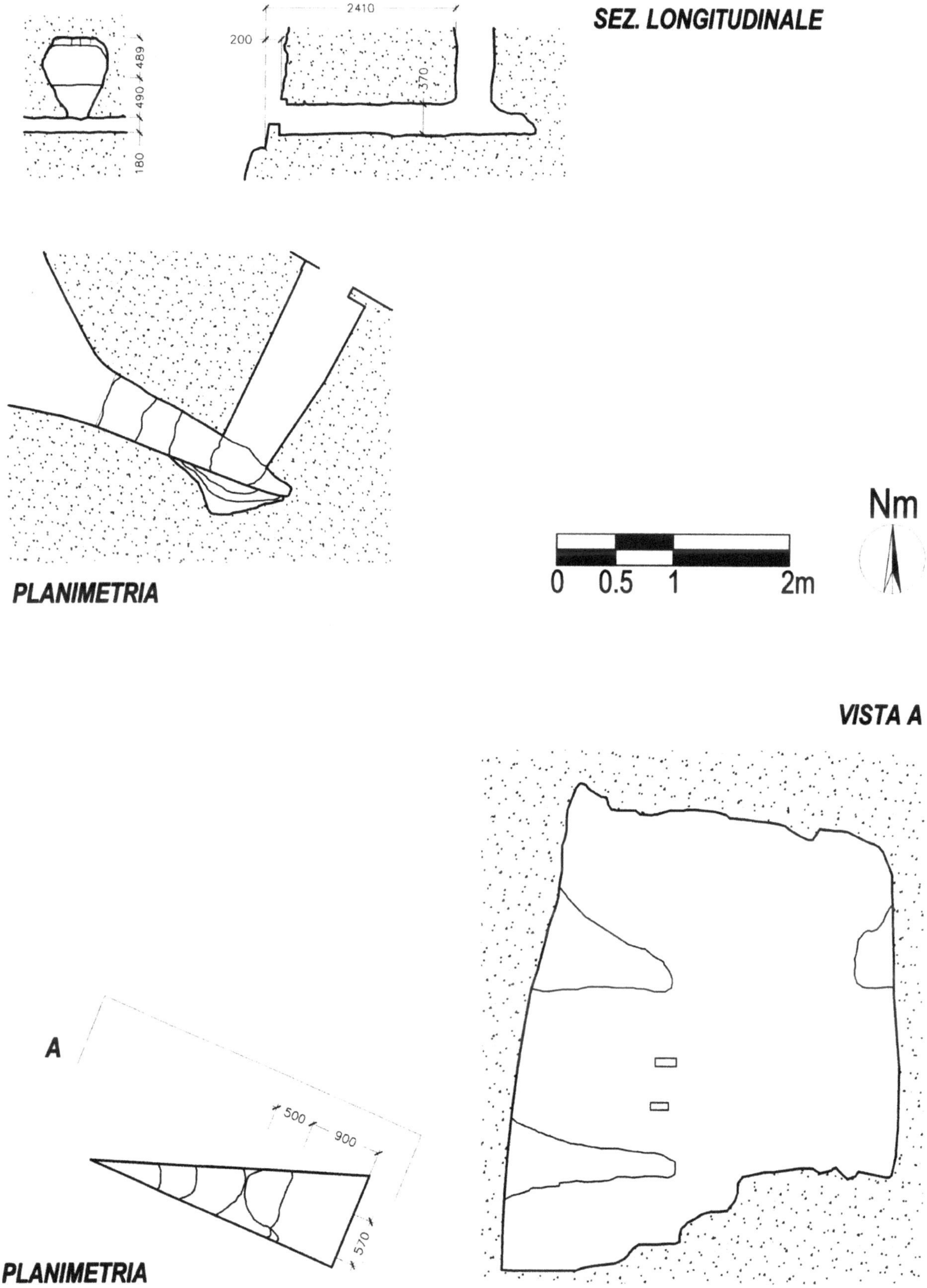

Fig. III.58. *Second Demolition Tunnel (CA 00021 PI TO) (Ass.ne S.C.A.M. Archive).*

Typology: 6 - demolition tunnel
Description: this is an "L-shaped" tunnel, the terminal branch of which would have contained a demolition chamber. It is accessed through a broken wall at 0.4 m above a wall off-set and said entrance measures 0.66 x 0.58 m. It is 6.9 m long and the section measures 0.67 x 0.37 m at the end of the first tunnel section. The structure is excavated through a wall at right angles to the bastion facade. The external coating is 0.3 m thick. There is then a 2.6 m breach and a second brick wall, with a 0.45 m opening. The tunnel extends along the right hand side of the second wall and continues through the breach. The bastion's external façade is positioned on a 112°-292° axis, while the tunnel proceeds in a 26° direction and the chamber in a 150° direction.
Interpretation: it is highly unlikely that the structure was excavated by Austrio-Piedmontese troops during the 1704-1705 sieges: providing the French besiegers with ready-made demolition chambers, ready to be charged, makes no sense. The tunnel may have been built by the French soldiers who partially demolished the Fortress during its French occupation, between 1705 and 1706. However, it is more plausible that it was built later, at the same time as the other demolition tunnels (First Demolition Tunnel CA 00020 PI TO and Third Demolition Tunnel CA 00022 PI TO). Following the War in the Alps which ended in 1796, with the Savoia being conquered by the French and the signing of the Peace Treaty in Paris in May of the same year, it was decided that several Savoia fortresses, such as Demonte Fort in Valle Stura and San Vittorio fort in Tortona, would be demolished. It is also possible that the demolition chambers were created at this time but never utilised.
Dating: end of the XVIII century or start of the XIX.
Notes: a clear view of the bastion can be seen from within the structure, which is not intercepted by the rock. The brick and cement fill along the southwest side, at the external south corner of the Bastion, could conceal the entrance to a further demolition tunnel (as yet uncovered).
Bibliography: none.
Data ownership: Associazione Speleologia Cavità Artificiali Milano (S.C.A.M.).
Compiled by: Gianluca Padovan (Ass.ne S.C.A.M.).

III.3.29 - *CA 00022 PI TO; Third Demolition Tunnel (Gian Domenico Cella Tunnel)*

Cadastral number: CA 00022 PI TO
Denomination: Third Demolition Tunnel (Gian Domenico Cella Tunnel) (fig. III.59)
Region-country: Piedmont, Italy
Province: Turin
Municipality: Verrua Savoia
Locality: Verrua Hill
Location: Terrace Bastion
Ownership: private
Cartography: CTR 1:10.000
Geological unit: marls, sandstones and Pliocene limestone resting on an Eocene Clay Shale Formation
Altitude: //
Position: //
Context: built into the northwest corner of the external façade of the Terrace Bastion.
Operations conducted: exploration, planimetric survey, photographic service
Work carried out by: Associazione Speleologia Cavità Artificiali Milano (S.C.A.M.) and Gruppo Grotte C.A.I. Novara; 2003
Warnings: possible collapse, especially in the internal section
Typology: 6 - demolition tunnel
Description: this is a "Y-shaped" tunnel, the terminal branches of which would have contained demolition chambers. The elliptic entrance measures 0.82 x 0.34 m; it was excavated through a wall and is currently situated just above the soil covering the base of the bastion. The structure extends for a total of 5.64 m and at the end of the first section at 2 m from the entrance; its section measures 0.67 x 0.41 m. The external coating measures 0.35 m. The tunnel then continues through the breach. The bastion's external façade is positioned on a 112°-292° axis, while the tunnel proceeds in a 207° direction and the two branches in a 155° and 244° direction.
Interpretation: it is highly unlikely that the structure was excavated by Austrio-Piedmontese troops during the 1704-1705 sieges: providing the French besiegers with ready-made demolition chambers, ready to be charged, would have made no sense. The tunnel may have been built by the French soldiers who partially demolished the Tunnel during its French occupation, between 1705 and 1706. However, it is more plausible that it was built later, at the same time as the other demolition tunnels (First Demolition Tunnel CA 00020 PI TO and Third Demolition Tunnel CA 00022 PI TO). Following the War in the Alps which ended in 1796, with the Savoia being conquered by the French and the signing of the Peace Treaty in Paris in May of the same year, it was decided that several Savoia fortresses, such as

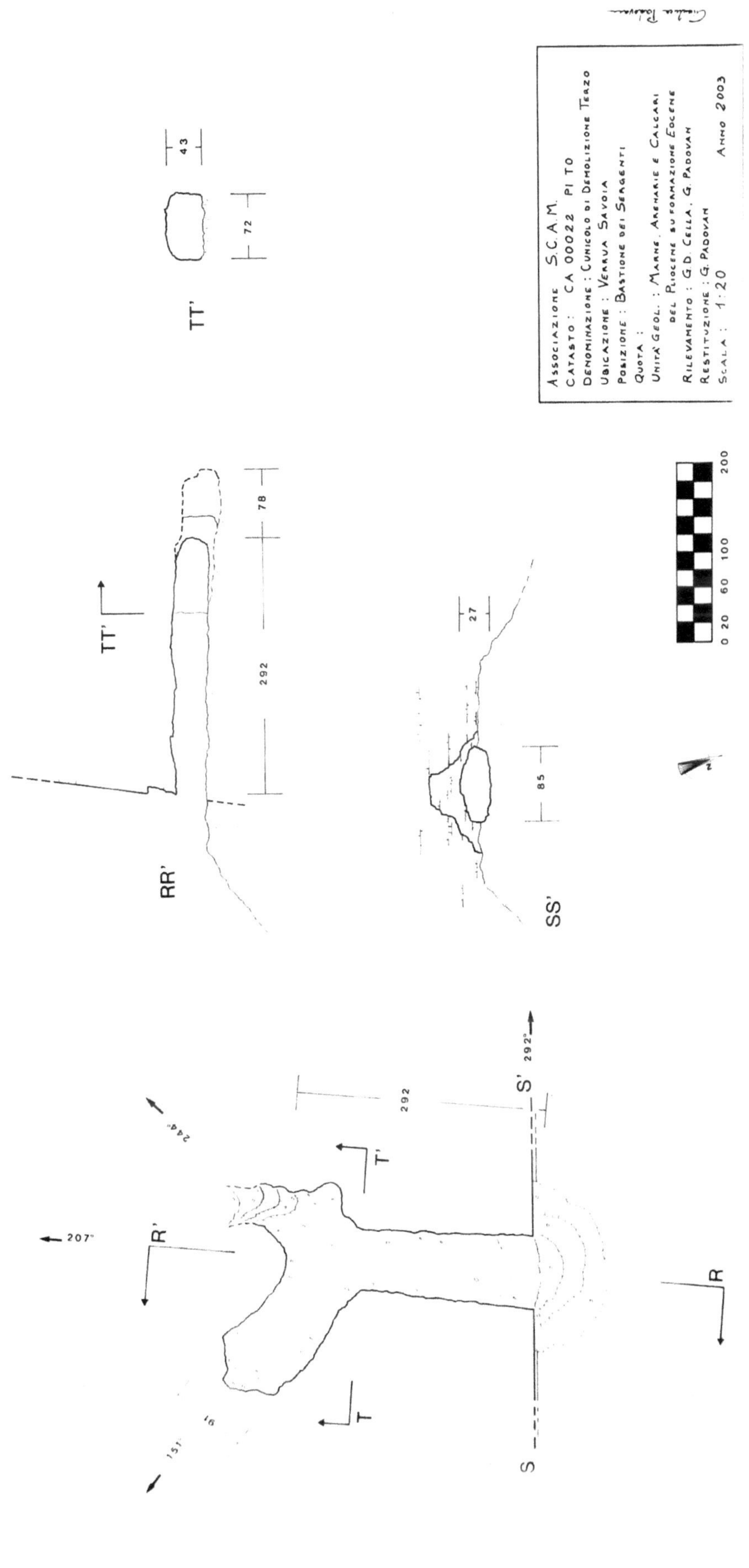

Fig. III.59. *Third Demolition Tunnel (CA 00022 PI TO) (Ass.ne S.C.A.M. Archive).*

Demonte Fort in Valle Stura and San Vittorio fort in Tortona, would be demolished. It is also possible that the demolition chambers were created at this time but never utilised.
Dating: end of the XVIII century or start of the XIX.
Notes: a clear view of the bastion can be seen from within the structure, which is not intercepted by the rock. The brick and cement fill along the east side, at the external south corner of the Bastion, could conceal the entrance to a further demolition tunnel (as yet uncovered).
Bibliography: none.
Data ownership: Associazione Speleologia Cavità Artificiali Milano (S.C.A.M.).
Compiled by: Gianluca Padovan (Ass.ne S.C.A.M.).

III.3.30 - *CA 00023 PI TO; Small Rampart Cistern*

Cadastral number: CA 00023 PI TO
Denomination: Small Rampart Cistern (fig. III.60)
Region-country: Piedmont, Italy
Province: Turin
Municipality: Verrua Savoia
Locality: Verrua Hill
Location: next to a shepherd's track and near a vineyard
Ownership: private
Cartography: CTR 1:10.000
Geological unit: marls, sandstones and Pliocene limestone resting on an Eocene Clay Shale Formation
Altitude: //
Position: //
Context: just outside the stronghold, in front of its south-east side
Operations conducted: exploration, planimetric survey, photographic service
Work carried out by: Federazione Nazionale Cavità Artificiali (F.N.C.A.); 2003
Warnings: none
Typology: 2c - cistern
Description: this is a circular concrete chamber with jack arch. The entrance consists of a 0.75 m masonry frame, just visible on the surface, with a 0.51 m manhole. The diameter of the environment is 2.18 m and it is just over 2 m deep; the base is concealed by a silt deposit.
Interpretation: this is a cistern for the storage of meteoric water, although it cannot be excluded that it may collect water from a small vein.
Dating: of undoubtedly recent construction, it may have been built on a pre-existing cistern.
Notes: the shepherd's track is what remains of a trench dug excavated by Franco-Spanish troops during the 1704-1705 sieges.
Bibliography: none.
Data ownership: Federazione Nazionale Cavità Artificiali Milano (F.N.C.A.).
Compiled by: Gianluca Padovan (Ass.ne S.C.A.M.).

III.3.31 - *CA 00024 PI TO; Po Quarry*

Cadastral number: CA 00024 PI TO
Denomination: Po Quarry (fig. III.61)
Region-country: Piedmont, Italy
Province: Turin
Municipality: Verrua Savoia
Locality: Verrua Hill
Location: situated laterally to the open-pit quarry
Ownership: private
Cartography: CTR 1:10.000
Geological unit: Upper Pliocene Valle Andona Sand
Altitude: 162 m a.s.l.
Position: coordinates Gauss Boaga coordinates 1429405 E; 5002863 N
Context: situated in proximity of the foot of the hill and close to Supply Bridge, the cistern is situated just to the side of the rockfall protection mesh, overlooking Provincial Road 111 to Sulpiano
Operations conducted: exploration, planimetric survey, photographic service

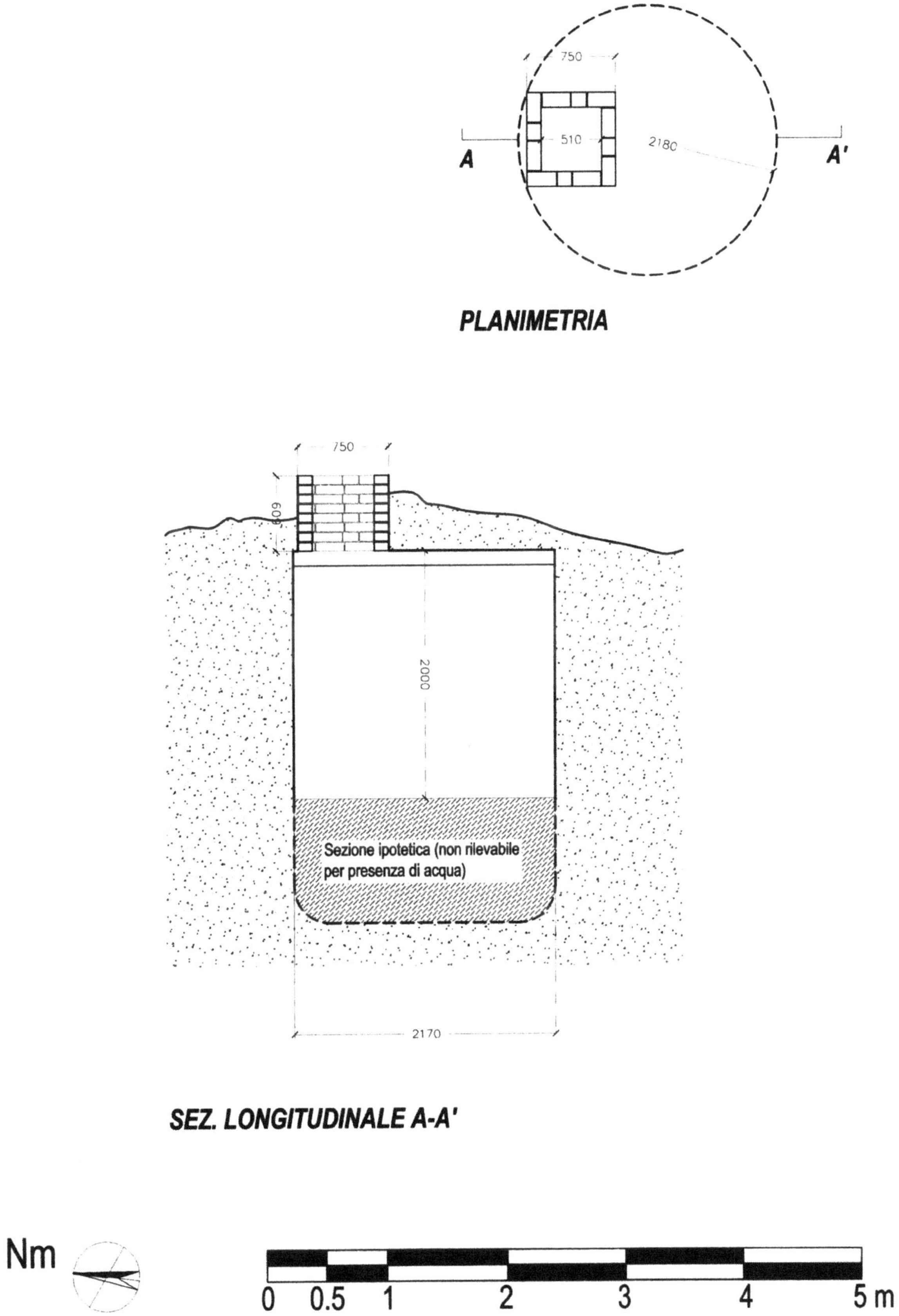

Fig. III.60. *Small Rampart Cistern (CA 000023 PI TO) (Federazione Nazionale Cavità Artificiali Archive).*

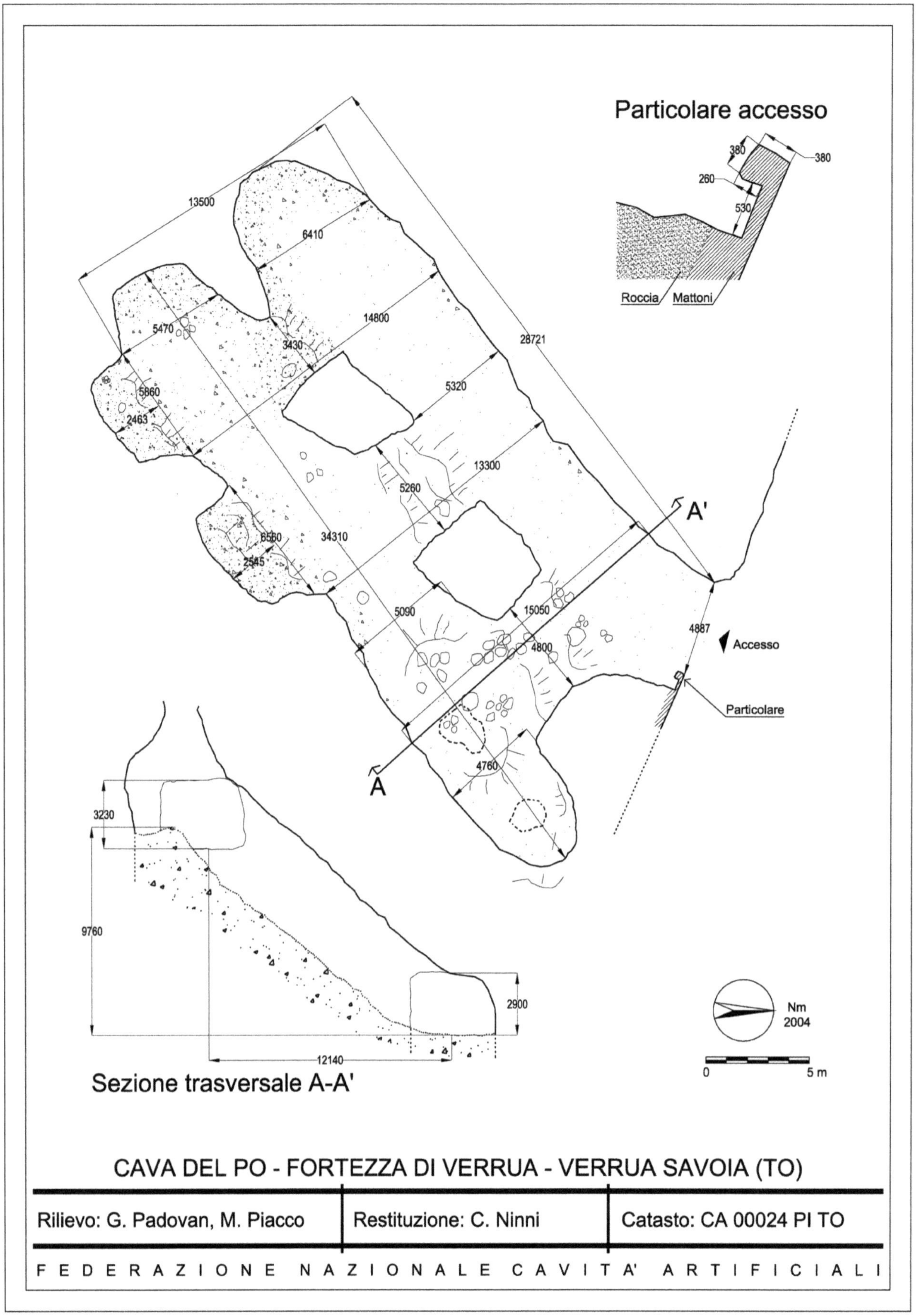

Fig. III.61. *Po Quarry (CA 00024 PI TO) (Federazione Nazionale Cavità Artificiali Archive).*

Work carried out by: Federazione Nazionale Cavità Artificiali (F.N.C.A.) and Gruppo Grotte C.A.I. Novara; 2003
Warnings: it is subject to structural collapse and collapse of material from the walls and vaulted roof
Typology: 1 - quarry
Description: the quarry consists of two level, rectilinear tunnels, which are not perfectly parallel and are situated on different levels; they are linked by short, sloping tunnels at right angles to the main tunnels; the upper one has two casting pits, which communicate with the surface. The quadrangular entrance is carved into the rocky hillside and was once closed by a gate; a large pile of sand fills the room and the remains of a small brick wall and a small wooden door can be seen at the right hand side of the room. The extraction base is full of fallen boulders. It has about 10 m of height-difference; the tunnels are on average 3-4 m wide and 3.5 m high. Numerous fossils can be seen on the roof and walls, primarily oysters, pectinids and lignitized vegetal remains; the signs left by drill-rods and excavation equipment can be seen on various parts of the walls.
Interpretation: multiple pillar extraction method was carried out here on a sloping surface, probably in order to follow the layer or useful material or to create lower panels. Following its abandonment, possibly following the 1957 cave-in, it was used as a garage for quarry trucks.
Dating: dating to approximately the 1950s, this pre-dates the 1957 collapse.
Notes: given its positioning close to the Supply Bridge and its precarious internal stability, it is recommended that the quarry be completely filled to prevent its structure from collapsing and compromising not only the Supply Bridge but also the underlying road and presenting a danger to road-users.
Bibliography: none.
Data ownership: Federazione Nazionale Cavità Artificiali (F.N.C.A.).
Compiled by: Gianluca Padovan (Ass.ne S.C.A.M.).

III.3.32 - *CA 00025 PI TO; Wasps' Static Tank*

Cadastral number: CA 00025 PI TO
Denomination: Wasps' Static Tank (fig. III.62)
Region-country: Piedmont, Italy
Province: Turin
Municipality: Verrua Savoia
Locality: Verrua Hill
Location: west curtain of the "bomb proof" Barracks
Ownership: private
Cartography: CTR 1:10.000
Geological unit: marls, sandstones and Pliocene limestone resting on an Eocene Clay Shale Formation
Altitude: (ellipsoid) 301.0408 m a.s.l.
Position: (topographic GPS) 45°10'27.64094" N, 8°06'02.00614" E
Context: within the Donjon structure, upper area
Operations conducted: exploration, planimetric survey, photographic service
Work carried out by: Associazione Speleologia Cavità Artificiali Milano (S.C.A.M.); 2003
Warnings: wasps' nests
Typology: 2d - static tank
Description: this underground structure, carved into sandstone has a brick drop arch. The roof in the north corner has collapsed. This was probably the site of an earthenware pipe that was later widened. In the opposite south angle, is a square brick shaft, covered by a fairly crude, thick sandstone cover. A layer of soil covers the sandstone floor. The excavation is regular and the walls show vague signs left by the excavation equipment. Access is currently gained through what appears to be a collapsed roof, while a small shaft with sandstone cover survives at the opposite corner. The environment is 2.1 x 1.5 x 1.4 x 1.6 m with a maximum height of 1.65 m.
Interpretation: this is a static tank for the collection of waste water from a "bomb-proof" Barracks toilet.
Dating: the tank predates the upper floor of the "bomb-proof" Barracks.
Notes: an earthenware pipe runs through the wall of the "bomb-proof" Barracks directly into the septic tank. Space for the pipe was created by breaking though the wall. After laying the pipe, the gap was quickly sealed with hydraulic mortar and fragments of stone. It should not be excluded that there may originally have been other similar environments, now destroyed or eliminated by building transformation or collapse.
Bibliography: none.
Data ownership: Associazione Speleologia Cavità Artificiali Milano (S.C.A.M.).
Compiled by: Gianluca Padovan (Ass.ne S.C.A.M.).

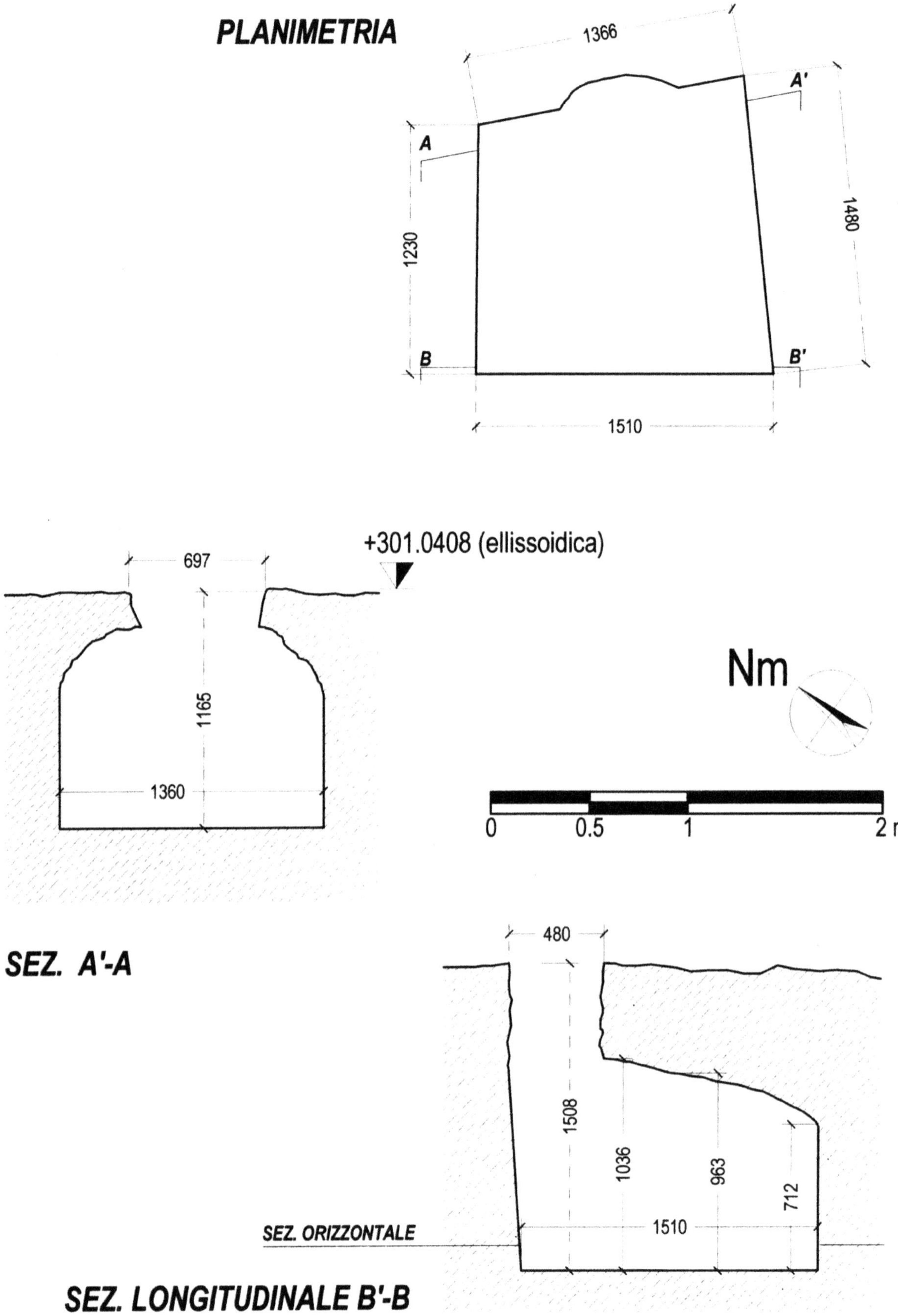

Fig. III.62. *Wasps' Static Tank (CA 00025 PI TO) (Ass.ne S.C.A.M. Archive).*

III.3.33 - *CA 00026 PI TO; Stairway Static Tank*

Cadastral number: CA 00026 PI TO
Denomination: Stairway Static Tank (fig. III.63.a and III.63.b)
Region-country: Piedmont, Italy
Province: Turin
Municipality: Verrua Savoia
Locality: Verrua Hill
Location: stairwell to the "Bomb-proof" Barracks
Ownership: private
Cartography: CTR 1:10.000
Geological unit: marls, sandstones and Pliocene limestone resting on an Eocene Clay Shale Formation
Altitude: (ellipsoid) 300.048 m a.s.l.
Position: UTM 429348.2008 E; 5002701.2139 N
Context: connected to the stairwell from the square behind the "bomb proof" Barracks to the top of the Donjon
Operations conducted: exploration, planimetric survey, photographic service
Work carried out by: Associazione Speleologia Cavità Artificiali Milano (S.C.A.M.); 2003
Warnings: poor air circulation
Typology: 2d - static tank
Description: the trapezoidal chamber is almost completely filled by soil and detritus. The access is a semi-circular opening, consisting of bricks placed in the stairwell access balcony floor, next to the wall which separates the building from the adjacent "bomb-proof" Barracks. It measures 1.79 x 0.82 x 1.87 x 1.32 and its current internal depth is of 1 m. Its brick vault is arched, suggestive of the rampart and the walls are also made of brick while the fill consists of regular rows of pebbles and broken rocks staggered by rows of bricks. The soil build-up probably conceals the piping as none is visible; the visible walls show no signs of hydraulic mortar.
Interpretation: it is probable that this was the static tank of an overlying "communal area", that is a troop latrine. This "communal area" is currently bricked-up and is almost completely filled by detritus. A few metres away is the last, second floor room of the "bomb-proof" Barracks, used as a toilet, with an underlying collection system, perfectly recognisable until 1996 when the room was literally demolished by unknown looters in search of unlikely treasure.
Dating: built at the same time as the Donjon service stairs and thus dates back to the third and fourth quarters of the XVII century.
Notes: as there is no sewage system, the Stronghold's "communal areas" would all have followed the same structure. If not for the presence of piping for inorganic waste and the fact that the walls were water-proofed, this could well have been something other than a static tank. Clearing of the room as well as the restoration of the overlying building would shed light on how toilets of that period were structured.
Bibliography: none.
Data ownership: Associazione Speleologia Cavità Artificiali Milano (S.C.A.M.).
Compiled by: Gianluca Padovan (Ass.ne S.C.A.M.).

III.3.34 - *CA 00027 PI TO; TE.S.E.S. Tunnel*

Cadastral number: CA 00027 PI TO
Denomination: TE.S.E.S. Tunnel (TE.S.E.S.: Experimental Underground Exploration Team) (figs III.45, III.46, III.64 and III.65)
Region-country: Piedmont, Italy
Province: Turin
Municipality: Verrua Savoia
Locality: Verrua Hill
Location: situated at the edge of the current quarry, by which it has now been intercepted
Ownership: private
Cartography: CTR 1:10.000
Geological unit: marls, sandstones and Pliocene limestone resting on an Eocene Clay Shale Formation
Altitude: 167 m a.l.m. (corresponding to the nearest entrance to the Room of the Opening, CA 00012 PI TO)
Position: (GPS investigation) east 1429481, north 5002835 (relative to the nearby entrance to the Room of the Opening)
Context: within the stronghold, situated in the barracks area of the Supply Gate, in the north-eastern section between the Gate and the enceinte wall; accessed via an open-air masonry corridor
Operations conducted: exploration, planimetric survey, photographic service

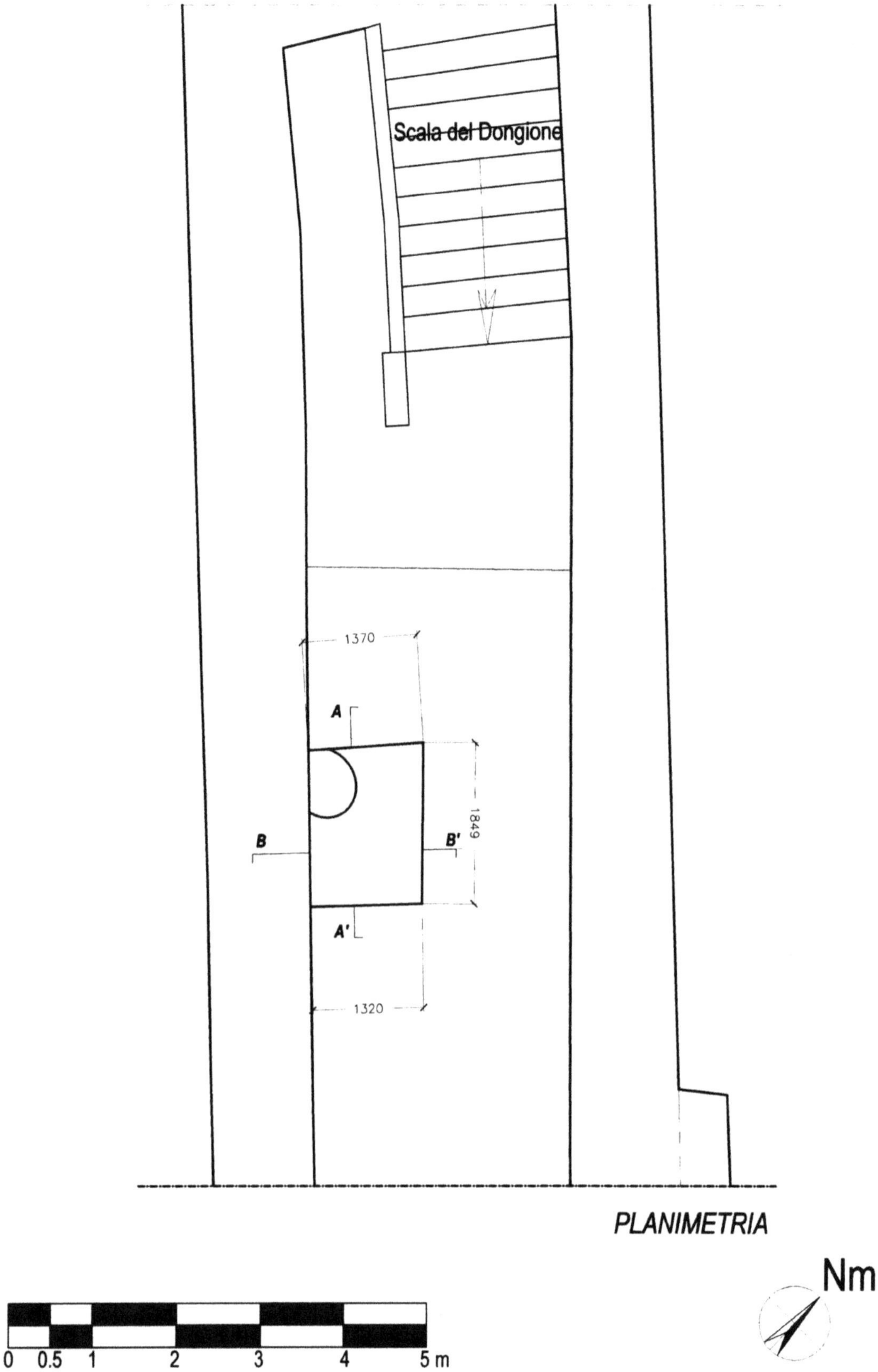

Fig. III.63.a. *Stairway Static Tank (CA 00026 PI TO) (Ass.ne S.C.A.M. Archive).*

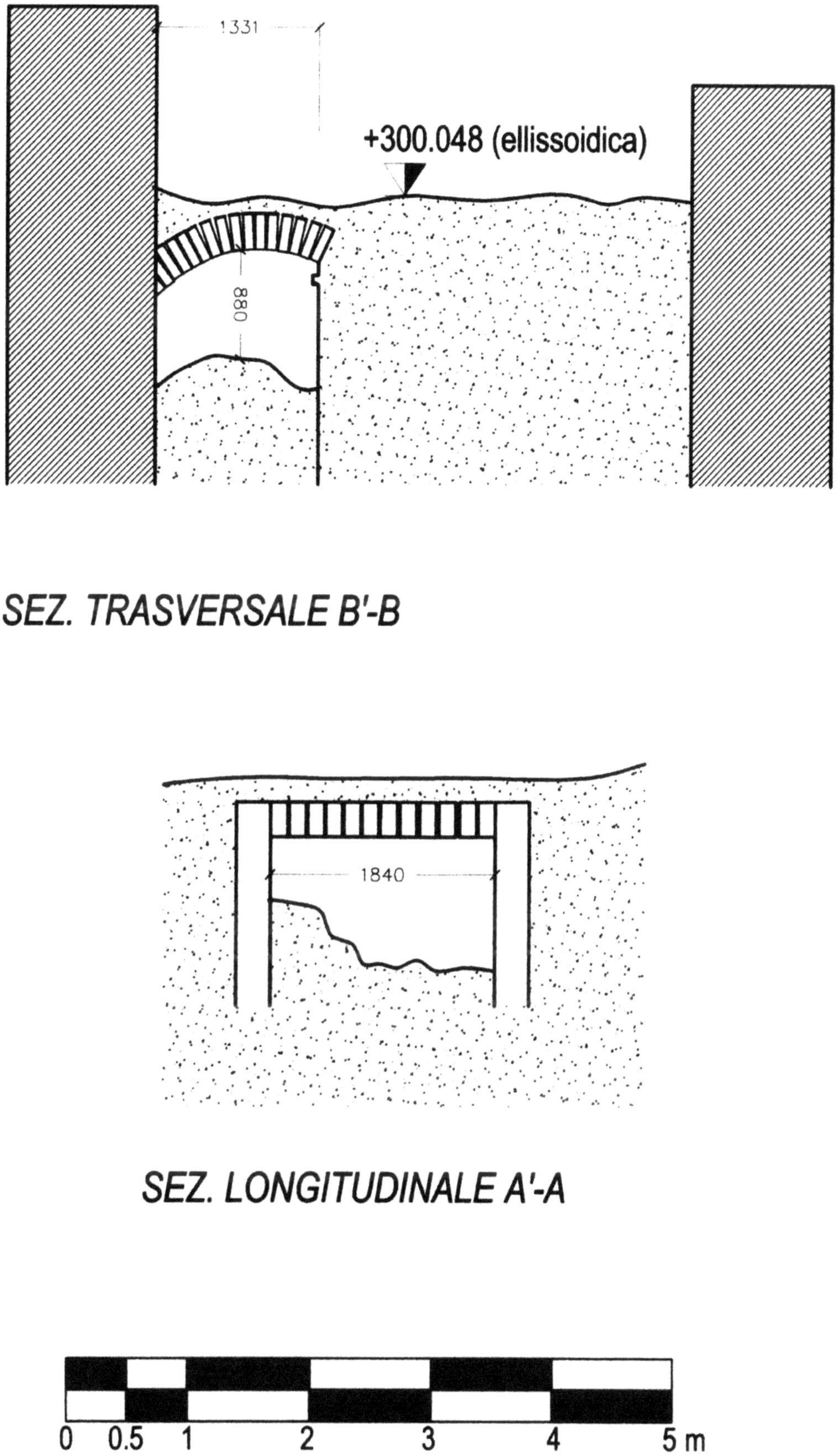

Fig. III.63.b. *Stairway Static Tank (CA 00026 PI TO) (Ass.ne S.C.A.M. Archive).*

Fig. III.64. *TE.S.E.S. Tunnel (CA 00027 PI TO). TE.S.E.S. Tunnel quarry face entrance. The rear wall extends a further 22 m to the left. The tunnel entrance can be seen at the left corner. The sector is larger that would appear from the old maps (photo L. Bavagnoli).*

Fig. III.65. *TE.S.E.S. Tunnel (CA 00027 PI TO). On the left is an opening, which leads to the adjacent Stanza del Foro (CA 00012 PI TO). The brick structure was recently demolished by explosives from the quarry works (photo L. Bavagnoli).*

Work carried out by: Associazione Speleologia Cavità Artificiali Milano (S.C.A.M.); 2003
Warnings: possible collapse of the final tract of tunnel
Typology: 7
Description: under the current ground surface and consisting of an additional quarry terrace amid the vegetation lies a well conserved 2.52 m structure with parallel brick walls, reminiscent of a dromos. The base of the dromos is not visible due to the debris, which given the sharp corners of masonry blocks and broken bricks, is undoubtedly all that remains of recently demolished structures. The head wall continues leftwards, orthogonal to the corridor and is visible for at least 22 m before it is again covered by vegetation. Descending into the structure, the entrance to the TE.S.E.S. Tunnel, framed by a brick arch, is visible on the left. 4.3 m of the tunnel, buried to as far as its piers, are practicable; the tunnel communicates with the adjacent Room of the Opening (CA 00012 PI TO) by means of a 0.24 x 0.18 opening with a depth of 0.38 m. The structure has been intercepted by quarry extraction works.
Interpretation: this could well be a walkway or as is more likely, a drainage and transport system for meteoric and/or infiltration water.
Dating: the structure should date back to the fortification's restoration following the 1625 siege. More precisely, given its position, it may date back to the third quarter of the XVII century when the triple bastion stronghold was built.
Notes: the opening, from which the name Room of the Opening airsides, would have been used for the purposes of air exchange as the environment presented no ventilation shafts and its only air inlet was a corridor, which communicated with the tunnel. Clearing the quarry face would uncover the large and mighty brick structure that is the tunnel- Room of the Opening -*dromos*- TE.S.E.S. Tunnel complex and provide clear information for its correct interpretation and possible restoration. The use of explosives in the demolition of historic works is attested by the surviving tunnel section, which is now interrupted: the uncovered section (the partially practicable section) is slightly misaligned in an upwards direction and points several mm toward the internal section of the hill. The fracture line clearly interrupts the internal speleothem covering and there are no calcareous deposits whatsoever within such fractures. Keeping in mind that the structure dates to no earlier than the latter part of the XVII century and that the Fortress was demilitarized no more than half a century ago, it is thought that mine demolition cannot be ascribed to this period, despite the fact that speleothems have been known to form quickly. The fissures are tangible proof that the tunnel remained intact and abandoned for sufficient time to allow water infiltration water to produce stalactites and flowstone formations. The tunnel was only recently demolished with an explosive charge.
Bibliography: none.
Data ownership: Associazione Speleologia Cavità Artificiali Milano (S.C.A.M.).
Compiled by: Gianluca Padovan (Ass.ne S.C.A.M.).

III.3.35 - *CA 00028 PI TO; Mini-Nymphaeum*

Cadastral number: CA 00028 PI TO
Denomination: Mini-Nymphaeum
Region-country: Piedmont, Italy
Province: Turin
Municipality: Verrua Savoia
Locality: Verrua Hill
Location: north sector of the Fortress
Ownership: private
Cartography: CTR 1:10.000
Geological unit: marls, sandstones and Pliocene limestone resting on an Eocene Clay Shale Formation
Altitude: //
Position: //
Context: situated within the curtain wall between Tour St Joseph e la Tour du Bienhereux
Operations conducted: exploration, survey, photographic service
Work carried out by: Associazione Speleologia Cavità Artificiali Milano (S.C.A.M.); 2003
Warnings: possible internal collapse
Typology: 6, transformed in small nymphaeum
Description: there is a niche, almost 2 m deep near the bottom of the surviving curtain wall. The niche was created by breaking the wall. The facing of the entire wall was restored relatively recently. The restoration was carried out using bricks and also cement upon which bricks were painted, to simulate the continuity of the rows. The breach however was retained, although it was rendered more aesthetically pleasing by the addition of a brick drop arch. A constant rivulet of water emerges from the breach.

Interpretation: it is difficult to believe that the enceinte wall had an external niche for the outflow of water as this would have presented an excellent point for placing a mine. It is far more probable that the wall structure included a water source and the construction of an appropriate water collection tank and underground overflow spillway channel, especially designed to pass under the defences and to thus discharge water further down the valley. The partial demolition of the system and area abandonment would have led to the deterioration of the discharge channel and the storage chamber itself and to the accumulation of debris, leaving the water to find its own outlet through cracks in the curtain wall. At some point someone repaired this particular section of city wall thus creating a niche for easier outflow of the water, by widening and deepening the breach.
Dating: dates to between the XIX century and the first part of the XX century.
Notes: it should be taken into account that the following maps indicate the presence of a fountain in or around this area: Anonymous, *Pianta della Fortezza di Verrua*, 1617 (?); Morello Carlo, *Planimetria della Fortezza*, 1651 (circa); Morello Carlo, *Planimetria della Fortezza*, 1653 (circa). The structure has not been surveyed.
Bibliography: Caramellino C., *Verrua Savoia, immagini di una Fortezza*, Verrua Savoia 1987.
Data ownership: Associazione Speleologia Cavità Artificiali Milano (S.C.A.M.).
Compiled by: Gianluca Padovan (Ass.ne S.C.A.M.).

III.3.36 - *CA 00029 PI TO; Postern Passage (Arcadio Corner)*

Cadastral number: CA 00029 PI TO
Denomination: Postern Passage (Arcadio Corner) (figs. III.66 and III.67)
Region-country: Piedmont, Italy
Province: Turin
Municipality: Verrua Savoia
Locality: Verrua Hill
Location: west sector of the Fortress (northwest of the Stronghold)
Ownership: private
Cartography: CTR 1:10.000
Geological unit: marls, sandstones and Pliocene limestone resting on an Eocene Clay Shale Formation
Altitude: //
Position: //
Context: situated in the hill summit area and at the foot of the city wall enclosing the northeast Stronghold area, between the Vineyard Bastion and the area where the "Verruca" once stood, in front of the Supply Gate detailed in the Cantoregio planimetry (1785). In the XVIII century, the area was subject to demolition and saw the collapse of the sandstone pinnacle known by the name of "Verruca".
Operations conducted: exploration, planimetric survey, photographic service
Work carried out by: Associazione Speleologia Cavità Artificiali Milano (S.C.A.M.); 2003
Warnings: possible internal collapse
Typology: 6 - tunnel
Description: this short passage is open on each side, is partly buried and inclines towards the opposite side of the city wall. Its entrance is just over 2 m from the city wall and is accessed by a well-like structure in the collapsed material. It is open and fully accessible from the other side. The passage is 2.7 m long, of which 2.19 m is still vaulted. Its parallel walls diverge slightly towards the base; towards the city wall it is 0.58 m wide at the height of the piers and 1.39 m high at the soil-level, while its opposite side measures 0.57 x 0.8 m. The distance between the lowest and highest tunnel sections is 0.86 m and the tunnel faces a 248° direction. The structure is covered by collapsed material and presents itself as a passage section. It is distinctly different from both the Celestino Tunnel (CA 00011 PI TO) and the four countermine tunnels (CA 00017 - 18 - 19 - 36 PI TO), in that its masonry workmanship is not as precise: the rows are not perfectly regular, the piers are not parallel and the poor quality mortar has been badly distributed; furthermore, the bricks utilised are not all of the same type and appear to have been reutilised. The bricks used to provide roof support protrude at regular intervals from the top of the breast walls.
Interpretation: in front of the passage section, the city walls contain two high and narrow adjacent rooms) of which only the left room is vaulted), sealed off by diverse walls. In this section, the Cantoregio planimetry indicates the existence of the Supply Gate and the interpretation of the situation becomes problematic: this could have been built following the destruction and subsequent restoration of the gate itself. The passage may have been built on top of collapsed material and this being the case, its purpose would be even more obscure.
Dating: the structure predates the Stronghold's northeast enceinte wall and thus predates the XVI century.
Notes: debris should be cleared from the entire section to clarify whether the passage once gave access to a lower room leading to the Supply Gate. The removal of material from the latter would further allow access to rear-lying

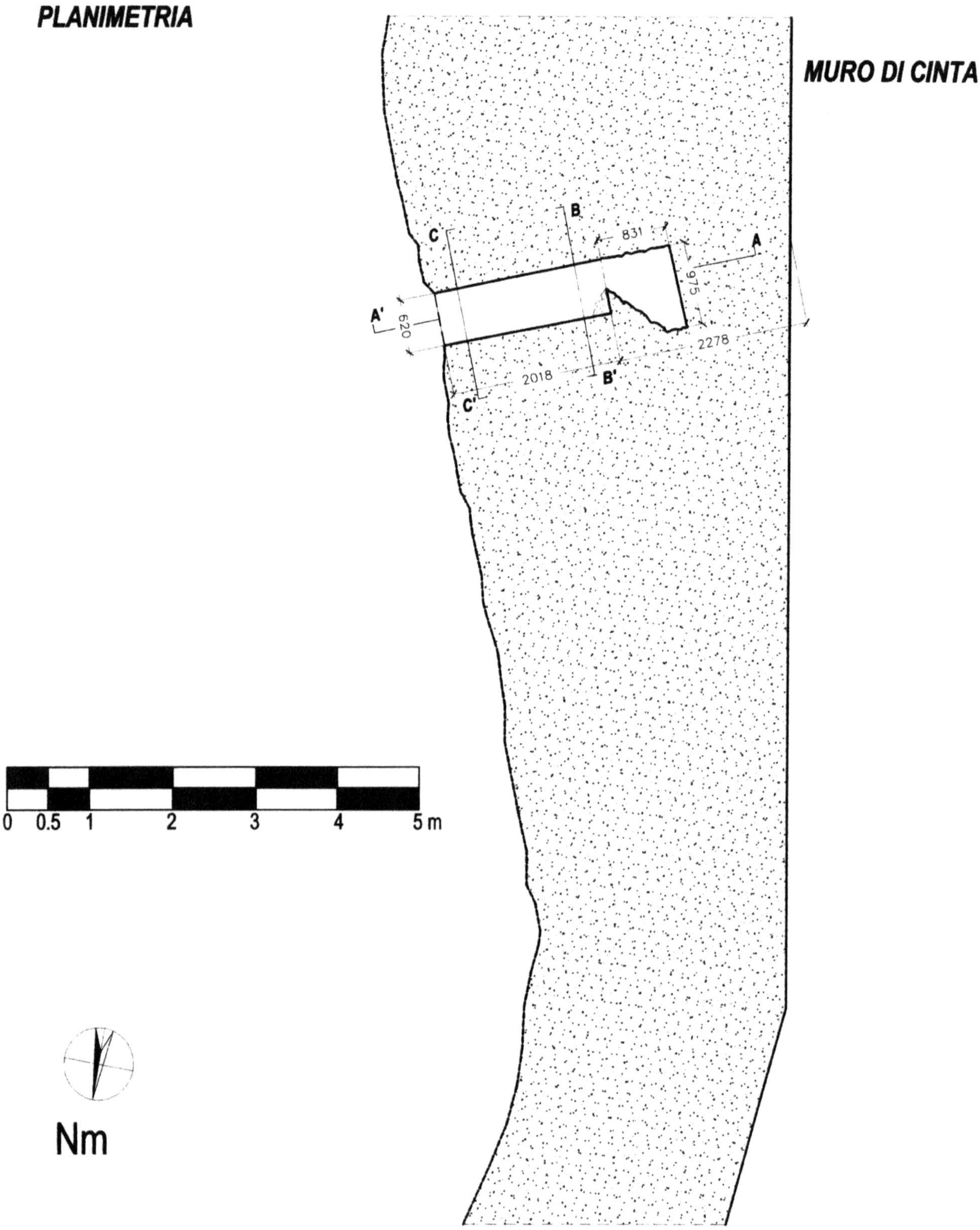

Fig. III.66. *Postern Passage (Arcadio Corner) (CA 00029 PI TO); plan (Ass.ne S.C.A.M. Archive).*

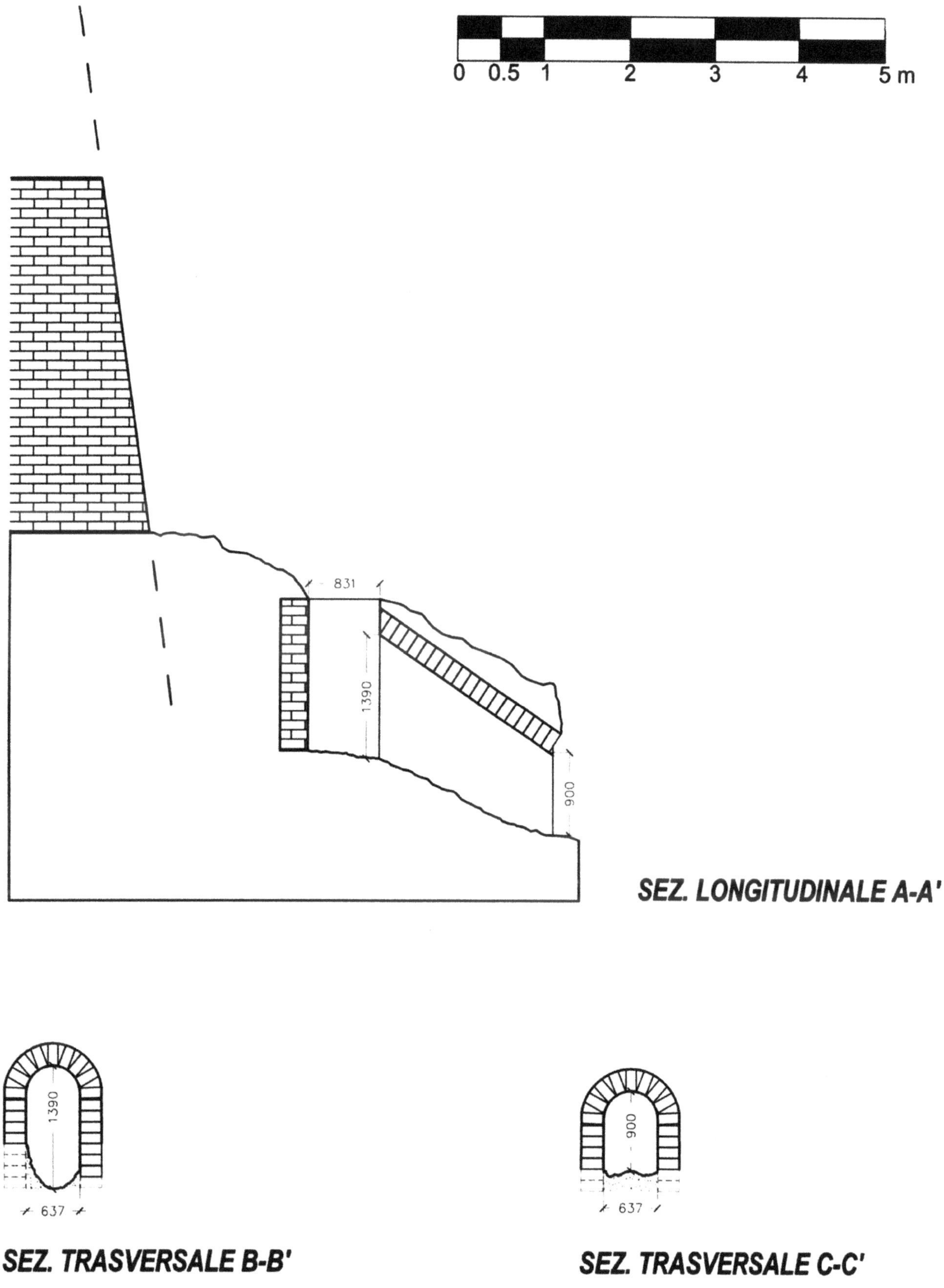

Fig. III.67. *Postern Passage (Arcadio Corner) (CA 00029 PI TO); sections (Ass.ne S.C.A.M. Archive).*

rooms and thus uncover the stairwell which led to the platform above, still essentially intact despite the collapse of the parapet.
Bibliography: none.
Data ownership: Associazione Speleologia Cavità Artificiali Milano (S.C.A.M.).
Compiled by: Gianluca Padovan (Ass.ne S.C.A.M.).

III.3.37 - *CA 00030 PI TO; Pavarolo Castle Well*

Cadastral number: 00030 PI TO
Denomination: Pavarolo Castle Well (figs. III.68, III.69 and III.70)
Region-country: Piedmont, Italy
Province: Turin
Municipality: Pavarolo
Locality: Pavarolo Castle
Location: upper Castle area, within the courtyard
Ownership: private
Cartography: CTR 156110
Geological unit: Fossiliferous sandstone from the Miocene period
Altitude: 386 m a.s.l.
Position: Gauss Boaga coordinates 1408308 E, 4991307N
Context: located in a brick building, on the internal façade of the central core that was once part of the mediaeval fortified structure
Operations conducted: exploration, planimetric survey, photographic service
Work carried out by: Associazione Speleologia Cavità Artificiali Milano (S.C.A.M.); 2003
Warnings: the building's earthing system and lightning conductor are located within the well
Typology: 2b - ordinary well
Description: the system consists of a brick room, a pail lifting system (which probably dates to the latter part of the XVIII century), a parapet (puteal) and a vertical, cylindrical soil perforation. The puteal consists of a brick parapet, which protects the well's opening on two sides and its base rests on the perimeter walls of the low construction mentioned above, at the external corner opposite the entrance (as shown in the 1:20 scale planimetry in section AA'). The well is 64.8 m deep (58.54 m at the water level) and has a variable circular section. The well shaft has a diameter of 1.38 m. The well is brick-lined to a depth of 14.6 m and in the terminal section the bricks are covered by calcium carbonate deposits. The well is not lined below this depth and the rock, covered by calcium carbonate deposits to a depth of almost 40 m, becomes visible. Further down, between the layers of clay and sandstone, the rock is clearly visible and presents clear signs left by excavation equipment. At a depth of -10.6 m, the structure widens and assumes a tronco-conical shape; there are four putlog holes in the last ring of bricks, just before the structure widens. This is again observed at the point where the diameter is 1.85 m, towards the area where the lining ends. At -39.9 m, the diameter of the well is a steady 3 m; almost all the formations have been removed and a new layer of calcium carbonate lines the well. Furthermore, the distribution of signs left on the rock by the excavation equipment is different to that which appears on the overlying rock. This indicates that the excavations were conducted over two distinct phases. At -56.64 m the well's diameter is 3.08 m. As the water table moved to a lower level, or simply to access larger quantities of water, it may have become necessary to deepen the well. There are no footholds along the unlined section; no pipes or drainpipes were found.
Interpretation: this well was used to capture the waters of a small aquifer.
Dating: unknown, but predates the XVIII century.
Notes: the fact that the well's opening is closed against the corner of a building and its parapet is not perfectly regular would indicate that the building was built later than the well. Although the well was excavated in alternating clay and sandstone strata, there is no structural collapse.
Bibliography: Bianchi S., Basilico R., Ninni C., Padovan G., *Il pozzo del castello di Pavarolo*, in Anzanello E., Dal Cin F., Gasparetto P., Gava S. (edited by), *Atti Montello 2002. Conglomeriamoci*, 21° Incontro Internazionale di Speleologia. *Nervesa della Battaglia 1-3 Novembre 2002*, Villorba 2003, pgs. 277-292.
Data ownership: Associazione Speleologia Cavità Artificiali Milano (S.C.A.M.).
Compiled by: Gianluca Padovan (Ass.ne S.C.A.M.).

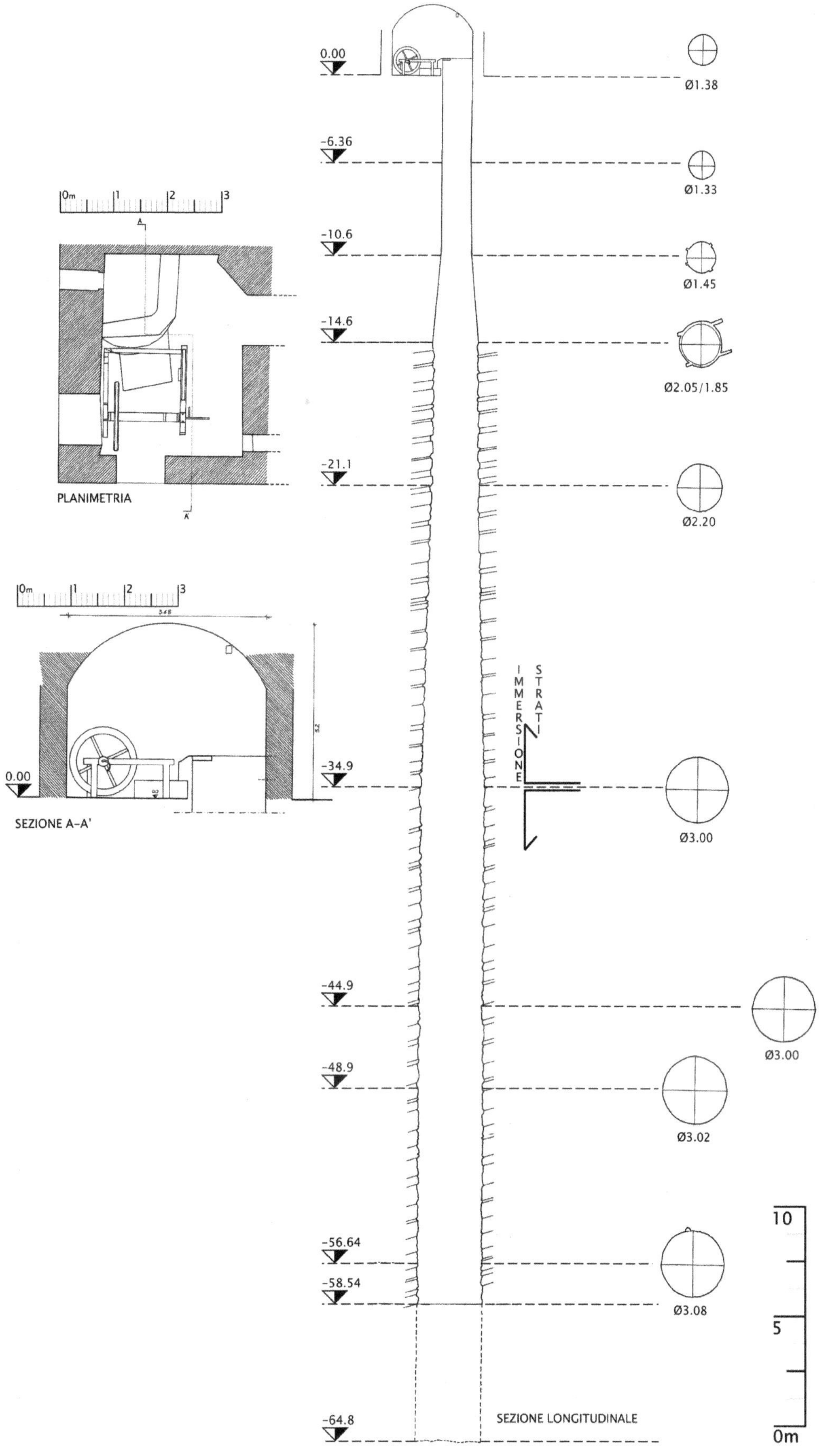

Fig. III.68. *Pavarolo Castle Well with plan and section of the room created for well-cover and of the pail-lifting system (winch) (CA 00030 PI TO) (Ass.ne S.C.A.M. Archive).*

Fig. III.69. *Speleologist at the Pavarolo Castle (photo G. Padovan).*

Fig. III.70. *Internal section of the first part of Pavarolo Castle Well (CA 00030 PI TO). The thick copper cable used to earth the lightning conductor, the rubber pipe linked to the motor pump and speleological ropes can be seen in the photograph (photo G. Padovan).*

III.3.38 - *CA 00031 PI TO; Garden Sump*

Cadastral number: 00031 PI TO
Denomination: Garden Sump (figs. III.71 and III.72)
Region-country: Piedmont, Italy
Province: Turin
Municipality: Pavarolo
Locality: Pavarolo Castle
Location: upper Castle area, within the courtyard
Ownership: private
Cartography: CTR 156110
Geological unit: Fossiliferous sandstone from the Miocene period
Altitude: 385.6 m a.s.l.
Position: //
Context: situated in the Castle gardens, the cistern would once have been situated within the courtyard itself
Operations conducted: exploration, photography, detritus clearance
Work carried out by: Associazione Speleologia Cavità Artificiali Milano (S.C.A.M.); 2003
Warnings: none
Typology: 2d - cesspit
Description: this vertical perforation of the ground is approximately 2 m deep and is brick-lined. Its base, also brick-lined, is slightly concave; mortar has rendered the entire structure water-proof from its base to the vaulted ceiling. The brick vault has a lateral sump, which is closed by a manhole cover; in line with this is a protruding masonry conduit. It should be noted that several perforations have been made to the base of the structure to allow the outflow of water and that these were subsequently blocked by silt and detritus.
Interpretation: the vaulted roof appears to pre-date the well's shaft. There are putlog holes a few centimetres from the base of the structure, indicating that the structure probably originated as an ordinary well and was subsequently filled with debris and equipped with a brick floor. In any case, its current purpose is that of collecting and discharging meteoric water, while its previous purpose was that of a cesspit.
Dating: unknown
Notes: a planimetric survey is required.
Bibliography: Bianchi S., Basilico R., Ninni C., Padovan G., *Il pozzo del castello di Pavarolo*, in Anzanello E., Dal Cin F., Gasparetto P., Gava S. (edited by), *Atti Montello 2002. Conglomeriamoci*, 21° Incontro Internazionale di Speleologia. *Nervesa della Battaglia 1-3 Novembre 2002*, Villorba 2003, pgs. 277-292.
Data ownership: Associazione Speleologia Cavità Artificiali Milano (S.C.A.M.).
Compiled by: Gianluca Padovan (Ass.ne S.C.A.M.).

III.3.39 - *CA 00032 PI TO; Garden Cistern*

Cadastral number: 00032 PI TO
Denomination: Garden Cistern (fig. III.73)
Region-country: Piedmont, Italy
Province: Turin
Municipality: Pavarolo
Locality: Pavarolo Castle
Location: upper Castle area, within the courtyard
Ownership: private
Cartography: CTR 156110
Geological unit: Fossiliferous sandstone from the Miocene period
Altitude: 385.7 m a.s.l.
Position: //
Context: situated in the Castle gardens, the cistern would once have been situated within the courtyard itself
Operations conducted: exploration, photography
Work carried out by: Associazione Speleologia Cavità Artificiali Milano (S.C.A.M.); 2003
Warnings: none
Typology: 2c - cistern
Description: this is a masonry, cement-lined tank, which is approximately 3 m long, one and a half m wide and two metres deep. There are two short branches at its end. It has a cement roof and a man-hole cover blocks its entrance.
Interpretation: a cistern for the collection of meteoric water for the purposes of watering the garden.

Fig. III.71. *Garden Sump (CA 00032 PI TO). Detail of the wall facing of the small, cylindrical cistern (photo G. Padovan).*

Fig. III.72. *Garden Sump (CA 00032 PI TO). Access well to the cylindrical cistern, faced with bricks (photo G. Padovan).*

Fig. III.73. *Garden Sump (CA 00031 PI TO). A speleologist prepares to come out of the cistern (photo G. Padovan).*

Dating: XX century
Notes: a planimetric survey is required.
Bibliography: Bianchi S., Basilico R., Ninni C., Padovan G., *Il pozzo del castello di Pavarolo*, in Anzanello E., Dal Cin F., Gasparetto P., Gava S. (edited by), *Atti Montello 2002. Conglomeriamoci*, 21° Incontro Internazionale di Speleologia. *Nervesa della Battaglia 1-3 Novembre 2002*, Villorba 2003, pgs. 277-292.
Data ownership: Associazione Speleologia Cavità Artificiali Milano (S.C.A.M.).
Compiled by: Gianluca Padovan (Ass.ne S.C.A.M.).

III.3.40 - *CA 00033 PI TO; Pavarolo Castle Icehouse*

Cadastral number: 00033 PI TO
Denomination: Pavarolo Castle Icehouse
Region-country: Piedmont, Italy
Province: Turin
Municipality: Pavarolo
Locality: Pavarolo Castle
Location: situated below the upper Castle area
Ownership: private
Cartography: CTR 156110
Geological unit: Fossiliferous sandstone from the Miocene period
Altitude: //
Position: //
Context: located half way up the crag beneath the upper castle area
Operations conducted: exploration, photography
Work carried out by: Associazione Speleologia Cavità Artificiali Milano (S.C.A.M.); 2003
Warnings: none
Typology: 2c - icehouse
Description: a masonry corridor leads to a small circular chamber with brick conch vault. Approximately two metres high with diameter of three metres, it still has the original overhanging channel for expelling ice or snow. The floor is covered in sediment.
Dating: unknown.
Notes: the icehouse was reutilised as a cesspit and is now in disuse.
Bibliography: Bianchi S., Basilico R., Ninni C., Padovan G., *Il pozzo del castello di Pavarolo*, in Anzanello E., Dal Cin F., Gasparetto P., Gava S. (edited by), *Atti Montello 2002. Conglomeriamoci*, 21° Incontro Internazionale di Speleologia. *Nervesa della Battaglia 1-3 Novembre 2002*, Villorba 2003, pgs. 277-292.
Data ownership: Associazione Speleologia Cavità Artificiali Milano (S.C.A.M.).
Compiled by: Gianluca Padovan (Ass.ne S.C.A.M.).

III.3.41 - *CA 00034 PI TO; TE.S.E.S. Room*

Cadastral number: CA 00034 PI TO
Denomination: TE.S.E.S. Room (TE.S.E.S.: Experimental Underground Exploration Team) (fig. III.74)
Region-country: Piedmont, Italy
Province: Turin
Municipality: Verrua Savoia
Locality: Verrua Hill
Location: in the Stronghold's northwest curtain
Ownership: private
Cartography: CTR 1:10.000
Geological unit: marls, sandstones and Pliocene limestone resting on an Eocene Clay Shale Formation
Altitude: (ellipsoid) 296.12 m
Position: //
Context: situated within the confines of the Stronghold's northwest curtain, which collapsed in 1957
Operations conducted: exploration, planimetric survey, photographic service
Work carried out by: Federazione Nazionale Cavità Artificiali (F.N.C.A.); 2003
Warnings: possible collapse of the entrance archway
Typology: 6 - room within the curtain wall

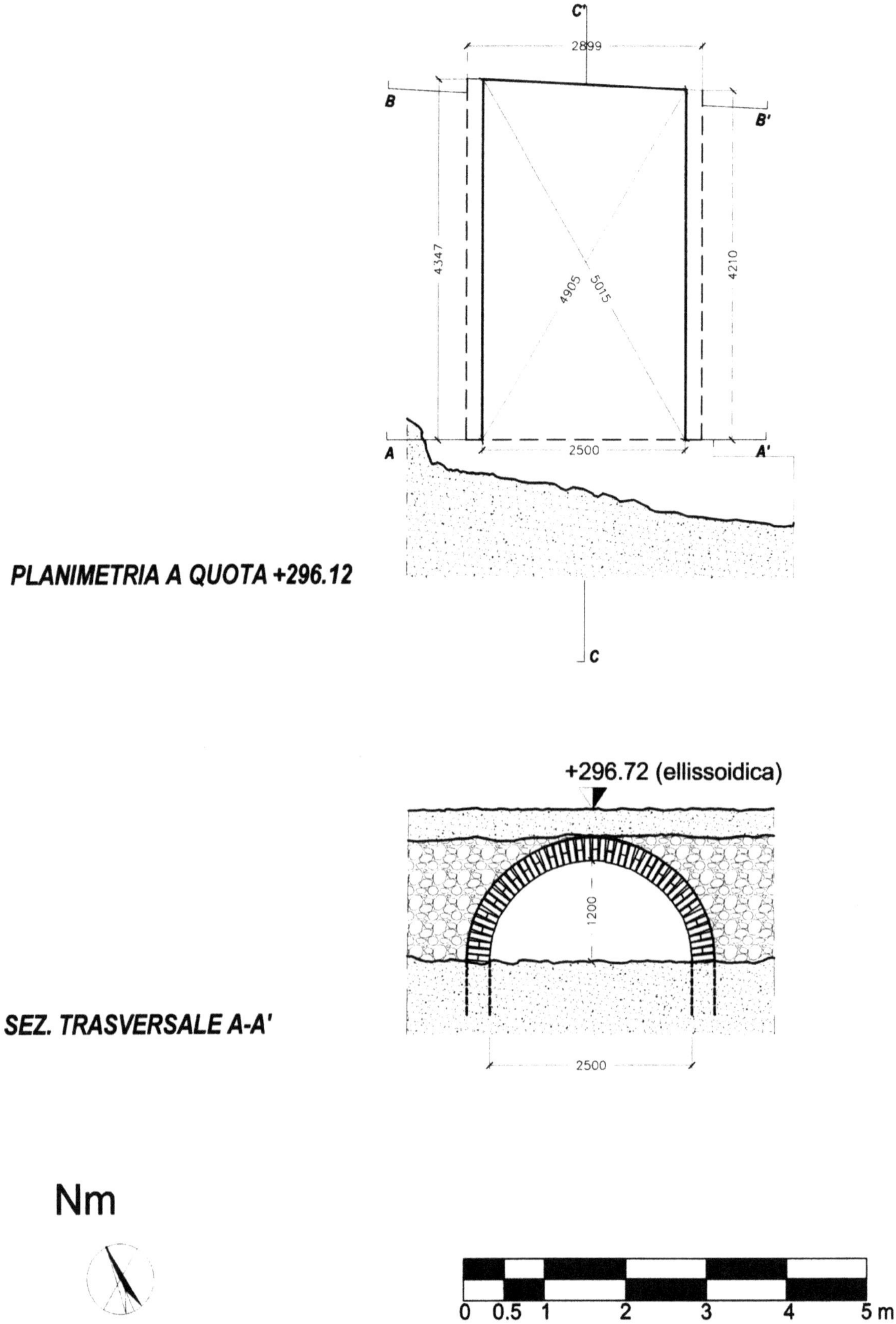

Fig. III.74. *TE.S.E.S. Room (CA 00034 PI TO) (Federazione Nazionale Cavità Artificiali Archive).*

Description: a rectangular room, with soil and detritus covering the piers reveals 1.2 m of the top section of a large brick archway, at right angles to the extern curtain wall. It is accessed by a trench created in the soil and debris. It central length is 4.32 m, while it has a reduced internal width of just 2.9 m; it has a brick barrel vault, while the fill material consists of broken stones interrupted by regular rows of bricks.
Interpretation: appears to be one of the curtain wall rooms.
Dating: given the material used to seal off the room, the environment could have been built at the same time as the Donjon. Furthermore, the wall-seals within the rooms of the Donjon were erected in the same way.
Notes: the collapse of the "Verruca" has undoubtedly determined the collapse of the structures built on the northwest curtain platform, while the later restructure of the area flattened and levelled off that which remained above the grade plane. It is also true that there are various environments and the necessary communication tunnels within the mighty curtain wall, which the Cantoregio tables bear no reference to (Caramellino 1987, pgs. 100-105). These were sealed following both the demilitarisation of the Stronghold and the collapse of the wall.
Bibliography: Caramellino C. (edited by), *Verrua Savoia immagini di una fortezza*, Verrua Savoia 1987.
Data ownership: Federazione Nazionale Cavità Artificiali (F.N.C.A.).
Compiled by: Gianluca Padovan (Ass.ne S.C.A.M.).

III.3.42 - *CA 00035 PI TO; Small Quadrivium Room*

Cadastral number: CA 00035 PI TO
Denomination: Small Quadrivium Room (figs. III.54, III.54.a, III.55 and III.55.a)
Region-country: Piedmont, Italy
Province: Turin
Municipality: Verrua Savoia
Locality: Verrua Hill
Location: situated outwith the west re-entrant flank of the Fortress
Ownership: private
Cartography: CTR 1:10.000
Geological unit: marls, sandstones and Pliocene limestone resting on an Eocene Clay Shale Formation
Altitude: 214 m a.s.l.
Position: (GPS investigation) east 1429488, west 5002496
Context: the room is situated on the plain overlooking the west flank of the Fortress and gives access to the Countermine Tunnels
Operations conducted: exploration, planimetric survey, photographic service
Work carried out by: Associazione Speleologia Cavità Artificiali Milano (S.C.A.M.); 2003
Warnings: none
Typology: 6 - countermine system
Description: a small room at the site of convergence of the three countermine system branches (CA 00017 PI TO, CA 00018 PI TO, CA 00019 PI TO) and the point where branch CA 00036 PI TO branches off. The roof is missing and an accumulation of debris obstructs the base.
Interpretation: part of the countermine system.
Dating: this structure predates the building phase of the second and third quarters of the XVII century. It may date back to between the end of XVI century and the first quarter of the XV century.
Notes: it appears to have been uncovered in the mid 1990s when excavators eradicated the remaining tree stumps from the coppice, to be used as a vineyard.
Bibliography: none.
Data ownership: Associazione Speleologia Cavità Artificiali Milano (S.C.A.M.).
Compiled by: Gianluca Padovan (Ass.ne S.C.A.M.).

III.3.43 - *CA 00036 PI TO; Fourth Countermine Tunnel*

Cadastral number: CA 00036 PI TO
Denomination: Fourth Countermine Tunnel (figs. III.54, III.54.a, III.55 and III.55.a)
Region-country: Piedmont, Italy
Province: Turin
Municipality: Verrua Savoia
Locality: Verrua Hill
Location: situated outwith the west re-entrant flank of the Fortress
Ownership: private

Cartography: CTR 1:10.000
Geological unit: marls, sandstones and Pliocene limestone resting on an Eocene Clay Shale Formation
Altitude: 214 m a.s.l.
Position: (GPS investigation) east 1429488, west 5002496
currently extends between the Major's Garden Bastion and the area where the Third Village once stood
Operations conducted: excavation, survey, photographic service
Work carried out by: Associazione Speleologia Cavità Artificiali Milano (S.C.A.M.); 2003
Warnings: poor air circulation and current structural collapse
Typology: 6 - countermine tunnel
Description: on the same axis as the Third Countermine Tunnel (CA 00019 PI TO) and appears to be its natural continuation in a 293° direction. The tunnel is almost completely buried and at just under 2 m, its roof has collapsed.
Interpretation: this is a perfect example of a countermine tunnel, which is probably part of the system used during the 1625 siege.
Dating: this structure predates the building phase of the second and third quarters of the XVII century. It may date back to the period of time between the end of XVI century and the first quarter of the XV century.
Notes: it appears to have been uncovered in the mid 1990s when excavators eradicated the remaining tree stumps from the coppice, to be used as a vineyard. Evidence uncovered during surface prospecting reveal that there are two rectilinear tunnel sections, approximately 30 m south of this "intersection". One of the tunnel sections ends in a T, indicating that it may have had two shot holes.
Bibliography: none.
Data ownership: Associazione Speleologia Cavità Artificiali Milano (S.C.A.M.).
Compiled by: Gianluca Padovan (Ass.ne S.C.A.M.).

III.4 - *Tuscany*

Fig. III. 75. *Geographic framework: Tuscan Region.*

Campiglia Marittima (Livorno). The first documented reference to Campiglia Marittima Castle dates back to 1004 (figs. III.75 and III.76). A document refers to the: «*castello de Campilia meditate cum ecclesia et curte*», which results to be one of the properties given to the Santa Maria di Serena Monastery by a Della Gherardesca Count. The results of the archaeological dig, carried out in the upper castle area, reveal that the castle system is IX century and thus predate the first known reference (Bianchi 1997. Ceccarelli Lemut 1985, pgs. 322-341). From the XII century onwards, the area was greatly transformed by the construction of a stone, two storey, quadrangular building (known as Building A) and the Palace, with an elegant biforate window adorned by a small column and capital in local white marble on the main floor (Boccacci 1984). The end of the XIII century saw significant developments with the construction of new, larger city walls. The new Village, consisting of districts with Pisan-style house-towers and churches, was thus enclosed. At the beginning of the XV century, Campiglia came under Florentine rule. During this and the subsequent century there the wall structures were altered and defensive works were strengthened.

From 1994 to 1999, Siena University's Departments of Archaeology and History of Art undertook a research project for Municipal Administration purposes. Such research involved the archaeological investigation of the Stronghold and of the deposits to be found both internally and externally to the buildings as well as those village ruins situated in an elevated position. Between 2000 and 2001, the Associazione Speleologia Cavità Artificiali Milano (S.C.A.M.), undertook the investigation, study and registration of wells and cisterns situated within city walls and on public land. It should be noted that various hydraulic works open onto private property and were not registered. Furthermore, several additional cisterns were identified and registered through surface reconnaissance in the municipal area of Campiglia.

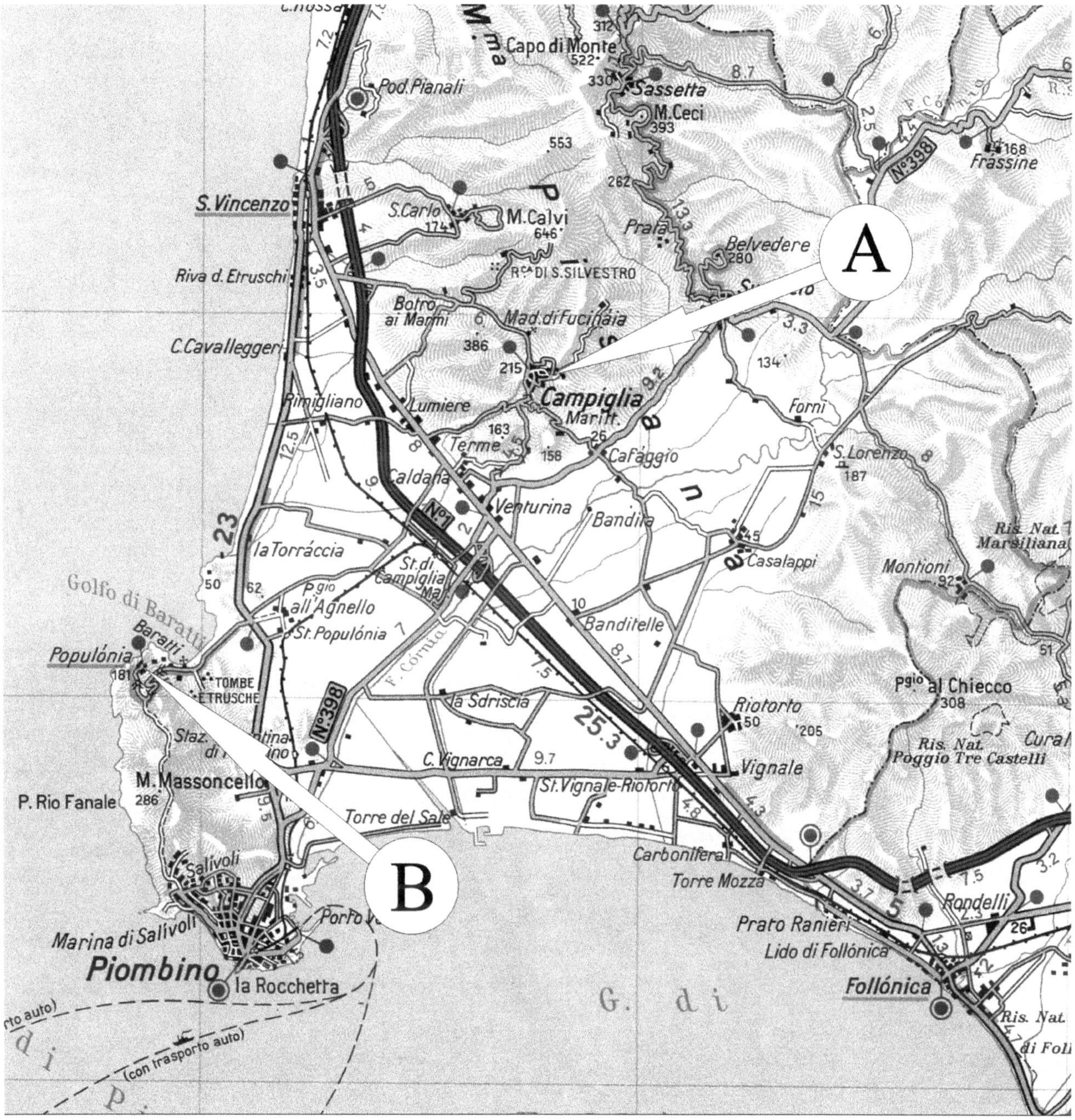

Fig. III.76. *Geographic framework: A: Campiglia Marittima (Livorno). B: Populonia (Livorno) (Touring Club Italiano 2002 b, table 12).*

Populonia (Livorno). The Etruscan city of Populonia dominates the Gulf of Baratti, overlooking the Isle of Elba (fig. III.76). Known for its necropolis, it primarily owes its fortune to its location. Overlooking a natural port, it became a sea port for the commerce of copper, bronze and iron. It also controls the Campigliese mines and benefits from economic commerce with the Isle of Elba. The acropolis was definitively abandoned following the Sillana repression and the fall of the city in around 80 B.C.

From 2001, exploration of the acropolis and the surrounding area has gradually revealed an urban network consisting of both monumental and private public buildings. Research is carried out by the Regional Board for Tuscan Archaeological Heritage in collaboration with Siena University's Department of Archaeology and Pisa University's Department of Archaeological Science (Mascione, Patera 2003, pg. 5). Please note that, in addition to the cistern in question, other hydraulic works, uncovered by recent archaeological digs, will be included in the register.

III.4.1 - *CA 01000 TO LI; Stronghold Cistern*

Cadastral number: CA 01000 TO LI
Denomination: Stronghold Cistern
Region-country: Tuscany, Italy
Province: Livorno
Municipality: Campiglia Marittima
Locality: Campiglia Marittima
Location: within monumental Rocca di Campiglia area
Ownership: Municipality of Campiglia Marittima
Cartography: C.T.R. 1:1000; C.T.R. 1: 2000, sheet 10G14
Geological unit: Eocene, Canetolo Formation (external unit)
Altitude: 280 m a.s.l.
Position: 4768800 N 1631825 E
Context: upper Stronghold area, between Palace and Tower B
Operations conducted: planimetric survey, archaeological dig, 1984, 1994-1999
Work carried out by: Siena University
Warnings: none
Typology: 2c - cistern
Description: consists of a storage chamber with surface communication well; this is a semi-subterranean, rectangular, vaulted structure; only part of the structure survives. The walls are made of alberese limestone blocks while the roof is made of travertine; the internal section shows traces of plaster with evident signs of humidity. The base is lined in hydraulic mortar. Entrance: consists of a small well, part of which has collapsed; however, two bardiglio marble blocks, which would once have framed the entrance, are visible. In the era following its construction, a second entrance was created, again on the roof, in line with the palace's south wall. Plan dimensions: 14 x 9 m. Footholds: none found. Piping: none found.
Interpretation: cistern serving the monumental complex.
Dating: XIII century.
Notes: survey published in Boccacci M., *La Rocca di Campiglia Marittima: studi e ricerche*, Firenze 1984.
Bibliography: Boccacci M., *La Rocca di Campiglia Marittima: studi e ricerche*, Firenze 1984.
Basilico R., Casini A., Padovan G., *Le cisterne del castello di Campiglia Marittima (LI)*, in Federazione Speleologica Toscana, *Atti del VII Congresso della Federazione Speleologica Toscana (Gavorrano, 31 Marzo - 1 Aprile 2001), Talp. Rivista della Federazione Speleologica Toscana*, No. 23, Lucca 2001, pgs. 139-147.
Basilico R., Casini A., Padovan G., *La storia dell'acqua in un castello della Maremma Toscana: Campiglia Marittima*, in Club Alpinistico Triestino - Gruppo Grotte Sezione Ricerche e Studi su Cavità Artificiali (edited by), *Atti del V Convegno Nazionale sulle Cavità Artificiali. Osoppo 28 Aprile - 1 Maggio 2001*, Trieste 2002, pgs. 41-68.
Data ownership: Siena University.
Compiled by: Gianluca Padovan (Ass.ne S.C.A.M.).

III.4.2 - *CA 01001 TO LI; San Jacopo and San Filippo Hospital Cistern*

Cadastral number: CA 01001 TO LI
Denomination: *San Jacopo and San Filippo Hospital Cistern* (figs. III.77.a, III.77.b, III.77.c and III.77.d)
Region-country: Tuscany, Italy
Province: Livorno
Municipality: Campiglia Marittima

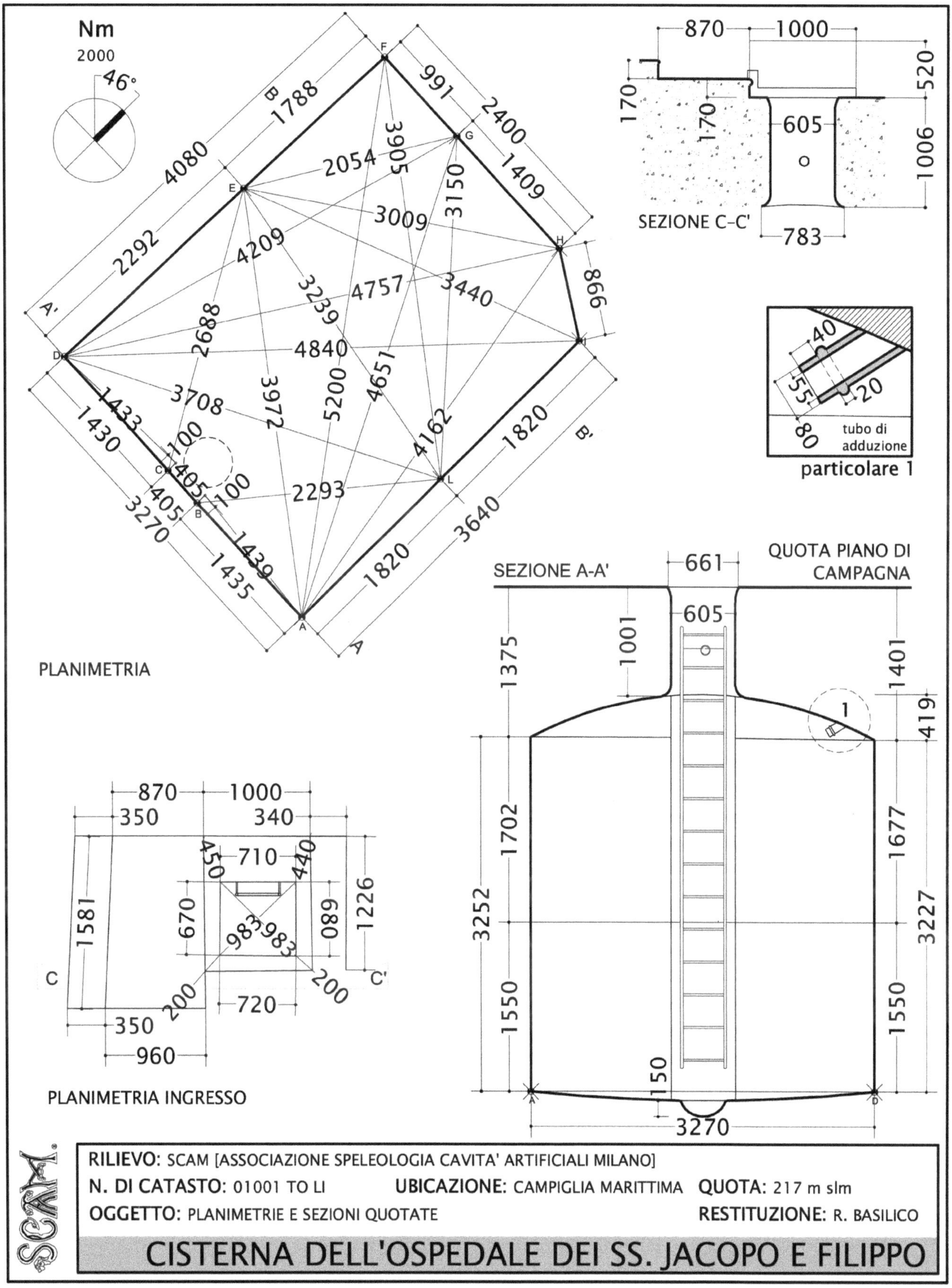

Fig. III.77.a. *San Jacopo and San Filippo Hospital Cistern (CA 01001 PI TO) (Ass.ne S.C.A.M. Archive).*

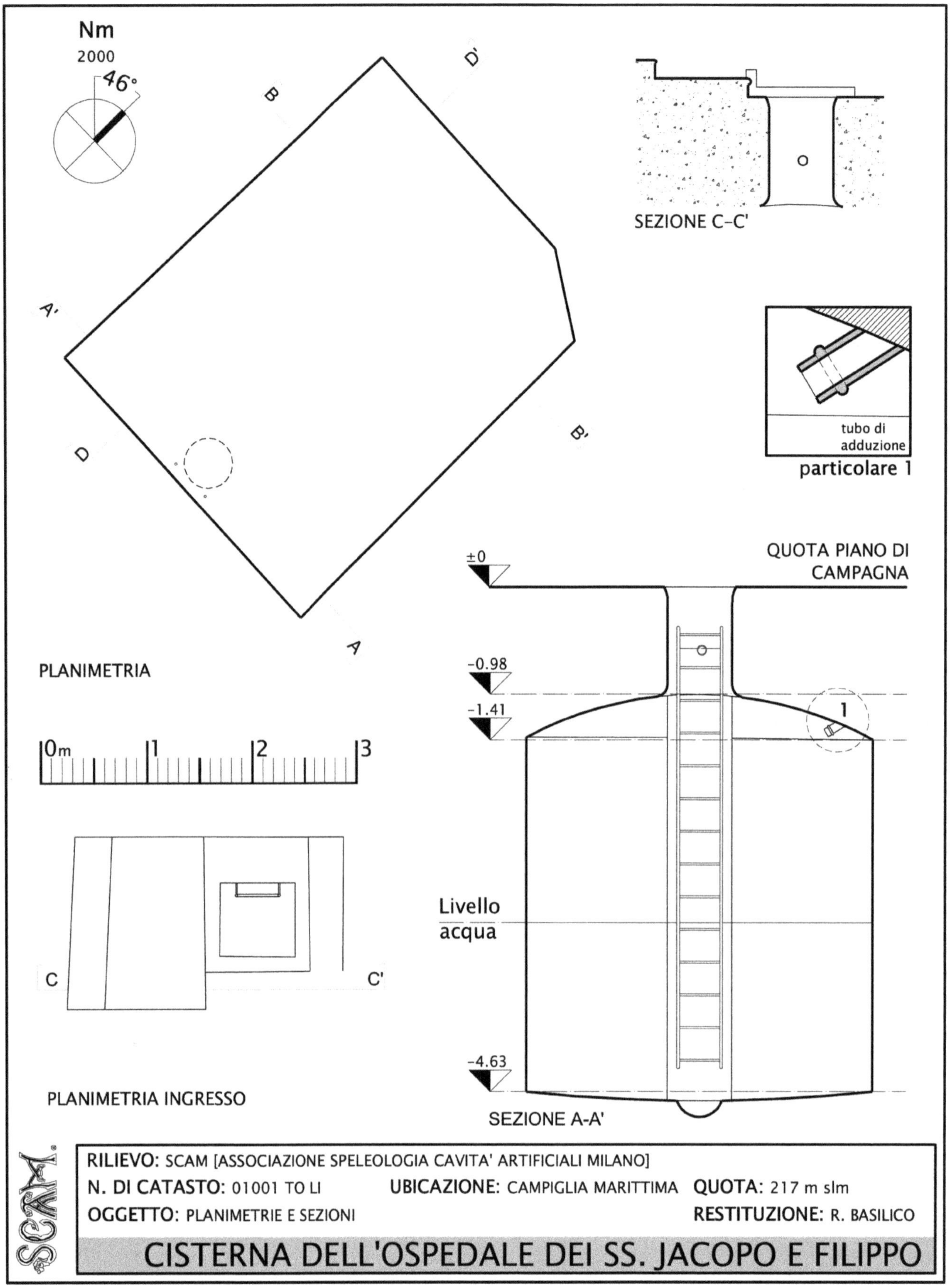

Fig. III.77.b. *San Jacopo and San Filippo Hospital Cistern (CA 01001 PI TO) (Ass.ne S.C.A.M. Archive).*

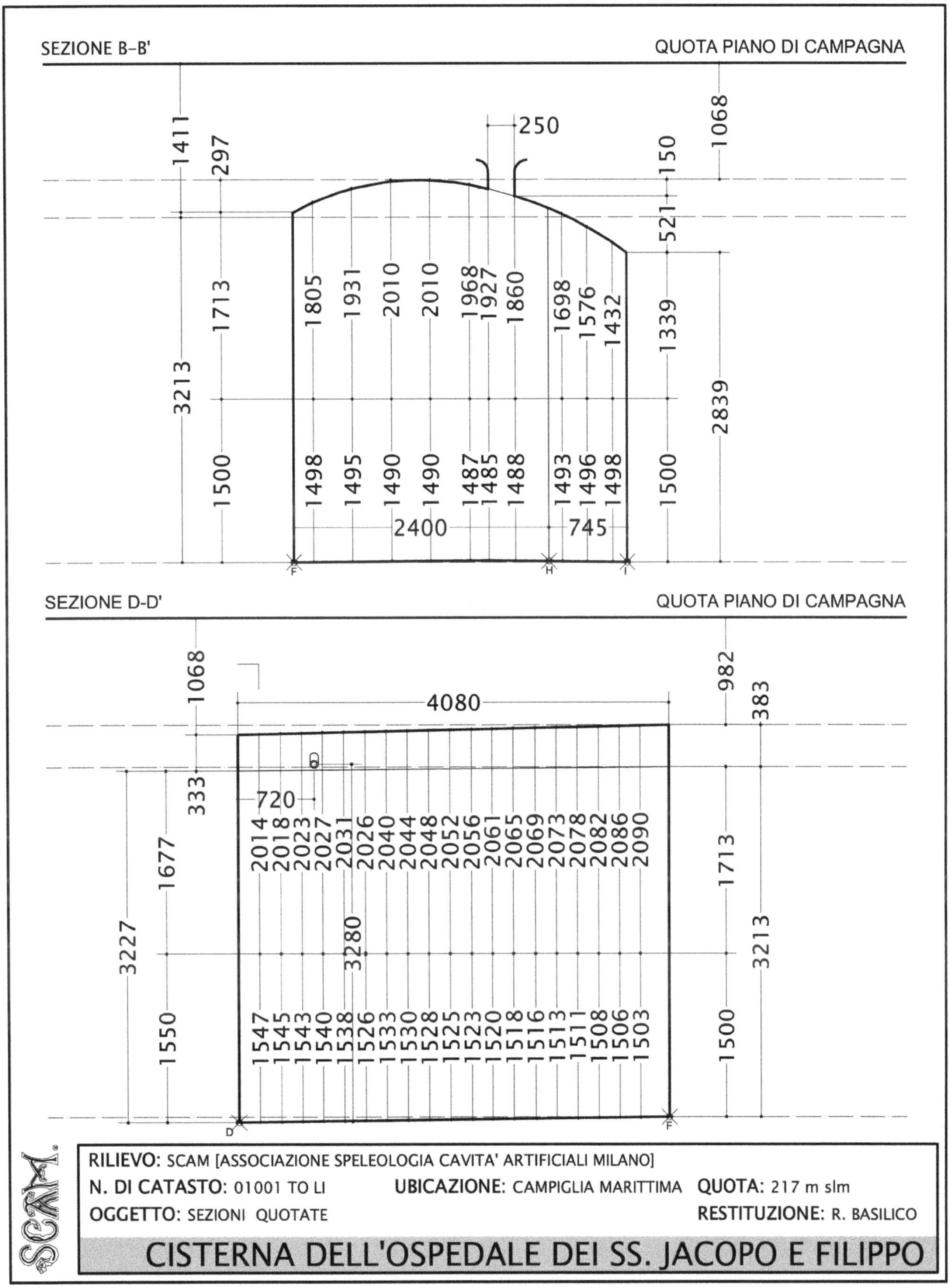

Fig. III.77.c. *San Jacopo and San Filippo Hospital Cistern (CA 01001 PI TO) (Ass.ne S.C.A.M. Archive).*

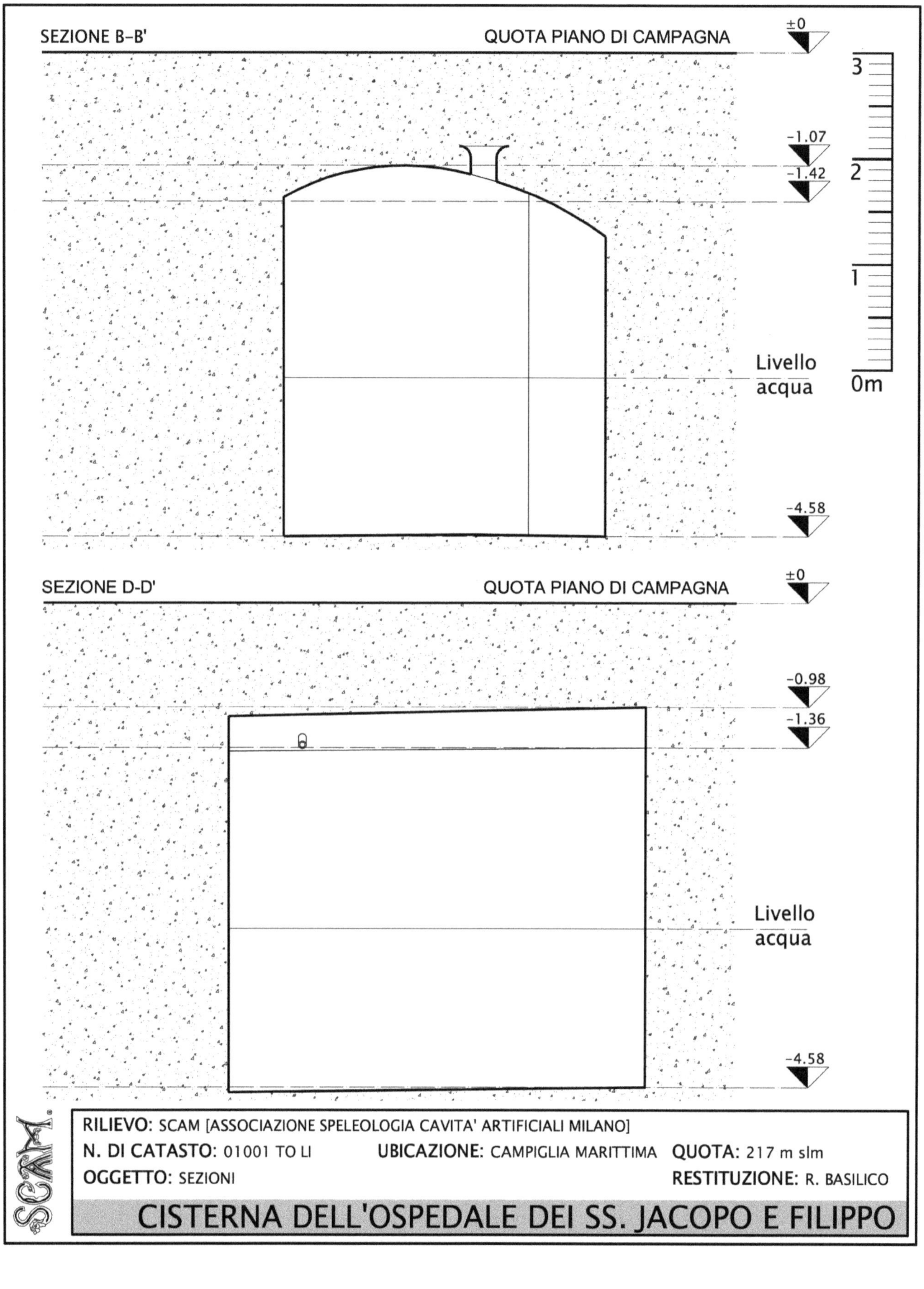

Fig. III.77.d. *San Jacopo and San Filippo Hospital Cistern (CA 01001 PI TO) (Ass.ne S.C.A.M. Archive).*

Locality: Campiglia Marittima
Location: inside the current Campiglia Marittima Municipal building
Ownership: Municipality of Campiglia Marittima
Cartography: C.T.R. 1:1000; C.T.R. 1: 2000, sheet 10G14
Geological unit: Oligocene-Cretaceous Eocene, varicoloured Argillite ("Scaglia Toscana")
Altitude: 217 m a.s.l.
Position: 4768750 N 1631750 E
Context: situated inside the old "San Jacopo and San Filippo Hospital" (as detailed in a XV century document, this structure shows evidence of XIII century walled structures); it is the currently site of the Municipal Building
Operations conducted: exploration, planimetric survey and photographic service
Work carried out by: Associazione Speleologia Cavità Artificiali Milano (S.C.A.M.); September 2000
Warnings: poor air circulation
Typology: 2c - cistern
Description: the storage system, consisting of an entrance shaft and a storage chamber, is located directly under a room. The well's opening is in the northeast corner of a room and is adjacent to the wall. It is positioned 0.52 m above the floor surface at the bottom of a short flight of steps. The bottom of the well slopes towards the basin under the entrance shaft. Dimensions: the quadrangular entrance measures approximately 0.7 m and descends for 1.05 m where it comes into contact with the roof of the storage chamber, its exposed area measuring 3.7 m and its other side measuring 3.5 m. The shorter side "closes" the east corner of the room, which would otherwise be rectangular. Footholds: none found. A metal wall ladder provides easy access to the well. Piping: an earthenware pipe with internal diameter of 9.5 cm lies just under the external edge of the well, almost on a level with the room's floor. Inside the chamber is an earthenware overhanging pipe with internal diameter of approximately 10 cm, which slopes inwards (CC' section); in the truncated corner at base of the chamber, is a large quadrangular opening in the vault (approximately 25 cm), which may once have housed another water supply inlet.
Interpretation: the external environment has clearly been re-structured and little evidence of its original state survives. Consequently, the cistern could have been built during the last restructuring.
Dating: XIII-XV century.
Notes: the wall lining appears homogenous as does the floor lining, although different materials have been used: this would suggest that they were built at different times or during a restructure.
Bibliography: Basilico R., Casini A., Padovan G., *Le cisterne del castello di Campiglia Marittima (LI)*, in Federazione Speleologica Toscana, *Atti del VII Congresso della Federazione Speleologica Toscana (Gavorrano, 31 Marzo - 1 Aprile 2001), Talp. Rivista della Federazione Speleologica Toscana*, No. 23, Lucca 2001, pgs. 139-147.
Basilico R., Casini A., Padovan G., *La storia dell'acqua in un castello della Maremma Toscana: Campiglia Marittima*, in Club Alpinistico Triestino - Gruppo Grotte Sezione Ricerche e Studi su Cavità Artificiali (edited by), *Atti del V Convegno Nazionale sulle Cavità Artificiali. Osoppo 28 Aprile - 1 Maggio 2001*, Trieste 2002, pgs. 41-68.
Data ownership: Associazione Speleologia Cavità Artificiali Milano (S.C.A.M.).
Compiled by: Gianluca Padovan (Ass.ne S.C.A.M.).

III.4.3 - *CA 01008 TO LI; Piazzetta Well*

Cadastral number: CA 01002
Denomination: Piazzetta Well (fig. III.78)
Region-country: Tuscany, Italy
Province: Livorno
Municipality: Campiglia Marittima
Locality: Campiglia Marittima
Location: situated at the north corner of Piazza Mazzini, near the Church of San Fiorenzo and next to the XIII century walls.
Ownership: Municipality of Campiglia Marittima
Cartography: C.T.R. 1:1000; C.T.R. 1: 2000, sheet 10G14
Geological unit: Eocene, Canetolo Formation (external unit)
Altitude: 225 m a.s.l.
Position: 4768750 N 1631750 E
Context: situated inside a square and without a puteal
Operations conducted: excavation, survey and partial photographic service
Work carried out by: Associazione Speleologia Cavità Artificiali Milano (S.C.A.M.); November 2000
Warnings: extremely poor air circulation
Typology: 2c - cistern

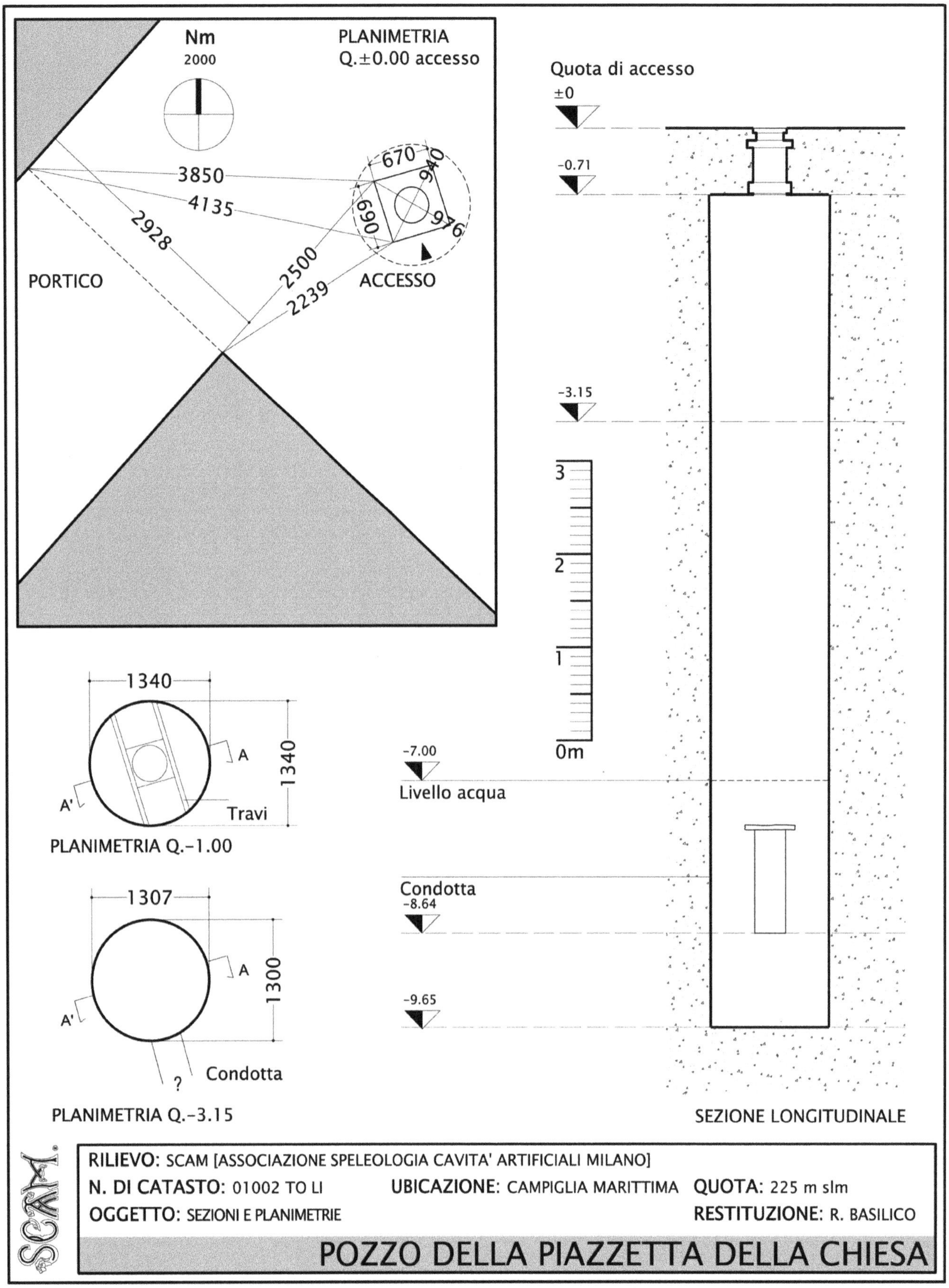

Fig. III.78. *Piazzetta Well (CA 01002 TO LI) (Ass.ne S.C.A.M. Archive).*

Description: a stone manhole covers the quadrangular entrance of a vertical shaft, consisting of a grey sandstone quadrangle with a circular cover of the same material, in the paving of the square. The vertical shaft was covered when the structure fell into disuse. The section between the cover and the lining is made of stone material and there are two iron beams, just under the street paving. The wall layout consists of rows of pale limestone, which are not always regular. The vaulted roof and jack arch of a southeast facing 1 m long tunnel is visible 7 m from the street entrance, just below the water level; this could well be the entrance to a hydraulic channel. It may also be linked to a cistern. The base appears to have been made with various materials and is covered by fine sediment. Dimensions: the entrance measures 0.32 m in diameter, widening to 1.23 m under the puteal, maintaining a fairly stable diameter until the point where it comes into contact with the water at a depth of -6.5 m. Footholds: none found. Piping: none found.
Interpretation: in all appearances a well, this is in fact a cistern. It will not be possible to ascertain whether the cistern was utilised for the storage of rainwater or for mains water until such a time as a full investigation is conducted.
Dating: XIII century.
Notes: the water level should be decreased to allow a clear view of the hypothetical tunnel and to remove the need for underwater speleological exploration. The well-curb is clearly 0.7 m higher that the lining: it appears that the covering was retained even after the square was restructured and paved. A lead pipe currently descends into the water from the roof.
Bibliography: Basilico R., Casini A., Padovan G., *Le cisterne del castello di Campiglia Marittima (LI)*, in Federazione Speleologica Toscana, *Atti del VII Congresso della Federazione Speleologica Toscana (Gavorrano, 31 Marzo - 1 Aprile 2001)*, *Talp. Rivista della Federazione Speleologica Toscana*, No. 23, Lucca 2001, pgs. 139-147.
Basilico R., Casini A., Padovan G., *La storia dell'acqua in un castello della Maremma Toscana: Campiglia Marittima*, in Club Alpinistico Triestino - Gruppo Grotte Sezione Ricerche e Studi su Cavità Artificiali (edited by), *Atti del V Convegno Nazionale sulle Cavità Artificiali. Osoppo 28 Aprile - 1 Maggio 2001*, Trieste 2002, pgs. 41-68.
Data ownership: Associazione Speleologia Cavità Artificiali Milano (S.C.A.M.).
Compiled by: Gianluca Padovan (Ass.ne S.C.A.M.).

III.4.4 - *CA 01003 TO LI; Church of San Fiorenzo Cistern*

Cadastral number: CA 01003 TO LI
Denominating: Church of San Fiorenzo Cistern (figs. III.79.a, III.79.b, III.79.c, III.79.d, III.79.e, III.80 and III.81)
Region-country: Tuscany, Italy
Province: Livorno
Municipality: Campiglia Marittima
Locality: Campiglia Marittima
Location: the storage work is situated behind the northwest wall of the XIII century church
Ownership: Municipality of Campiglia Marittima
Cartography: C.T.R. 1:1000; C.T.R. 1: 2000, sheet 10G14
Geological unit: Eocene, Canetolo Formation (external unit)
Altitude: 222 m a.s.l.
Position: 4768750 N 1631750 E
Context: located within a square
Operations conducted: exploration, survey, photographic and video service
Work carried out by: Associazione Speleologia Cavità Artificiali Milano (S.C.A.M.); November 2000
Warnings: poor air circulation
Typology: 2c - cistern
Description: consists of an entrance shaft, a storage chamber and a small room just below the road surface. The opening is in the southeast corner of the square, right next to the church wall. The quadrangular cistern is covered by a drop arch. It is made up of blocks covered in waterproof plaster and the previous water level is indicted at a height of 3.5 m from the base of the cistern. A small filtering chamber is situated behind the southeast wall at a similar depth to the springer. It was not surveyed as it is not accessible. A previous entrance, situated in the southeast corner of the chamber and visible from inside the cistern, is now blocked by a newspaper stand. Dimensions: the circular entrance has a diameter of 36 cm and descends for 0.7 m where it comes into contact with the roof of the storage chamber, measuring 7.14 m in height. Both piers measure 4.4 m. In a clockwise direction starting from the entrance side, the base measures 4.65 x 4.98 x 4.74 x 5.08 m. Footholds: none found. Piping: two overlapping pipes (with respective internal diameter of 5.5 cm and 5 cm), emerge from the filtering chamber.
Interpretation: this is undoubtedly a rainwater storage cistern
Dating: XIII century.
Note: a row of stones protrudes at a depth of -4.8 m. The cistern currently contains both water and hydrocarbons.

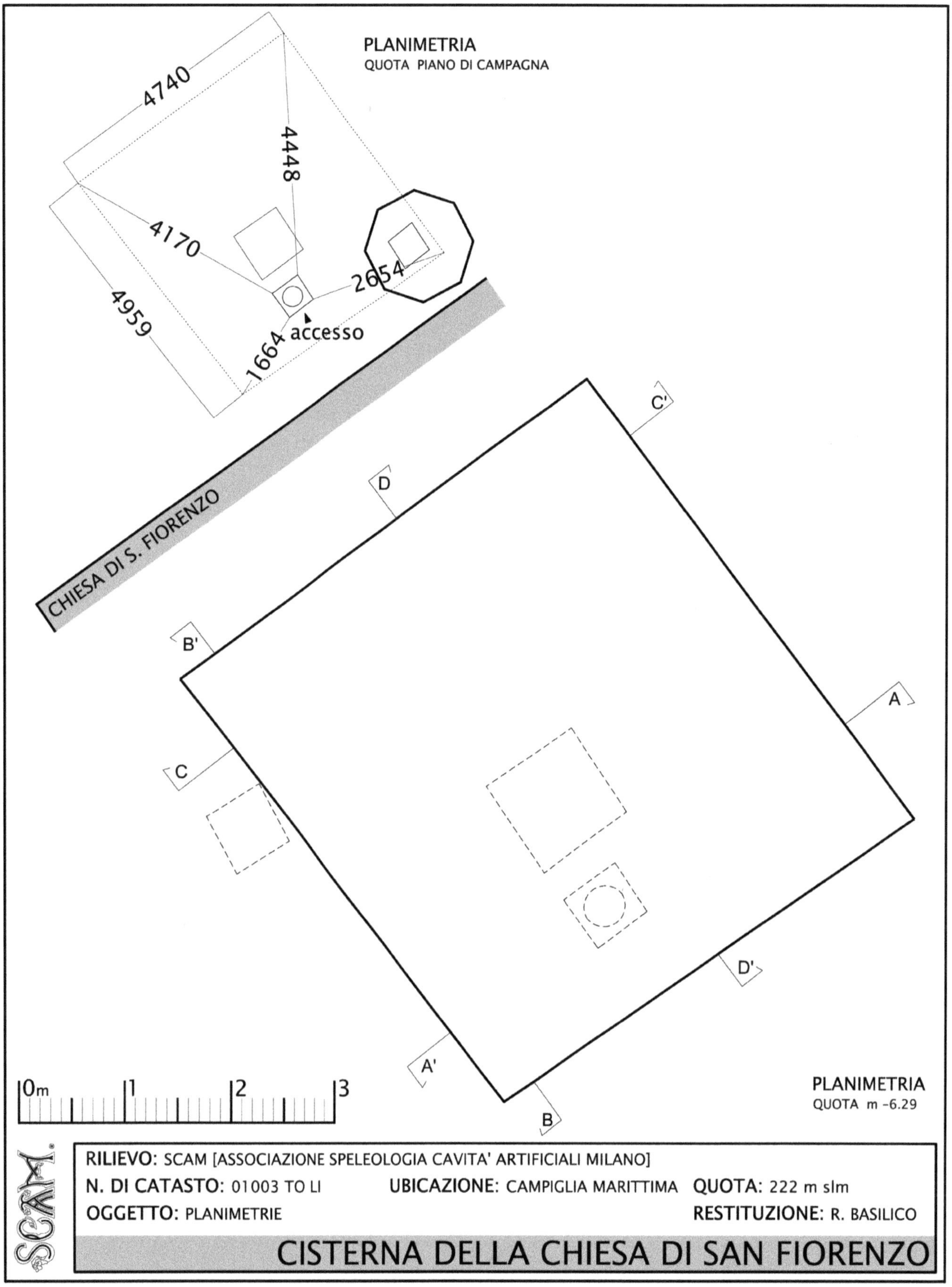

Fig. III.79.a. *Church of San Fiorenzo Cistern (CA 01003 TO LI) (Ass.ne S.C.A.M. Archive).*

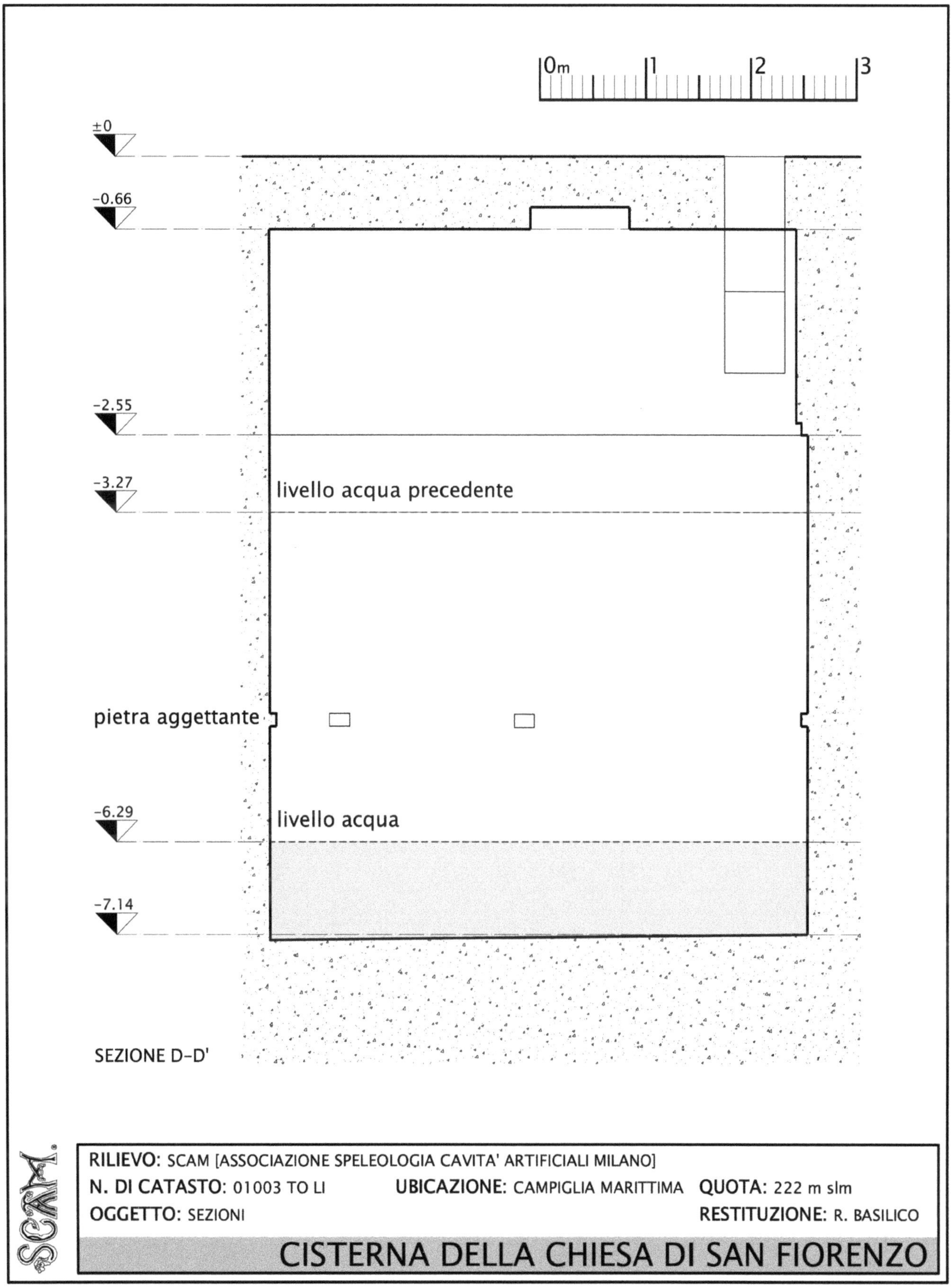

Fig. III.79.b. *Church of San Fiorenzo Cistern (CA 01003 TO LI) (Ass.ne S.C.A.M. Archive).*

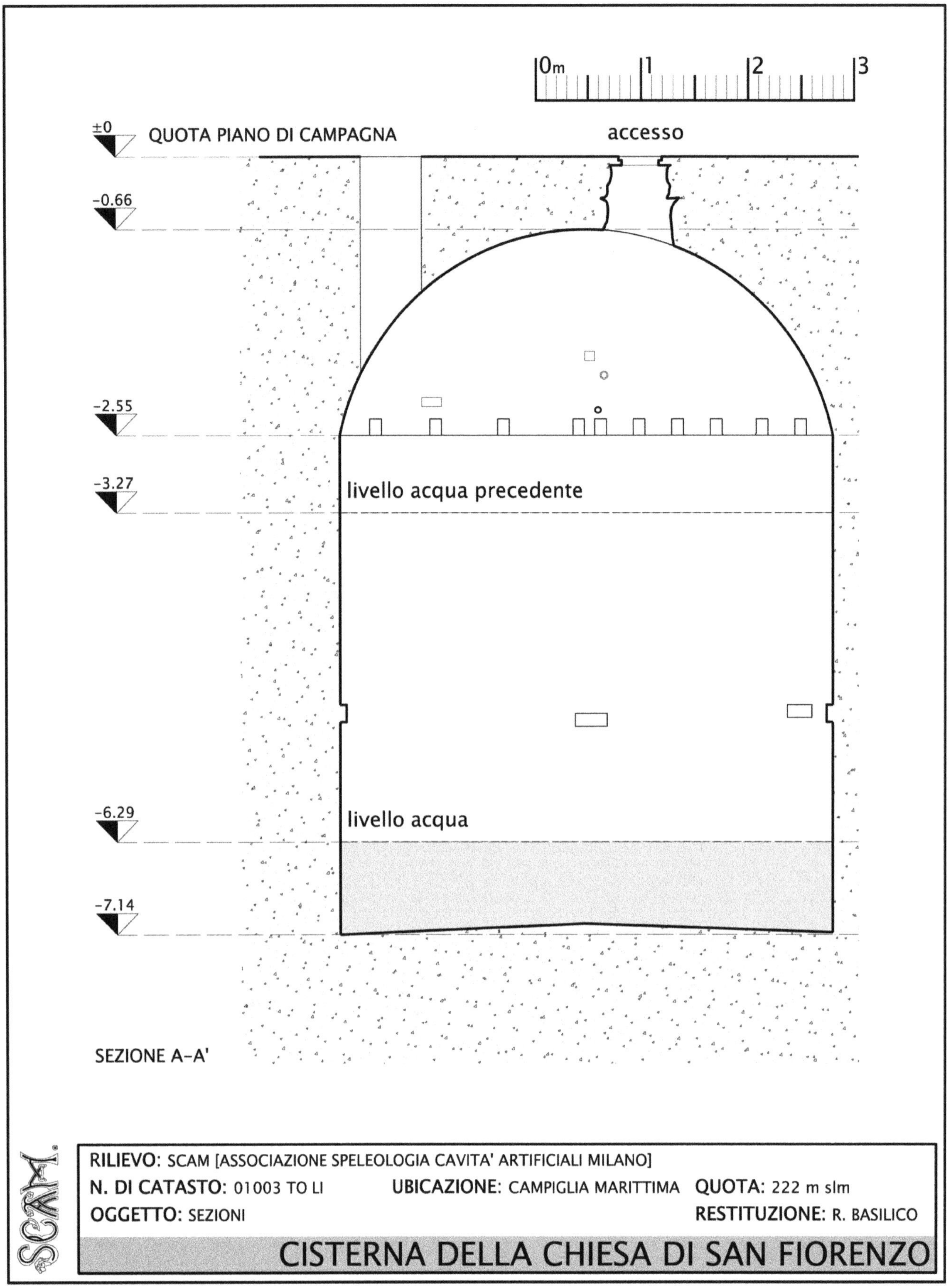

Fig. III.79.c. *Church of San Fiorenzo Cistern (CA 01003 TO LI) (Ass.ne S.C.A.M. Archive).*

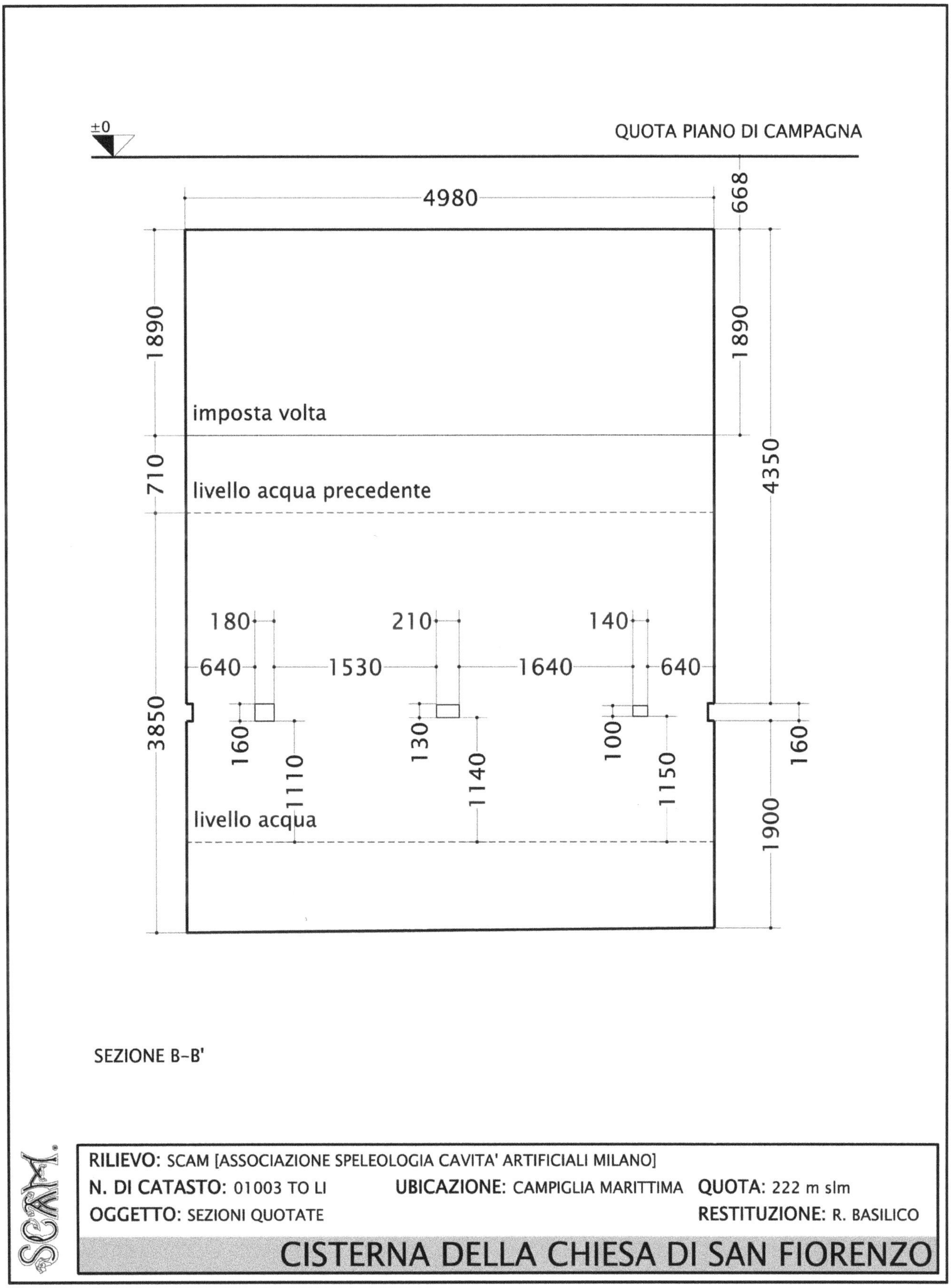

Fig. III.79.d. *Church of San Fiorenzo Cistern (CA 01003 TO LI) (Ass.ne S.C.A.M. Archive).*

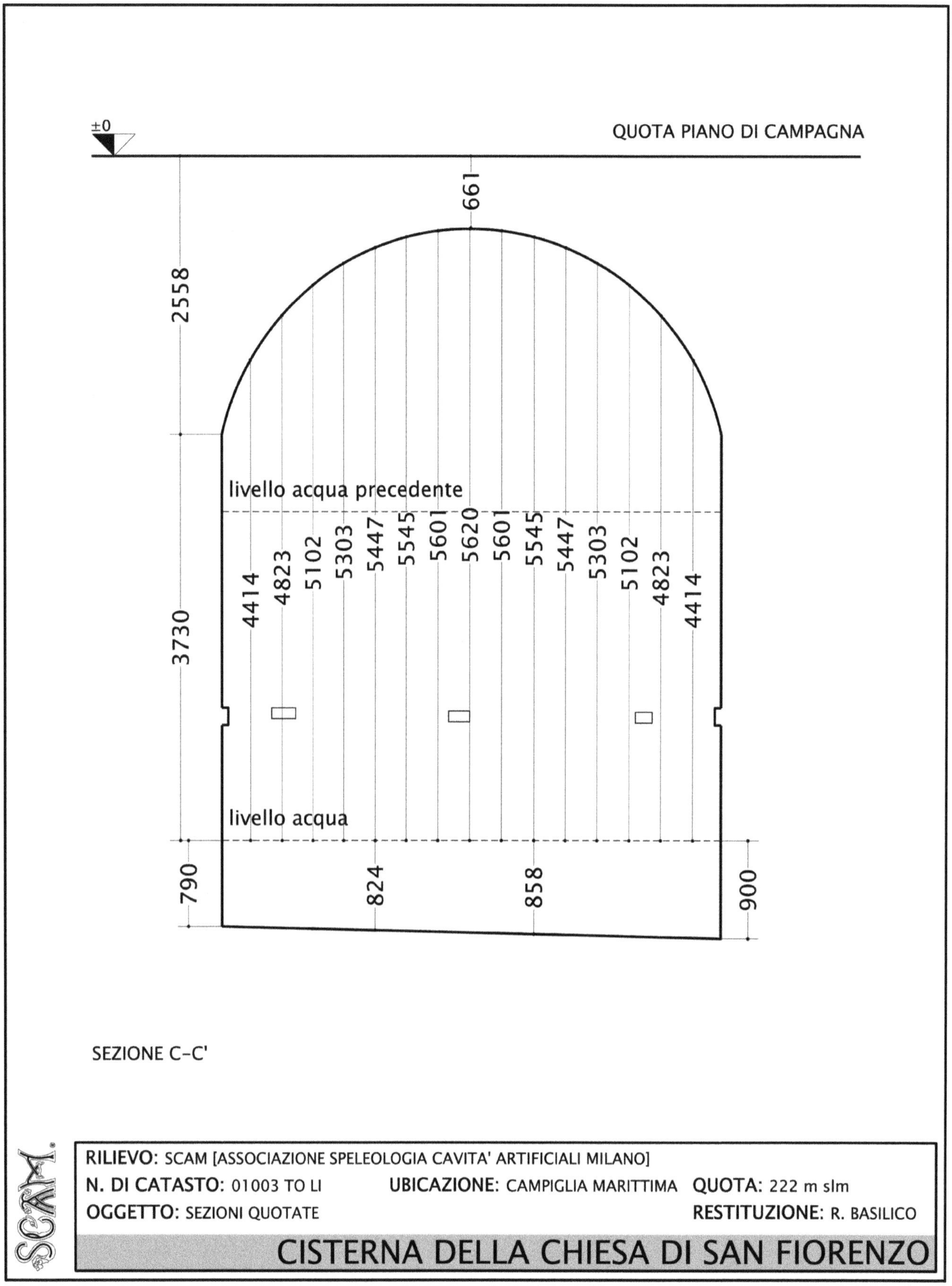

Fig. III.79.e. *Church of San Fiorenzo Cistern (CA 01003 TO LI) (Ass.ne S.C.A.M. Archive).*

Fig. III.80. Church of San Fiorenzo Cistern (CA 01003 TO LI). A speleologist at the structure's north corner (photo G. Padovan).

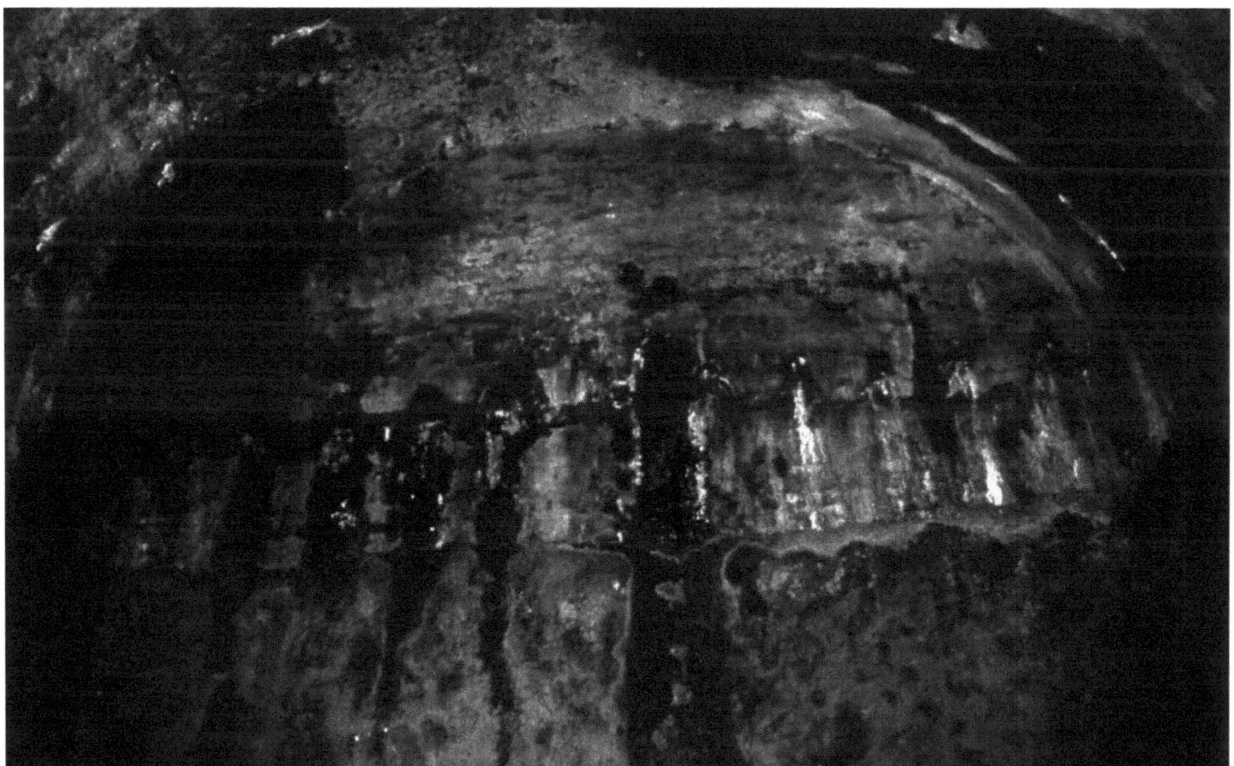

Fig. III.81. *Church of San Fiorenzo Cistern (CA 01003 TO LI). High section of east wall (section AA' in the survey) (photo G. Padovan).*

Bibliography: Basilico R., Casini A., Padovan G., *Le cisterne del castello di Campiglia Marittima (LI)*, in Federazione Speleologica Toscana, *Atti del VII Congresso della Federazione Speleologica Toscana (Gavorrano, 31 Marzo - 1 Aprile 2001), Talp. Rivista della Federazione Speleologica Toscana*, No. 23, Lucca 2001, pgs. 139-147. Basilico R., Casini A., Padovan G., *La storia dell'acqua in un castello della Maremma Toscana: Campiglia Marittima*, in Club Alpinistico Triestino - Gruppo Grotte Sezione Ricerche e Studi su Cavità Artificiali (edited by), *Atti del V Convegno Nazionale sulle Cavità Artificiali. Osoppo 28 Aprile - 1 Maggio 2001*, Trieste 2002, pgs. 41-68.
Data ownership: Associazione Speleologia Cavità Artificiali Milano (S.C.A.M.).
Compiled by: Gianluca Padovan (Ass.ne S.C.A.M.).

III.4.5 - *CA 01004 TO LI; Long Well*

Cadastral number: CA 01004 TO LI
Denomination: Long Well (fig. III.82)
Region-country: Tuscany, Italy
Province: Livorno
Municipality: Campiglia Marittima
Locality: Campiglia Marittima
Location: centrally positioned within a small square; until the early 1970s it was located inside a building, which was later demolished
Ownership: Municipality of Campiglia Marittima
Cartography: C.T.R. 1:1000; C.T.R. 1: 2000, sheet 10G14
Geological unit: Oligocene-Cretaceous Eocene, varicoloured Argillite ("Scaglia Toscana")
Altitude: 219 m a.s.l.
Position: 4768800 N 1631700 E
Context: centrally positioned within Piazza Martiri di via Fani
Operations conducted: exploration, survey, photographic service
Work carried out by: Associazione Speleologia Cavità Artificiali Milano (S.C.A.M.); September 2000
Warnings: poor air circulation, sewage
Typology: 2b - ordinary well (most probably)
Description: the well consists of a single cylindrical element; the wall, in local limestone rock, is set out in fairly regular rows. The niches carved into the sides of the well were probably used for beam support. Access is gained from by means of a manhole on the road surface, the first 0.82 m of which is partially obstructed. Dimensions: the depth of the well from the road surface is 12.7 m and the shape of the well is regular (1.7 m in diameter) throughout its course. The section is intercepted by niches carved at -1.75 m and -5.6 m. Footholds: none found. Piping: none found.
Interpretation: this appears to be an ordinary well, however it cannot be excluded that it could be a cistern.
Dating: XIII century.
Notes: at a depth of between 4 m and 8 m, the walls present evident signs of sewage; it is thought that a nearby sewage channel may present a leak.
Bibliography: Basilico R., Casini A., Padovan G., *Le cisterne del castello di Campiglia Marittima (LI)*, in Federazione Speleologica Toscana, *Atti del VII Congresso della Federazione Speleologica Toscana (Gavorrano, 31 Marzo - 1 Aprile 2001), Talp. Rivista della Federazione Speleologica Toscana*, No. 23, Lucca 2001, pgs. 139-147. Basilico R., Casini A., Padovan G., *La storia dell'acqua in un castello della Maremma Toscana: Campiglia Marittima*, in Club Alpinistico Triestino - Gruppo Grotte Sezione Ricerche e Studi su Cavità Artificiali (edited by), *Atti del V Convegno Nazionale sulle Cavità Artificiali. Osoppo 28 April - 1 May 2001*, Trieste 2002, pgs. 41-68.
Data ownership: Associazione Speleologia Cavità Artificiali Milano (S.C.A.M.).
Compiled by: Gianluca Padovan (Ass.ne S.C.A.M.).

III.4.6 - *CA 01005 TO LI; Pretorio Palace Cistern*

Cadastral number: CA 01005 TO LI
Denomination: Pretorio Palace Cistern (figs. III.83.a, III.83.b, III.83.c, III.83.d and III.84,)
Region-country: Tuscany, Italy
Province: Livorno
Municipality: Campiglia Marittima
Locality: Campiglia Marittima
Location: situated on the via Vecchio Pretorio side, in front of Pretorio Palace
Ownership: Municipality of Campiglia Marittima

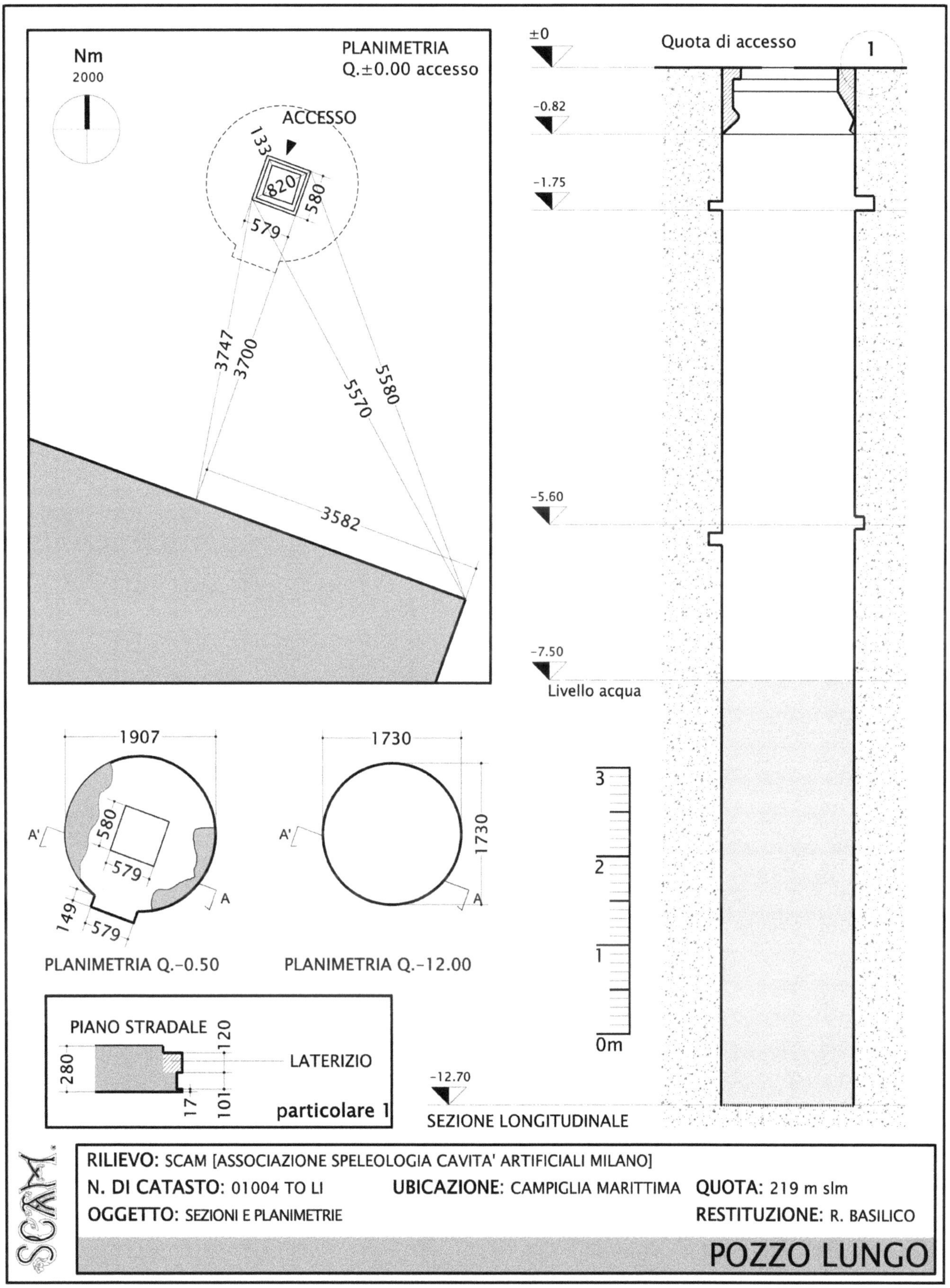

Fig. III.82. *Long Well (CA 01004 TO LI) (Ass.ne S.C.A.M. Archive).*

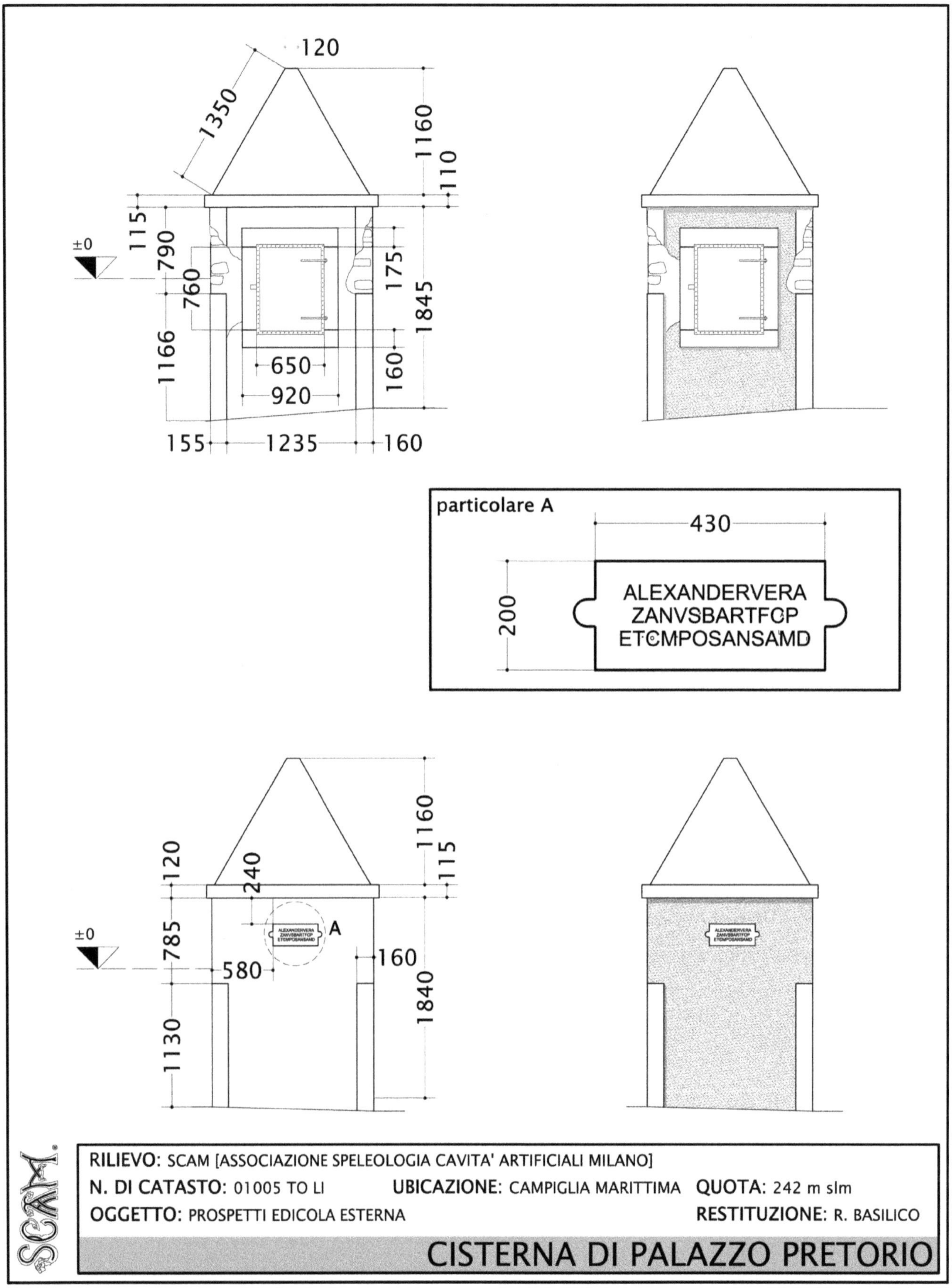

Fig. III.83.a. *Pretorio Palace Cistern (CA 01005 TO LI) (Ass.ne S.C.A.M. Archive).*

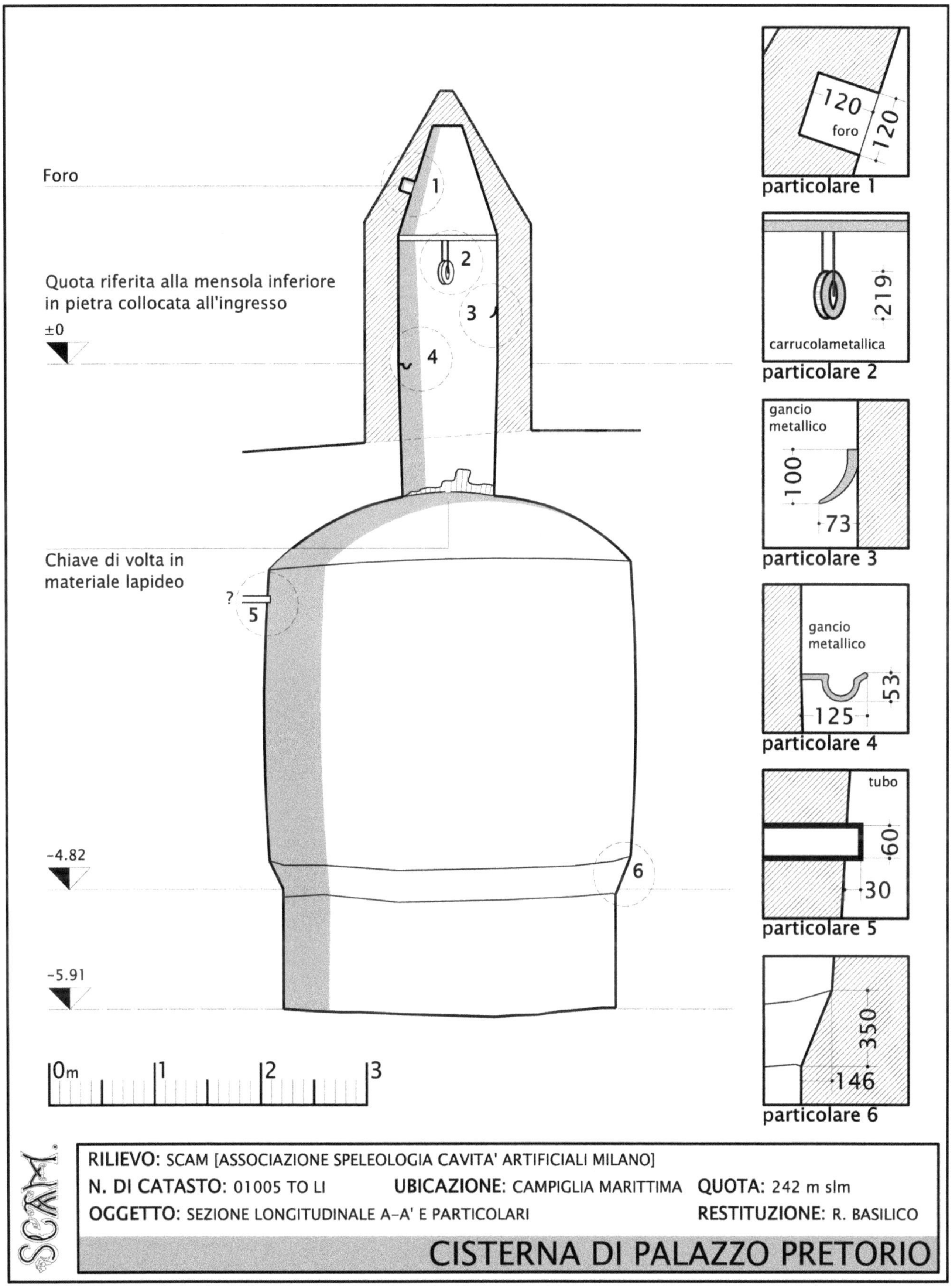

Fig. III.83.b. *Pretorio Palace Cistern (CA 01005 TO LI) (Ass.ne S.C.A.M. Archive).*

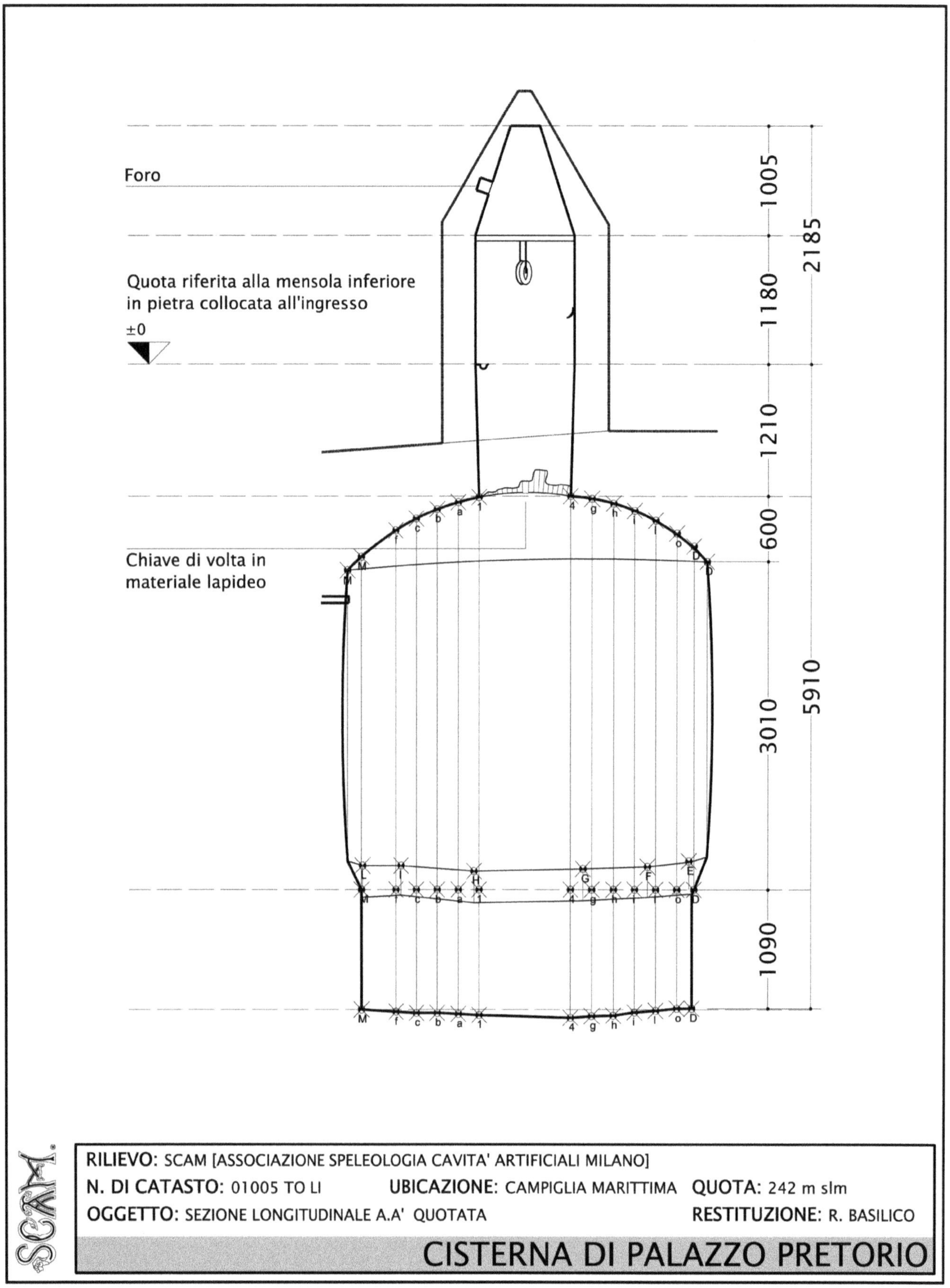

Fig. III.83.c. *Pretorio Palace Cistern (CA 01005 TO LI) (Ass.ne S.C.A.M. Archive).*

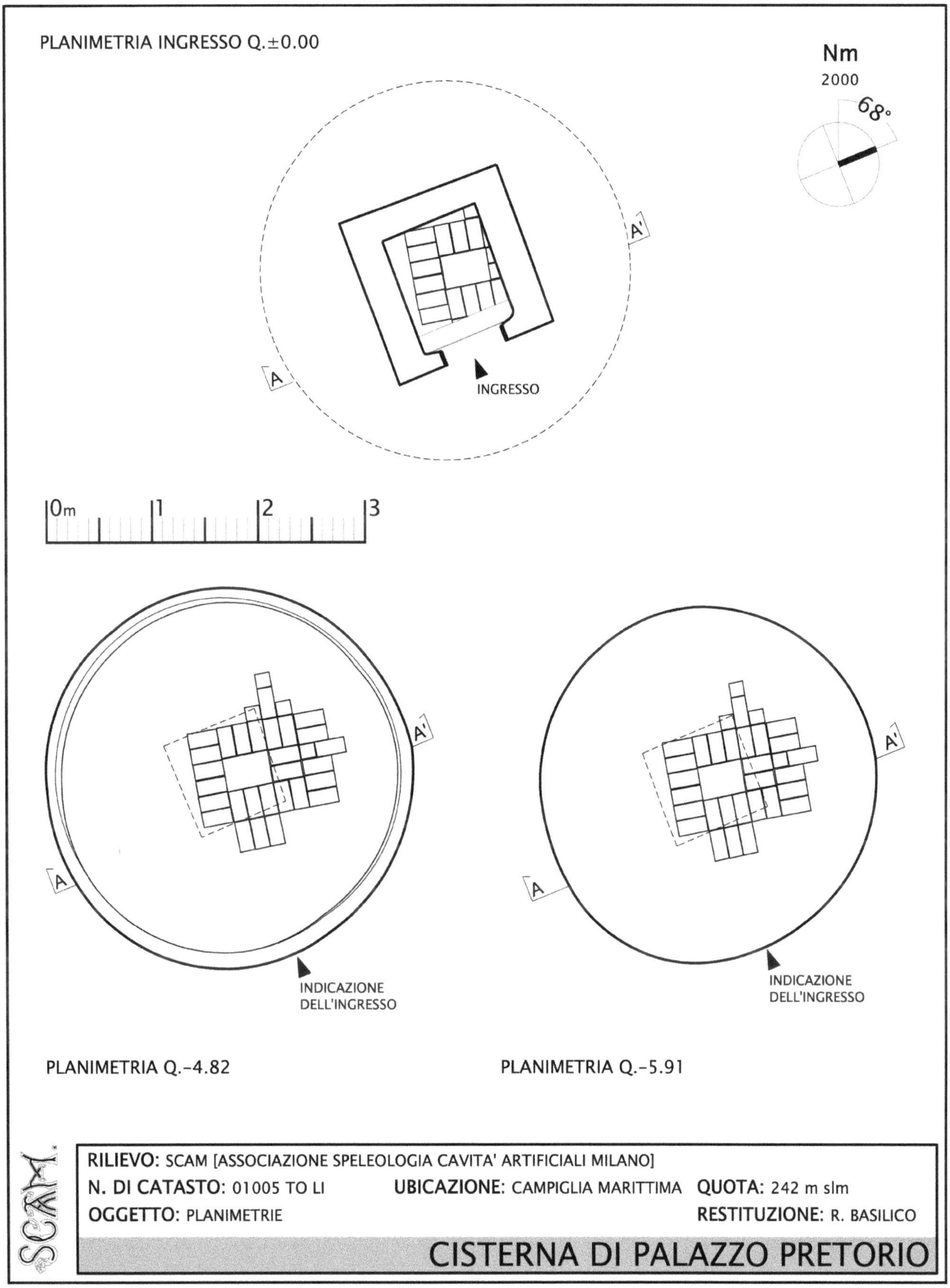

Fig. III.83.d. *Pretorio Palace Cistern (CA 01005 TO LI) (Ass.ne S.C.A.M. Archive).*

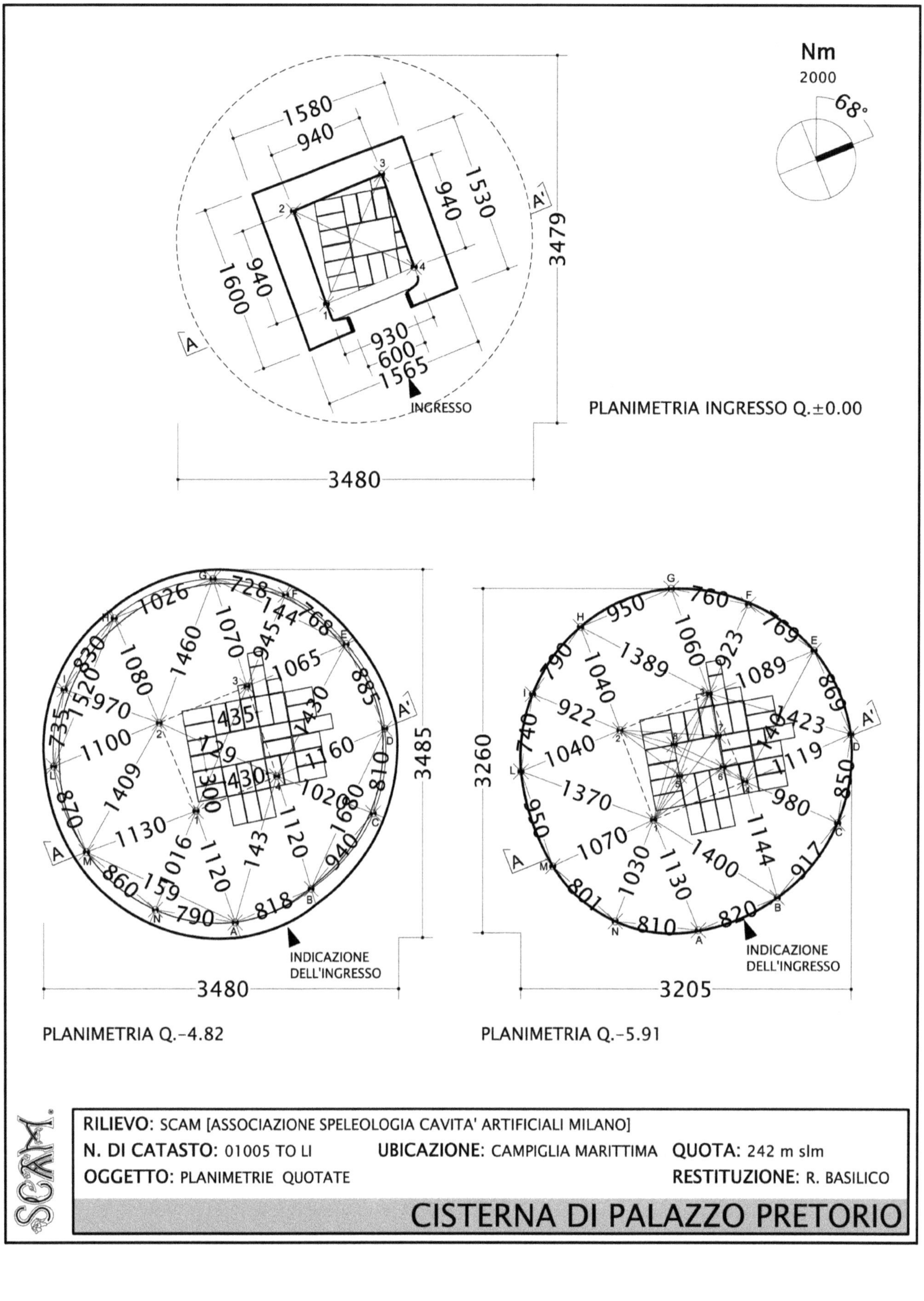

Fig. III.84. *Pretorio Palace Cistern (CA 01005 TO LI). Trilateration detail (Ass.ne S.C.A.M. Archive).*

Cartography: C.T.R. 1:1000; C.T.R. 1: 2000, sheet 10G14
Geological unit: Eocene, Canetolo Formation (external unit)
Altitude: 242 m a.s.l.
Position: 4768800 N 1631730 E
Context: the storage system is located next to a private building, at the top of a set of steps
Operations conducted: exploration, survey and photographic service
Work carried out by: Associazione Speleologia Cavità Artificiali Milano (S.C.A.M.); November 2000
Warnings: none
Typology: 2c - cistern
Description: the cistern consists of a circular chamber, accessed via a well located within an aedicule. The man-made structure consists of a quadrangular aedicule with a small pyramidal cupola, the upper part of which is missing. A horizontal central fascia emphasises the gap between the two, while the corners of the lower element are enhanced by four bardiglio marble pilasters, which extend to half the height of the prism. The base of the cylindrical chamber is inclined towards the collection well under the entrance shaft. The transversal section of the walls between the springer and the point in which it narrows are convex, thus reducing the diameter. Hydraulic mortar, missing in various places, has rendered the brick chamber impermeable. Dimensions: the 0.6 x 0.8 m entrance descends for 1.4 m where it comes into contact with the roof of the storage chamber, which is 4.9 m high, at the external surface. At a depth of -4.91 m, its diameter is of 2.9 m, while at -3.55 m, its diameter is of 3.2 m. Footholds: none found. Piping: there is a metal pipe 22 cm below the springer.
Interpretation: thanks to its easily recognizable shape, the cistern is an identifiable sign of the urban network. The entrance threshold presents grooving caused by the ropes or chains, indicative of the fact that a pulley was not always used to lift the buckets, which survive to this very day.
Dating: 1503.
Notes: the storage chamber appears to have been created during a later excavation, indicated by the point where the diameter narrows.
Bibliography: Basilico R., Casini A., Padovan G., *Le cisterne del castello di Campiglia Marittima (LI)*, in Federazione Speleologica Toscana, *Atti del VII Congresso della Federazione Speleologica Toscana (Gavorrano, 31 Marzo - 1 Aprile 2001)*, *Talp. Rivista della Federazione Speleologica Toscana*, No. 23, Lucca 2001, pgs. 139-147.
Basilico R., Casini A., Padovan G., *La storia dell'acqua in un castello della Maremma Toscana: Campiglia Marittima*, in Club Alpinistico Triestino - Gruppo Grotte Sezione Ricerche e Studi su Cavità Artificiali (edited by), *Atti del V Convegno Nazionale sulle Cavità Artificiali. Osoppo 28 Aprile - 1 Maggio 2001*, Trieste 2002, pgs. 41-68.
Data ownership: Associazione Speleologia Cavità Artificiali Milano (S.C.A.M.).
Compiled by: Roberto Basilico (Ass.ne S.C.A.M.).

III.4.7 - *CA 01006 TO LI; Wall Cistern (or Arch Cistern)*

Cadastral number: CA 01006 TO LI
Denomination: Wall Cistern (or Arch Cistern) (figs. III.85.a, III.85.b, III.85.c, III.85.d and III.86)
Region-country: Tuscany, Italy
Province: Livorno
Municipality: Campiglia Marittima
Locality: Campiglia Marittima
Location: situated in Via Vecchio Pretorio in the area next to Pretorio Palace and near the remains of a mediaeval building
Ownership: Municipality of Campiglia Marittima
Cartography: C.T.R. 1:1000; C.T.R. 1: 2000, sheet 10G14
Geological unit: Eocene, Canetolo Formation (external unit)
Altitude: 245 m a.s.l.
Position: 4768800 N 1631730 E
Context: the structure is situated within the pier of an archway
Operations conducted: exploration, survey and photographic service
Work carried out by: Associazione Speleologia Cavità Artificiali Milano (S.C.A.M.); September 2000
Warnings: none.
Typology: 2c - storage
Description: consisting of an entrance shaft and storage chamber, the layout of the local limestone masonry framework is fairly regular; the deep well reaches and extends beyond the top of the archway, which opens onto terraced earthwork. Entrance: located within the pier of an arch, the entrance is framed in limestone rock and is closed by a wooden door with a small bolt. The threshold presents numerous ridges, caused by the ropes or chains

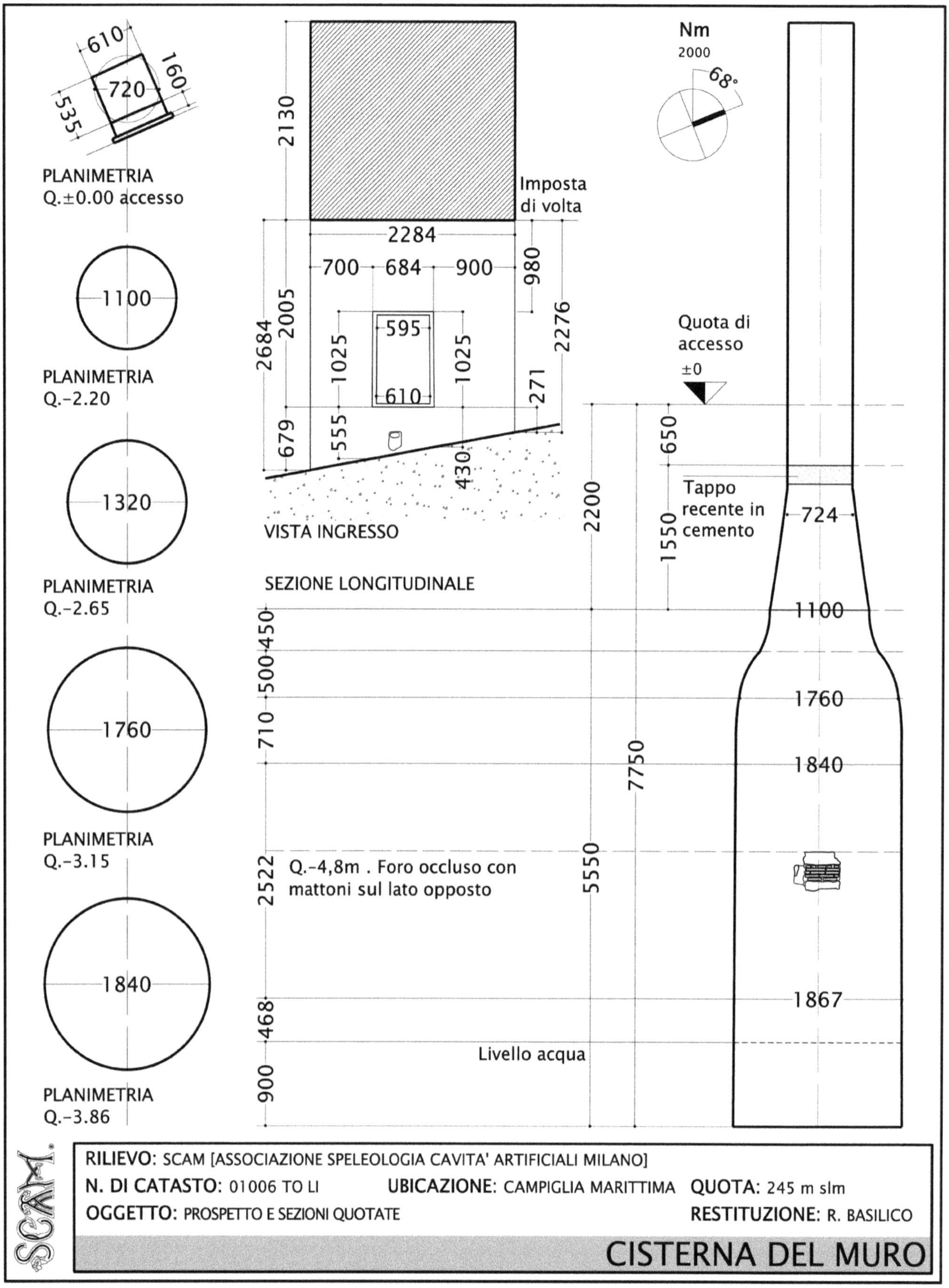

Fig. III.85. a. *Wall Cistern (or Arch Cistern) (CA 01006 TO LI) (Ass.ne S.C.A.M. Archive).*

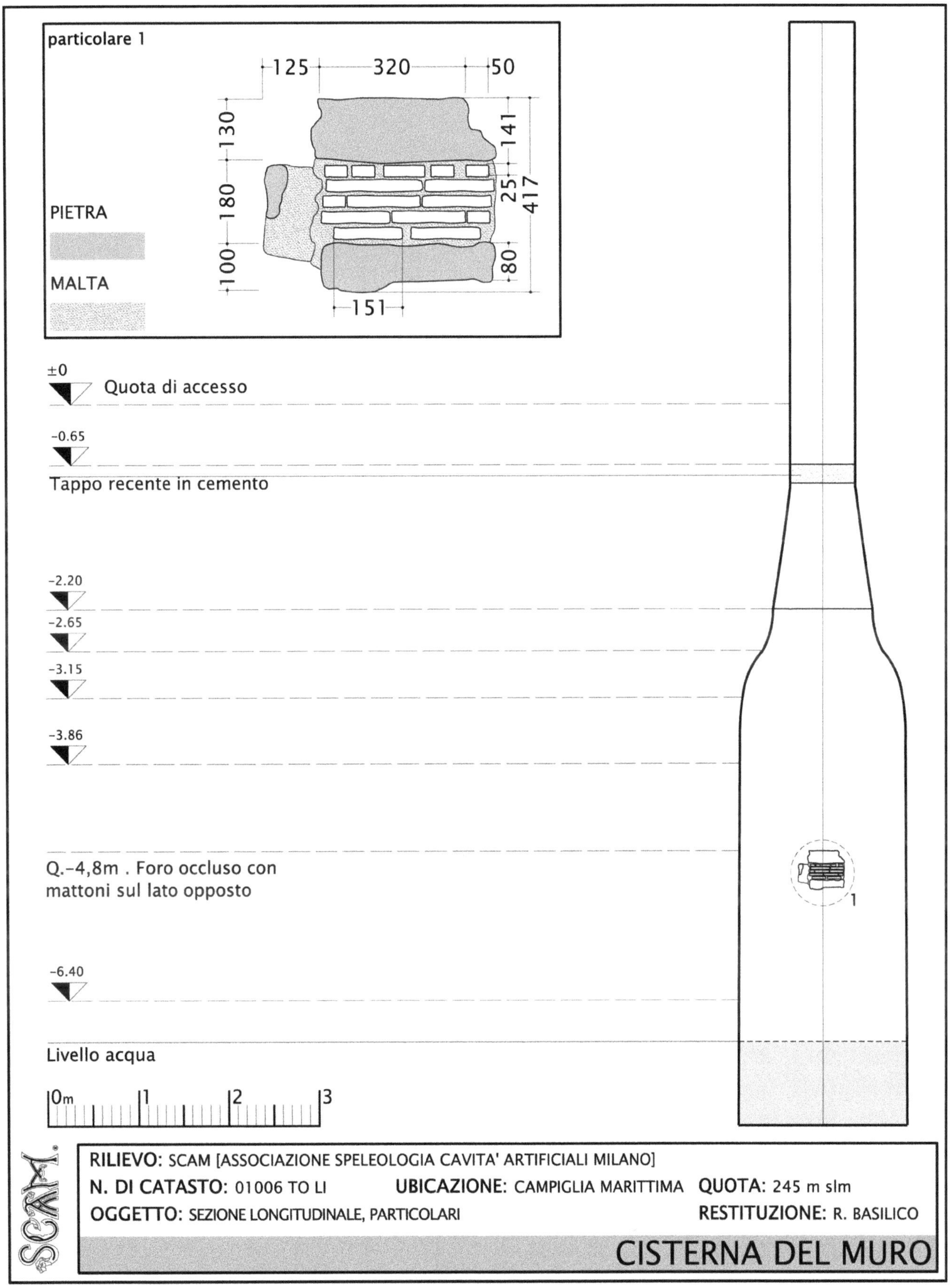

Fig. III.85.b. *Wall Cistern (or Arch Cistern) (CA 01006 TO LI) (Ass.ne S.C.A.M. Archive).*

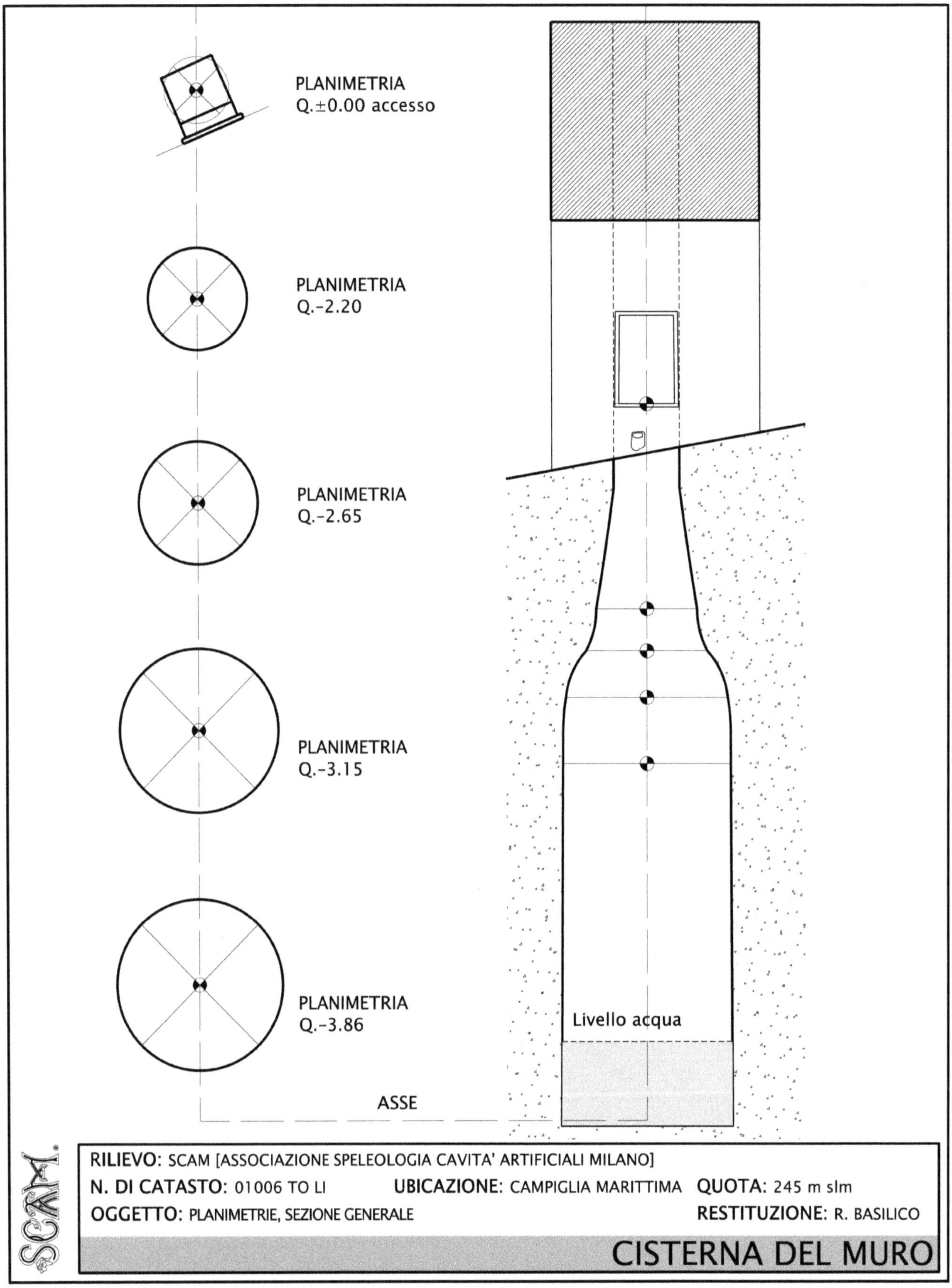

Fig. III.85.c. *Wall Cistern (or Arch Cistern) (CA 01006 TO LI) (Ass.ne S.C.A.M. Archive).*

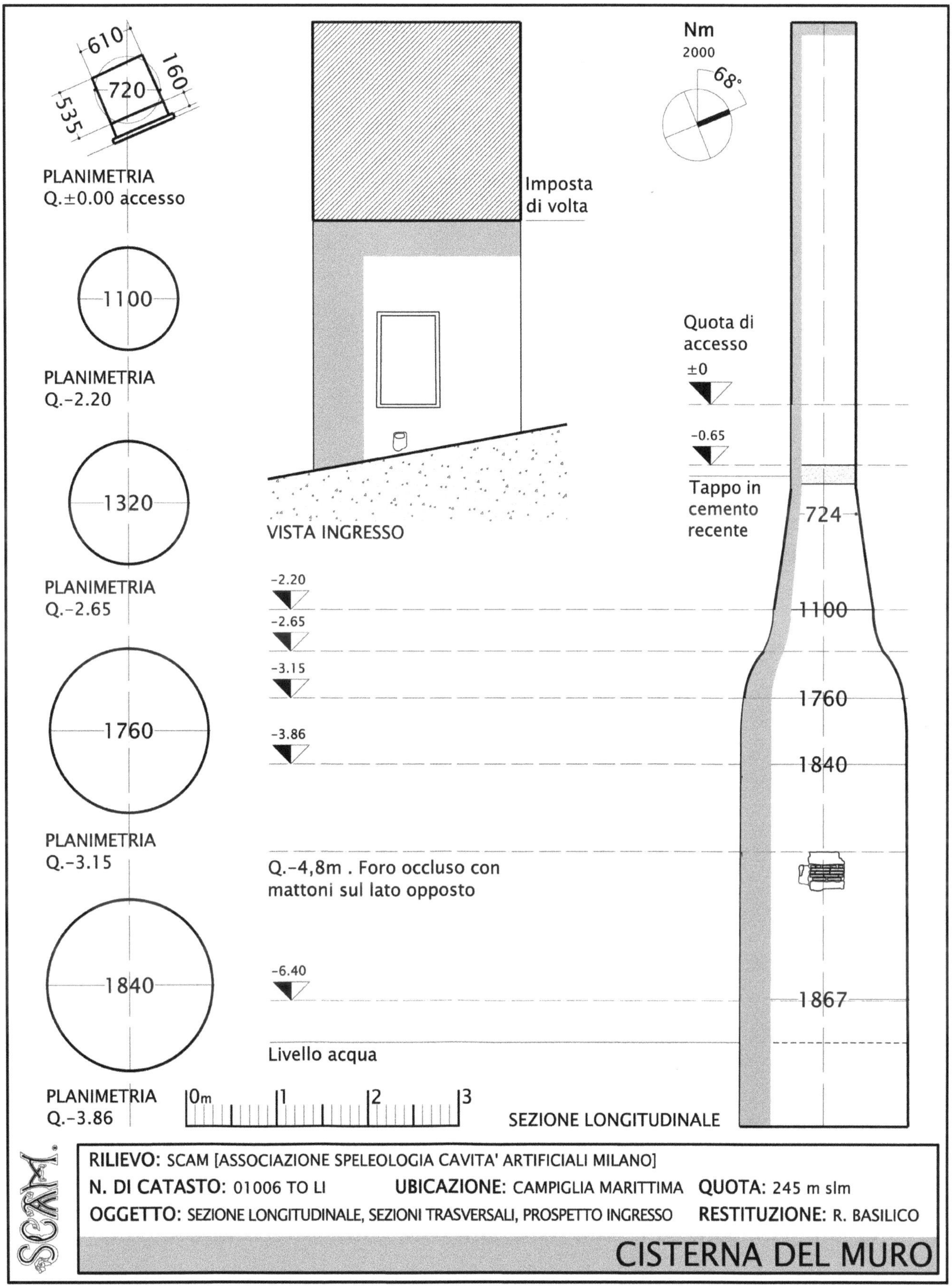

Fig. III.85.d. *Wall Cistern (or Arch Cistern) (CA 01006 TO LI) (Ass.ne S.C.A.M. Archive).*

Fig. III.86. *Stone Arch surmounted by an aedicule: inside is the Wall Cistern (or Arch Cistern) (CA 01006 TO LI) (photo K. P. Wilke).*

used to lift the pails. Dimensions: the entrance measures 0.6 x 0.95 m and 7.75 m at its base. The entrance shaft is 0.61 x 0.53 m and the internal diameter of the storage chamber varies between 1.84 m and 1.87 m. At the time of exploration the water was 0.9 m deep. Footholds: several quadrangular recesses are identifiable throughout the cistern; each and every one is blocked by brick fragments. Piping: there is a pipe at 0.6 m from the entrance threshold which has been blocked with cement.
Interpretation: its final purpose was that of a cistern however it cannot be excluded that it may have originated as an ordinary well.
Dating: due to the lack of elements, it has not been possible to date the structure; it could date back to the XIII century if not before.
Notes: given the homogeneity of the bricks used, analysis of both floor and wall bricks would be useful. Those in the floor may predate those used in the walls. Investigation of the upper section and private areas would be of assistance.
Bibliography: Basilico R., Casini A., Padovan G., *Le cisterne del castello di Campiglia Marittima (LI)*, in Federazione Speleologica Toscana, *Atti del VII Congresso della Federazione Speleologica Toscana (Gavorrano, 31 Marzo - 1 Aprile 2001), Talp. Rivista della Federazione Speleologica Toscana*, No. 23, Lucca 2001, pgs. 139-147.
Basilico R., Casini A., Padovan G., *La storia dell'acqua in un castello della Maremma Toscana: Campiglia Marittima*, in Club Alpinistico Triestino - Gruppo Grotte Sezione Ricerche e Studi su Cavità Artificiali (edited by), *Atti del V Convegno Nazionale sulle Cavità Artificiali. Osoppo 28 Aprile - 1 Maggio 2001*, Trieste 2002, pgs. 41-68.
Data ownership: Associazione Speleologia Cavità Artificiali Milano (S.C.A.M.).
Compiled by: Gianluca Padovan (Ass.ne S.C.A.M.).

III.4.8 - *CA 01007 TO LI; Villa Lanzi Cistern*

Cadastral number: CA 01007 TO LI
Denomination: Villa Lanzi Cistern (figs. III.87.a, III.87.b, III.87.c, III.87.d and III.88)
Region-country: Tuscany, Italy
Province: Livorno
Municipality: Campiglia Marittima
Locality: Valle dei Lanzi
Location: located in the central court of Villa Lanzi, built in 1556
Ownership: Parchi Val di Cornia S.p.A.
Cartography: C.T.R. 1: 2000, sheet 11G61
Geological unit: Lower Jurassic, "Calcare massiccio" (Tuscan series)
Altitude: 277 m a.s.l.
Position: 4771900 N 1630950 E
Context: the storage system is located in the centre of the court
Operations conducted: exploration, survey and photographic service
Work carried out by: Associazione Speleologia Cavità Artificiali Milano (S.C.A.M.); November 1998
Warnings: none
Typology: 2c - cistern
Description: the cistern consists of a circular chamber, accessed via a well located within an aedicule; next to this is a small filtering and settling chamber for meteoric water. The man-made structure consists of a quadrangular aedicule with a small pyramidal cupola, the upper part of which is missing. The inner horizontal fascia emphasises the gap between the two; its plaster is even and next to it is a metal pipe, which was part of the old water-pumping system. The base of the cylindrical chamber is flat and there is a collection basin under the entrance shaft. The transversal section of the cistern is vaguely bell-shaped. At the time of exploration the water level was 4.3 m thus special underwater speleology equipment was required. Dimensions: the 0.94 x 0.98 m entrance descends for 0.74 m where it comes into contact with the roof of the storage chamber, at a depth of 8.21 m. Footholds: none found. Piping: a water supply inlet from the settling and filtering chamber is located to the north; the aforementioned metal pipe descends into the structure. Description of the settling and filtering chamber: until a few months ago, a marble slab with a central 36 cm perforation was closed by a manhole of the same material. The chamber is quadrangular (1.96 x 1.2 x 1.94 x 1.24 m) and is 1.6 m high. It is internally divided into three sectors by rendered walls, which are 0.45 m in height. There are two earthenware pipes at the southeast and southwest corners, which are connected to the building's eavestroughs. Inside, the pipes lead to two small channels consisting of overturned tiles, which transported the water to the first sector. The third sector, on the southwest corner, was cleared in 2001; the deposit consisted of a layer of wood charcoal. It should be highlighted that, at an as yet undetermined time, the third sector was by-passed by an earthenware pipe, which directly connected the second sector to the storage chamber.
Interpretation: this is a cistern for the storage of rainwater collected from roofs and channelled into special settling and filtering cisterns before reaching the storage chamber.

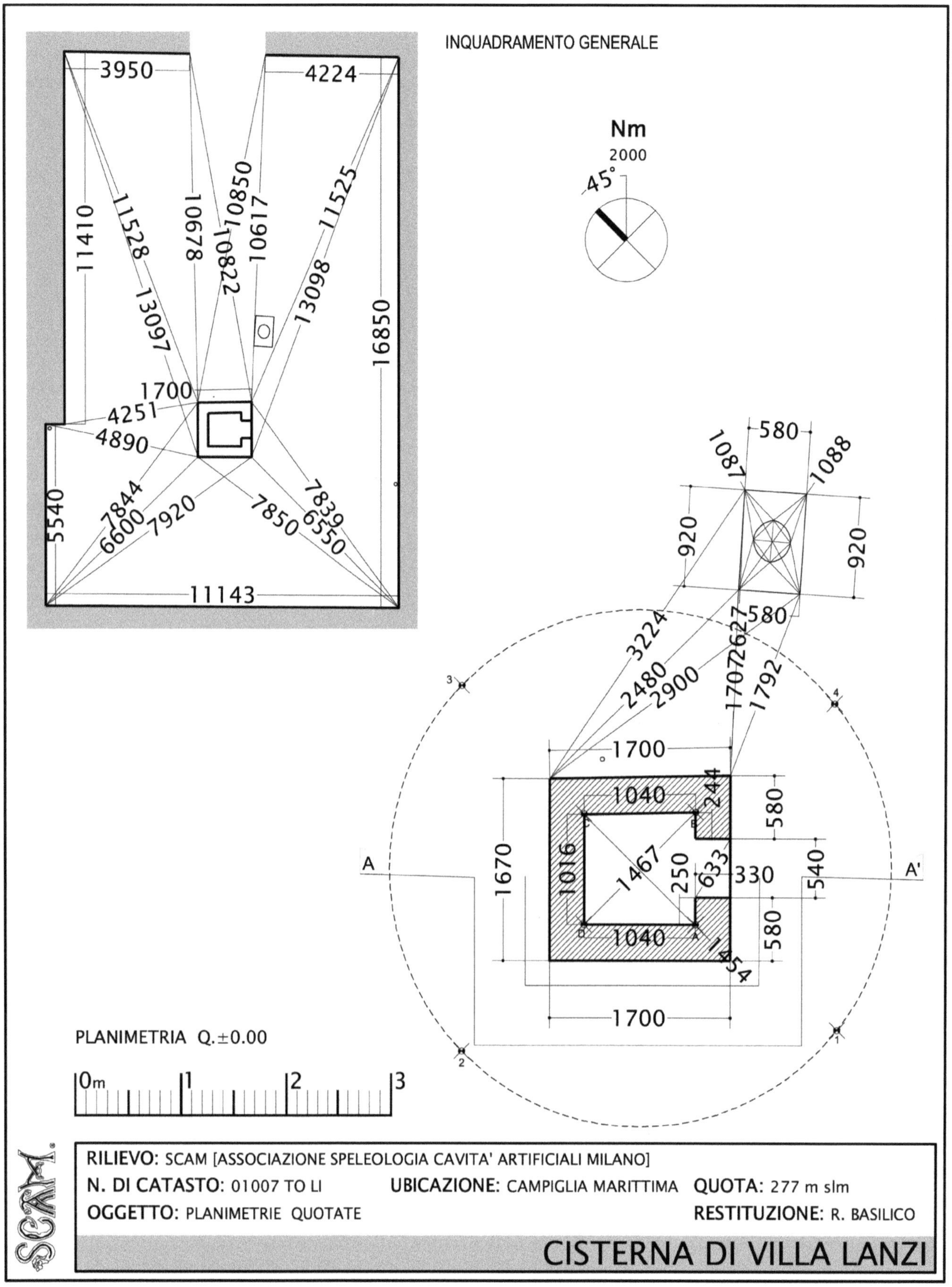

Fig. III.87.a. *Villa Lanzi Cistern (CA 01007 TO LI) (Ass.ne S.C.A.M. Archive).*

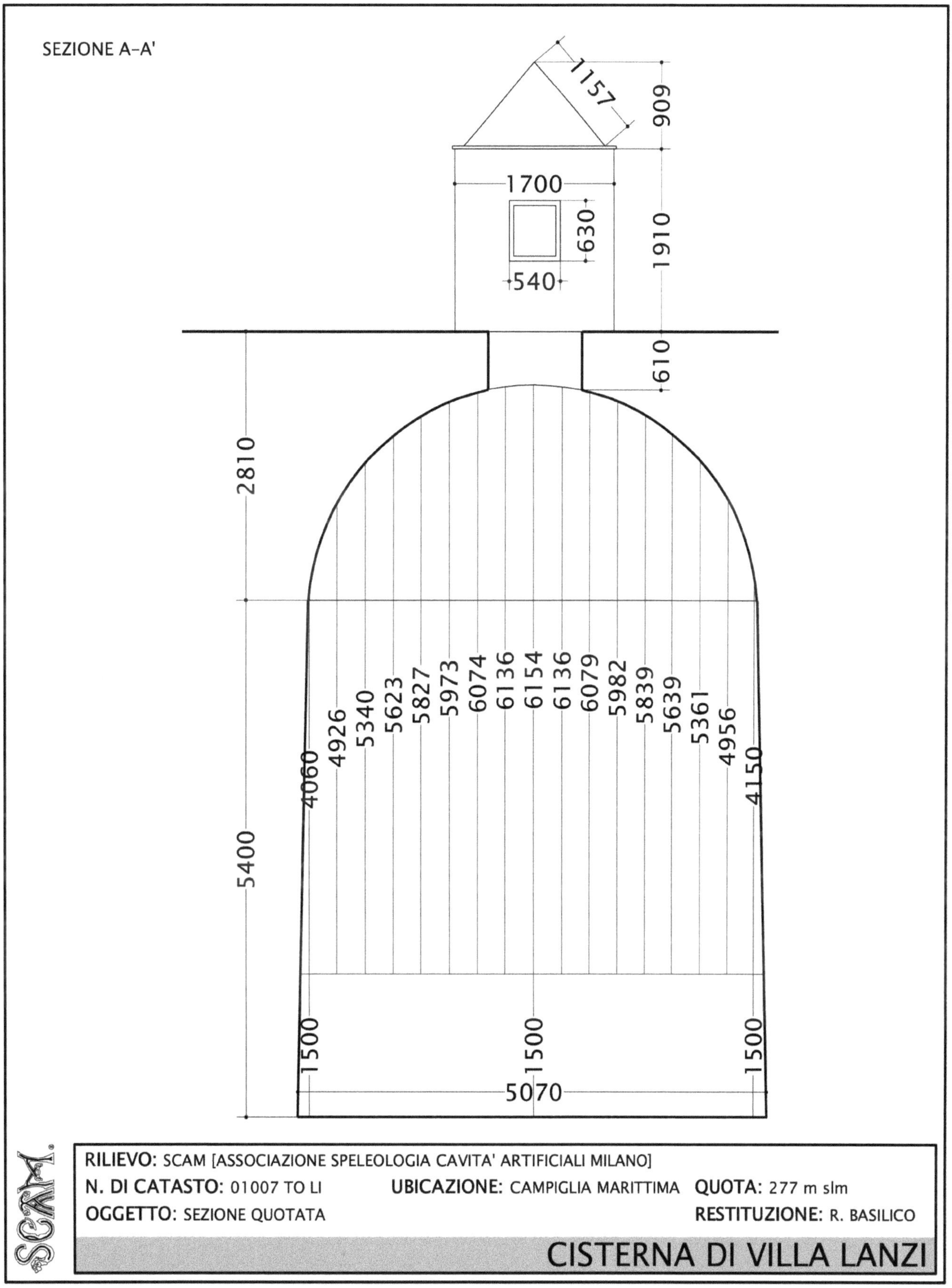

Fig. III.87.b. *Villa Lanzi Cistern (CA 01007 TO LI) (Ass.ne S.C.A.M. Archive).*

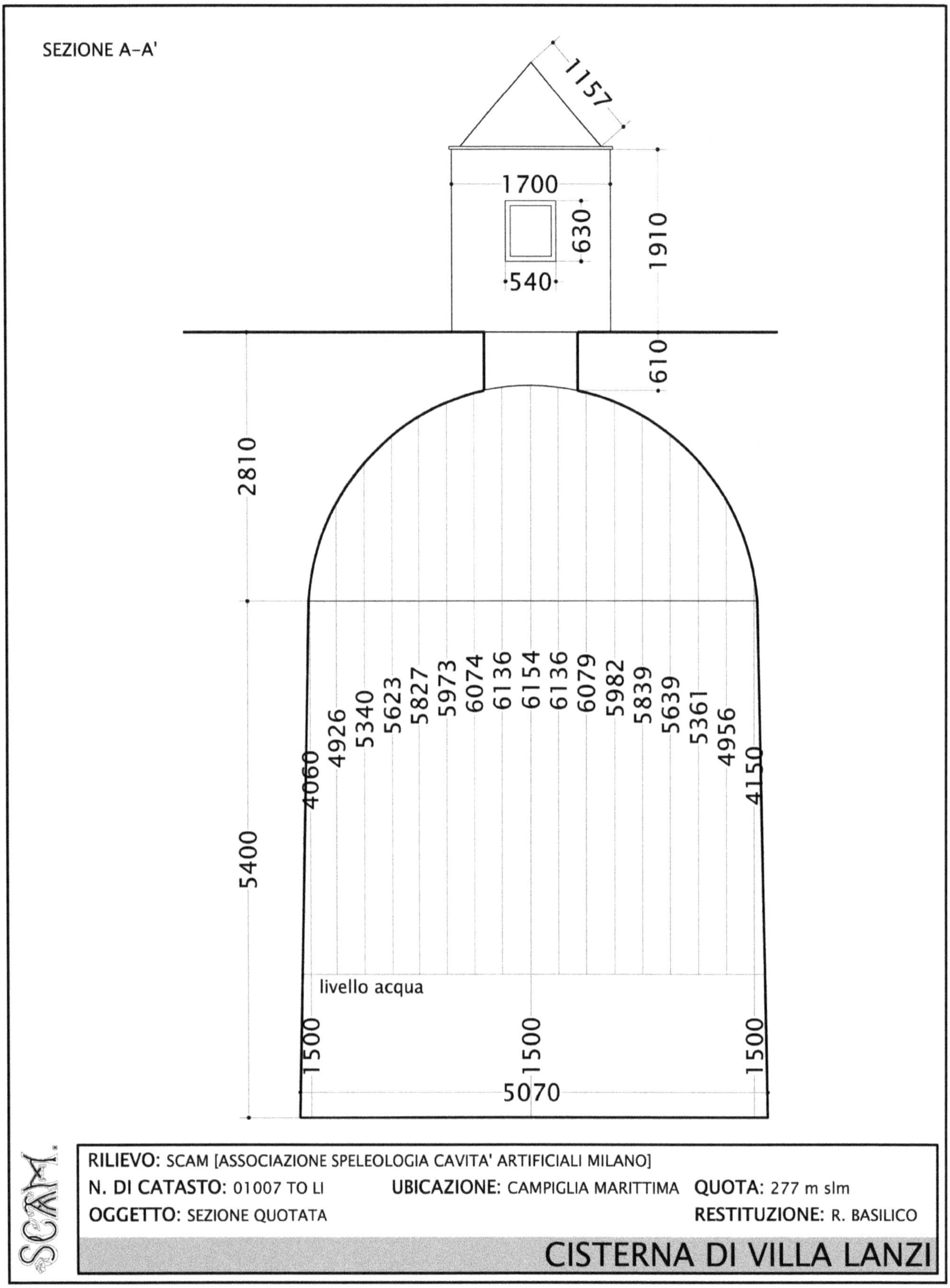

Fig. III.87.c. *Villa Lanzi Cistern (CA 01007 TO LI) (Ass.ne S.C.A.M. Archive).*

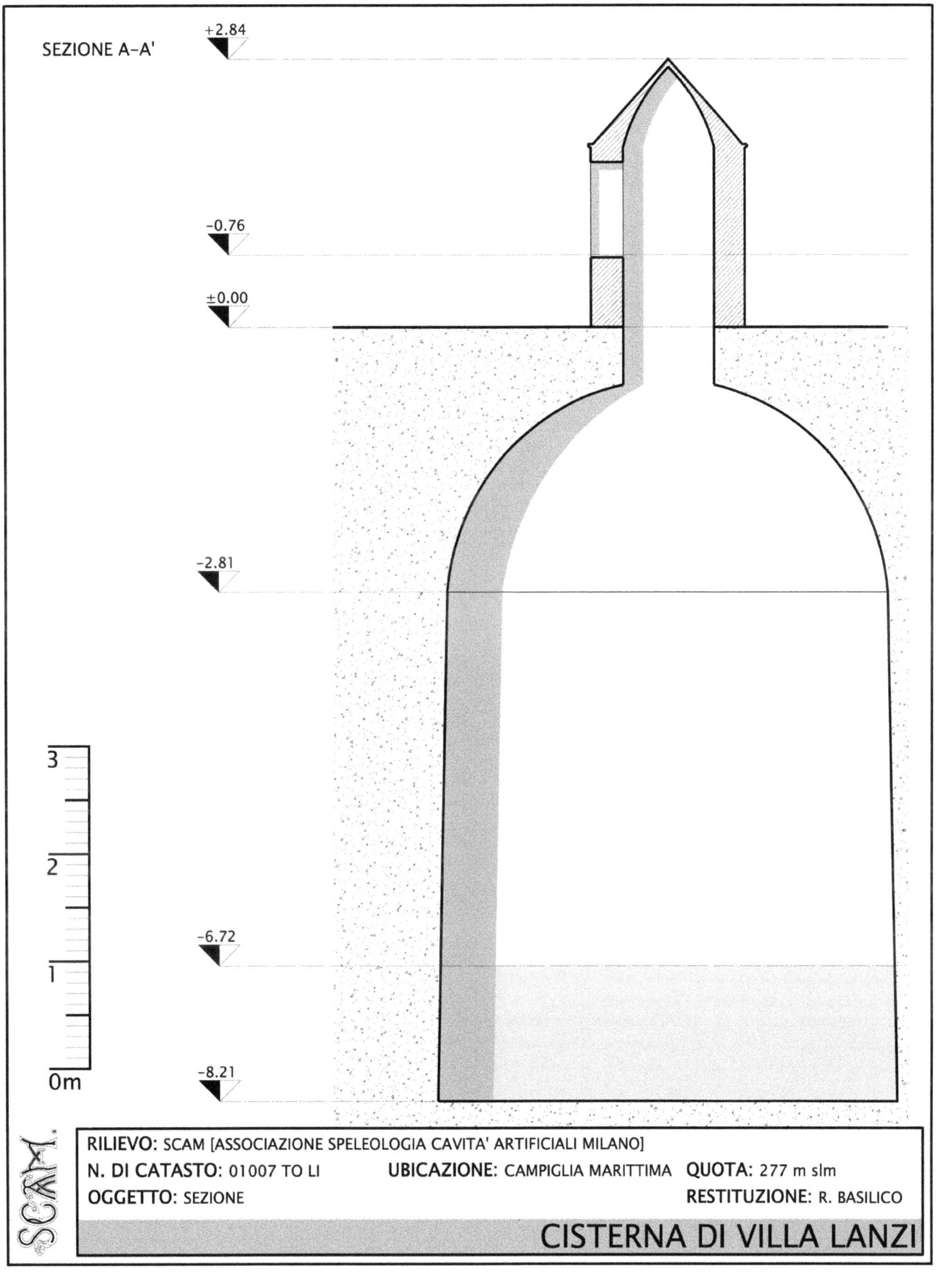

Fig. III.87.d. *Villa Lanzi Cistern (CA 01007 TO LI) (Ass.ne S.C.A.M. Archive).*

Fig. III.88. *Speleo-Diver in the Villa Lanzi Cistern (CA 01007 TO LI) (photo D. Padovan).*

Dating: latter part of the XVI century.
Notes: there is archive documentation relative to its construction (State Archives relating to Florence, the Medici family and Mines). The cistern is to be restored as part of the Medici building restoration programme. Villa Lanzi will probably become the headquarters of the Parks Documentation and Training Centre for Val di Cornia.
Bibliography: Falcini M., *Rapporto del Regio Ingegnere Architetto Mariano Falcini ai Signori proprietari della Miniera di piombo argentifero posta nell'agro campigliese nella maremma Toscana*, in *Progetto di una Società Anonima per l'escavazione e la lavorazione della Miniera di Piombo argentifero posta nell'agro campigliese della Maremma Toscana*, Firenze 1842.
Basilico R., Casini A., Padovan G., *La storia dell'acqua in un castello della Maremma Toscana: Campiglia Marittima*, in Club Alpinistico Triestino - Gruppo Grotte Sezione Ricerche e Studi su Cavità Artificiali (edited by), *Atti del V Convegno Nazionale sulle Cavità Artificiali. Osoppo 28 Aprile - 1 Maggio 2001*, Trieste 2002, pgs. 41-68.
Data ownership: Associazione Speleologia Cavità Artificiali Milano (S.C.A.M.).
Compiled by: Gianluca Padovan (Ass.ne S.C.A.M.).

III.4.9 - *CA 01008 TO LI; Caprareccia Cistern*

Cadastral number: CA 01008 TO LI
Denomination: Caprareccia Cistern (figs. III.89.a, III.89.b, III.89.c and III.89.d)
Region-country: Tuscany, Italy
Province: Livorno
Municipality: Campiglia Marittima
Locality: San Silvestro Archaeological Mining Park
Location: situated next to the Case Caprareccia complex, also known as "del Pozzo Re".
Ownership: Parchi Val di Cornia S.p.A.
Cartography: C.T.R. 1: 2000, sheet 11G61
Geological unit: Lower Jurassic, "Calcare massiccio" (Tuscan series)
Altitude: 235 m a.s.l.
Position: 771700 N 1631000 E
Context: situated near the XVI century buildings, which were restored in the XIX century
Operations conducted: exploration, survey and photographic service
Work carried out by: Associazione Speleologia Cavità Artificiali Milano (S.C.A.M.); November 1998
Warnings: poor air circulation
Typology: 2c - cistern
Description: the cistern consists of a circular chamber, accessed via an aedicule well. The man-made structure consists of a quadrangular aedicule, with a small pyramidal cupola, to the side of which are the remains of an external channel, which was possibly used as drinking trough for animals. The underlying storage chamber is circular. Its base converges towards the centre into a large detritic cone, directly below the entrance shaft. It has therefore been impossible to verify the existence of a basin. Dimensions: the circular entrance, measuring just under one metre, descends 0.5 m where it comes into contact with the roof of the storage chamber, which is 6.15 m high, at the external surface. Footholds: none found. Piping: to the northeast, the water supply inlet can be seen protruding from the metal pipe on the roof of a nearby building.
Interpretation: the cistern was also used to provide drinking water to livestock and/or for the washing of clothes as suggested by the adjacent structure.
Dating: uncertain.
Notes: the cistern is in disuse and the entrance to the settling and filtering chamber for meteoric waters has not been located; however, the eavestrough and the storage chamber are connected.
Bibliography: Basilico R., Casini A., Padovan G., *La storia dell'acqua in un castello della Maremma Toscana: Campiglia Marittima*, in Club Alpinistico Triestino - Gruppo Grotte Sezione Ricerche e Studi su Cavità Artificiali (edited by), *Atti del V Convegno Nazionale sulle Cavità Artificiali. Osoppo 28 Aprile - 1 Maggio 2001*, Trieste 2002, pgs. 41-68.
Data ownership: Associazione Speleologia Cavità Artificiali Milano (S.C.A.M.).
Compiled by: Gianluca Padovan (Ass.ne S.C.A.M.).

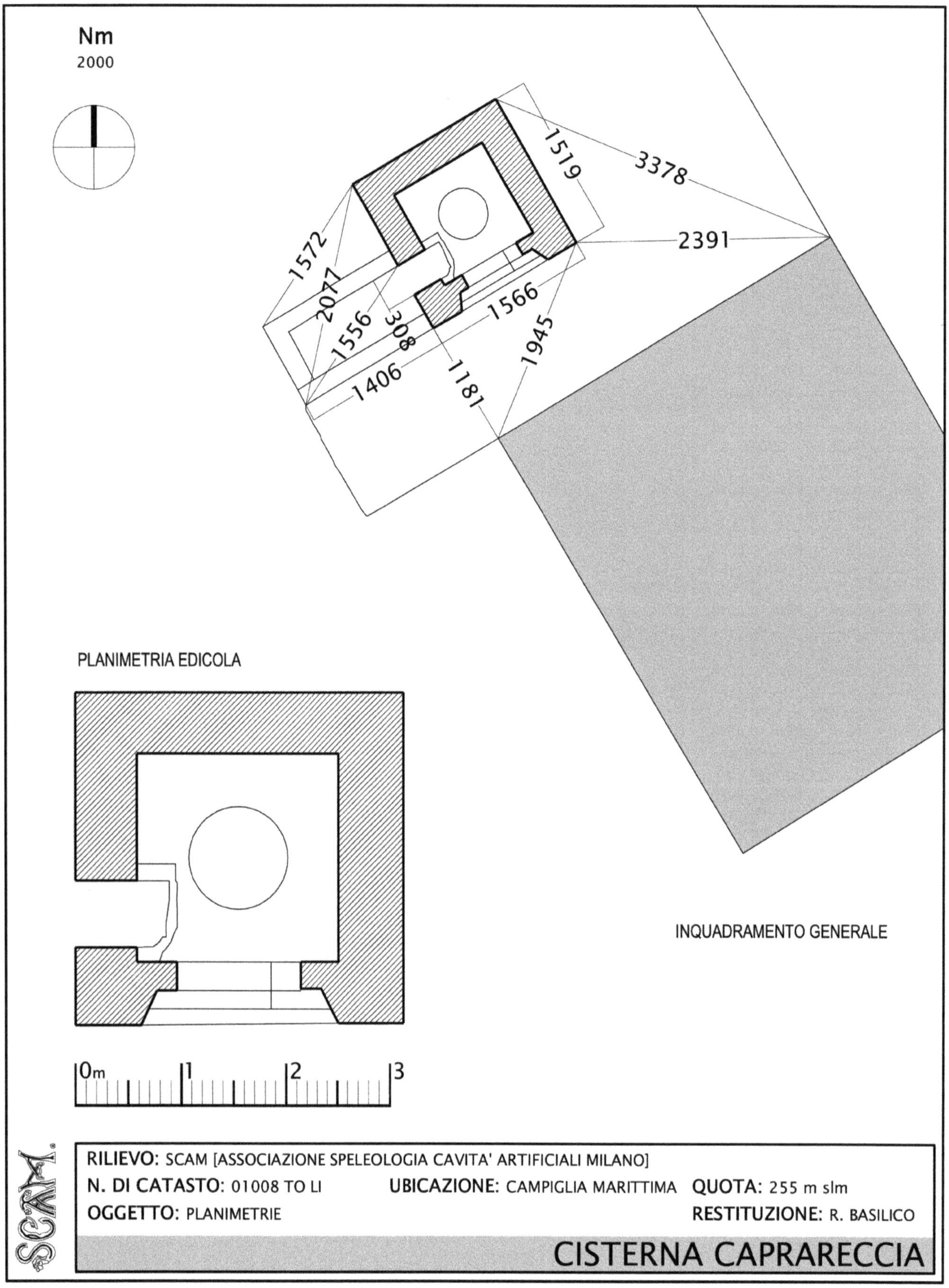

Fig. III.89.a. *Caprareccia Cistern (CA 01008 TO LI) (Ass.ne S.C.A.M. Archive).*

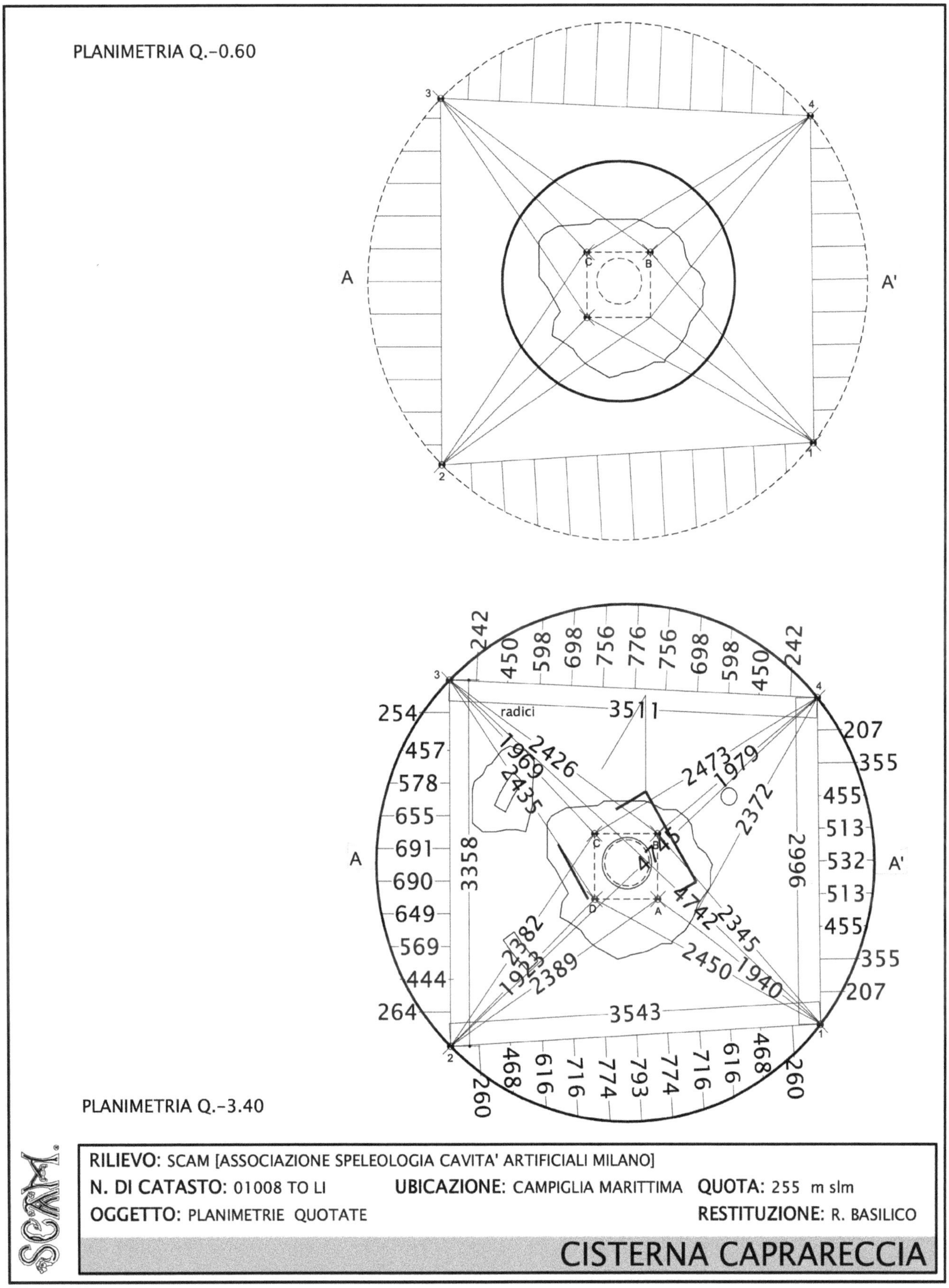

Fig. III.89.b. *Caprareccia Cistern (CA 01008 TO LI) (Ass.ne S.C.A.M. Archive).*

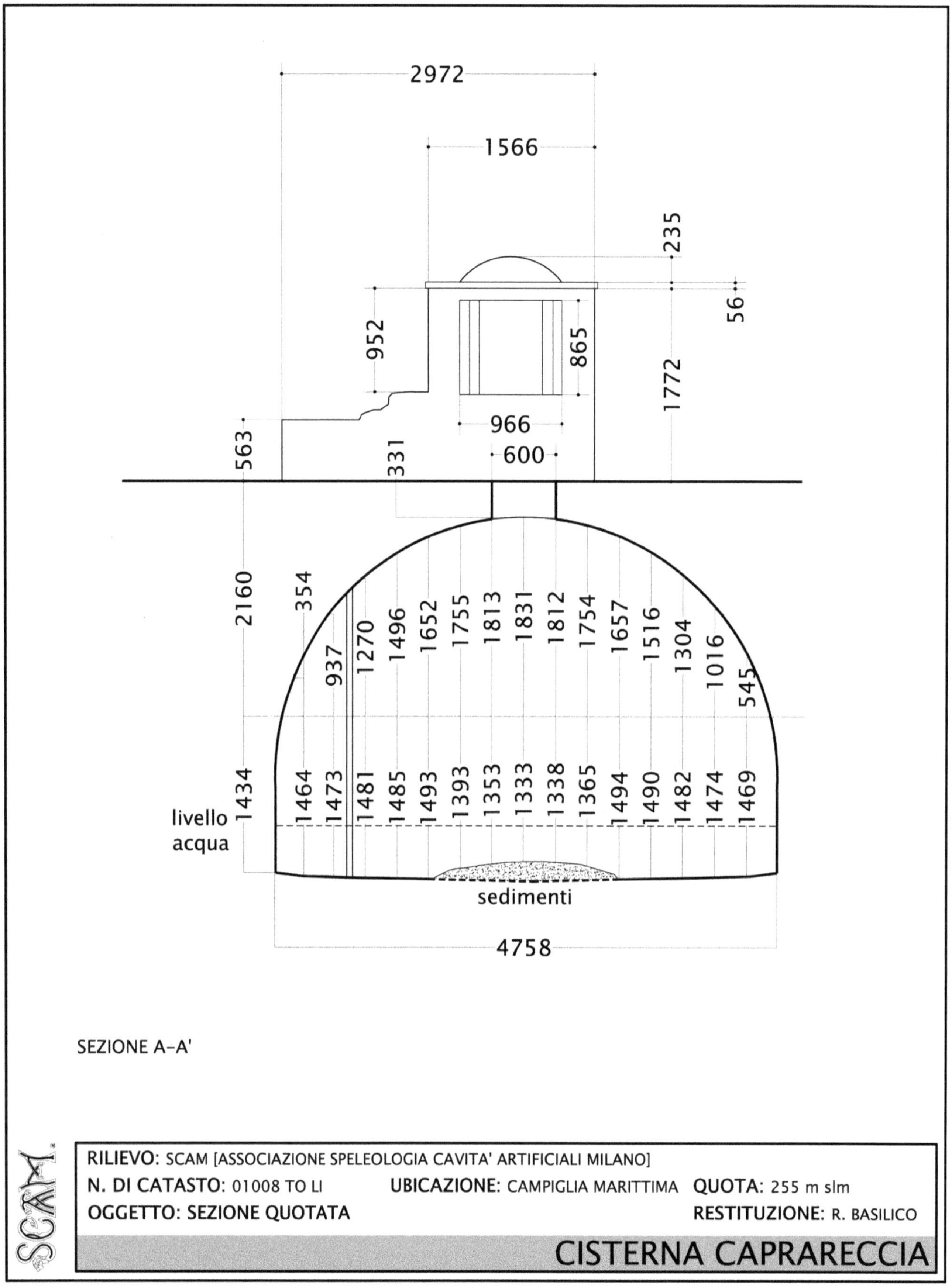

Fig. III.89.c. *Caprareccia Cistern (CA 01008 TO LI) (Ass.ne S.C.A.M. Archive).*

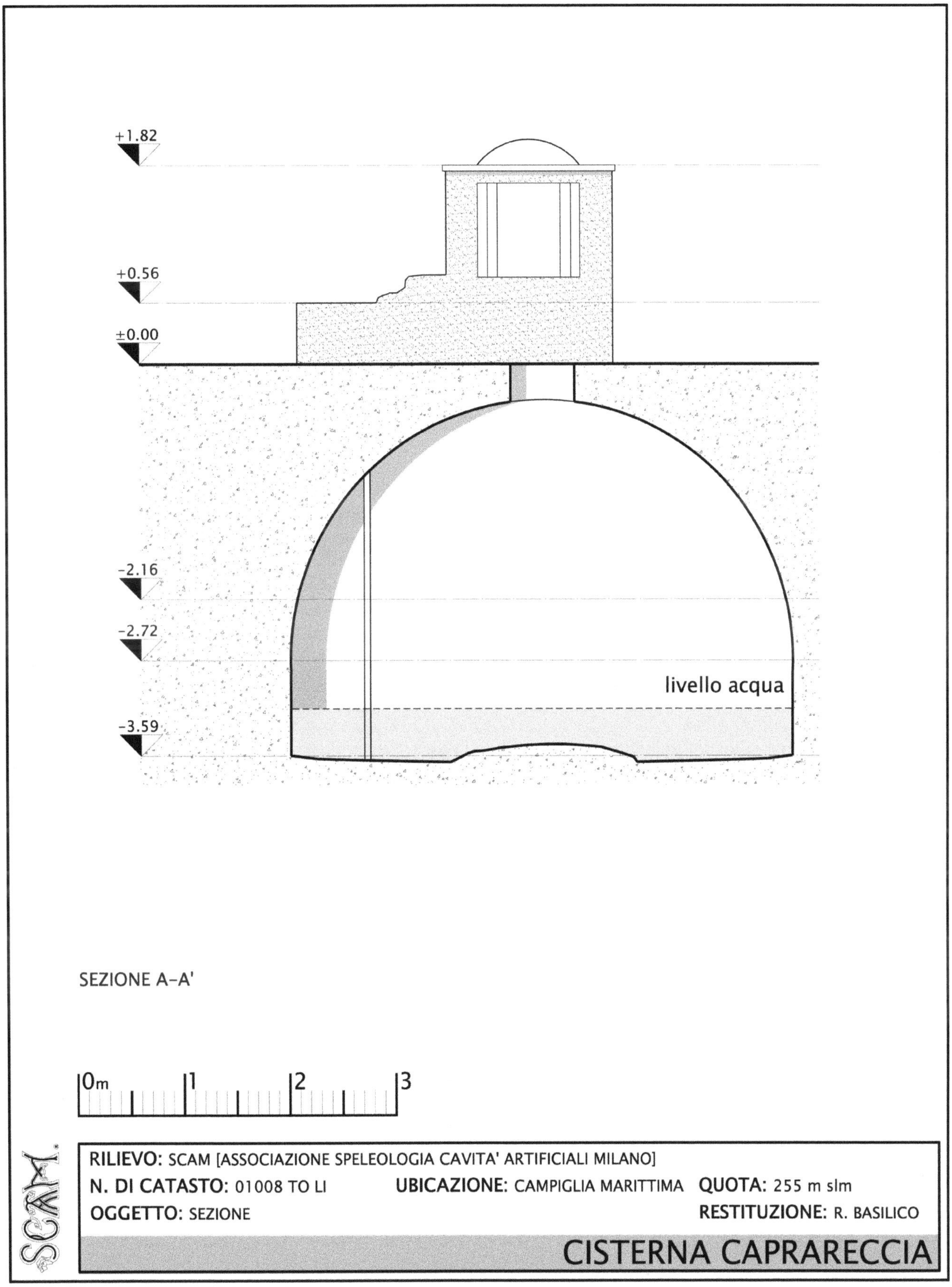

Fig. III.89.d. *Caprareccia Cistern (CA 01008 TO LI) (Ass.ne S.C.A.M. Archive).*

III.4.10 - *CA 01009 TO LI; Logge Cistern*

Cadastral number: CA 01009 TO LI
Denomination: Logge Cistern C1 (figs. III.90, III.91, III.92 and III.93)
Region-country: Tuscany, Italy
Province: Livorno
Municipality: Populonia
Locality: Poggio del Telegrafo – Populonia Acropolis
Location: Area called "Le Logge"
Ownership: Regional Board for the Protection of Monuments
Cartography: C.T.R. 1:10.000, F. 317020
Geological unit: "Scaglia Toscana"
Altitude: 186.46 m a.s.l.
Position: UTM north 4760635.9065, UTM east 621528.9309
Context: in the acropolis area of the ancient city of Populonia. The area is subject to archaeological investigation by the Universities of Pisa and Siena
Operations conducted: exploration, survey and photographic service
Work carried out by: Associazione Speleologia Cavità Artificiali Milano (S.C.A.M.); 2001
Warnings: the structure may completely collapse
Typology: 2c - cistern
Description: the cistern is situated in the monumental area known as Le Logge and is carved into the rock. It is of a composite, fairly standard shape. It may be the extension of a pre-existing larger building or it may be the point where two structures, built at different times, meet. Its shape is akin to a rectangular figure, the secondary side of which is placed next to trapezoidal side, upon the wide base of which is a third, quadrangular side. The irregular lines of part of the north wall result from partial collapse of the pier. The primary entrance, carved at the very top of the arch crown on the west side, is in a fairly central position in respect of the square base. Originally quadrangular, a missing stone block is apparent; no masonry structures appear to surround the structure on the surface. The edges of the stone blocks do not present furrows left by ropes or chains used to lift pails. The secondary opening is in the arch crown of the east wall but is slightly closer to the south side. It is not centrally positioned in respect of the rectangular base, but close to the fill-walls. Quadrangular in shape, the second opening is smaller in size (0.46 x 0.47 m) and like the first opening, is not protected by a surface puteal. Again, wear around the edges of the stone blocks does not indicate furrows left by ropes or chains used in the lifting of pails. The roof consists of two barrel vaults with drop arch, one larger than the other and one next to other in the point where the chamber widens, although not on the same axis. Both arches are made of segments of sedimentary rock (calcarenite). The dimensions are as follows: maximum depth 5.52 m at the soil fill; measurement at +2.95 m (relative to the base) are 5.66 m in length, 1.49 m in width for the east fill and 2.87 m in width for the west fill. No water supply conduits were found, although it would seem that there may have been one in the northeast corner. Unfortunately, a cave-in prevented any form of survey from being conducted. The wall facings present construction differences. There are numerous signs left by the excavation equipment in the parts carved into the rock. No niches or footholds were found.
Interpretation: cistern, probably for the storage of meteoric water.
Dating: this has not been allocated an exact chronological date, however it is related to the times of the ancient city of Populonia.
Notes: in order to prevent its total collapse, the structure should be restored with certain urgency.
Bibliography: Casini A., Padovan G., *La Cisterna C1 delle "Logge"*, in Mascione C. e Patera A. (edited by), *Materiali per Populonia 2*, Quaderni del Dipartimento di Archeologia e Storia delle Arti Sezione Archeologica – Università di Siena, Sesto Fiorentino (Firenze) 2003, pgs. 129-141.
Data ownership: Associazione Speleologia Cavità Artificiali Milano (S.C.A.M.).
Compiled by: Gianluca Padovan (Ass.ne S.C.A.M.).

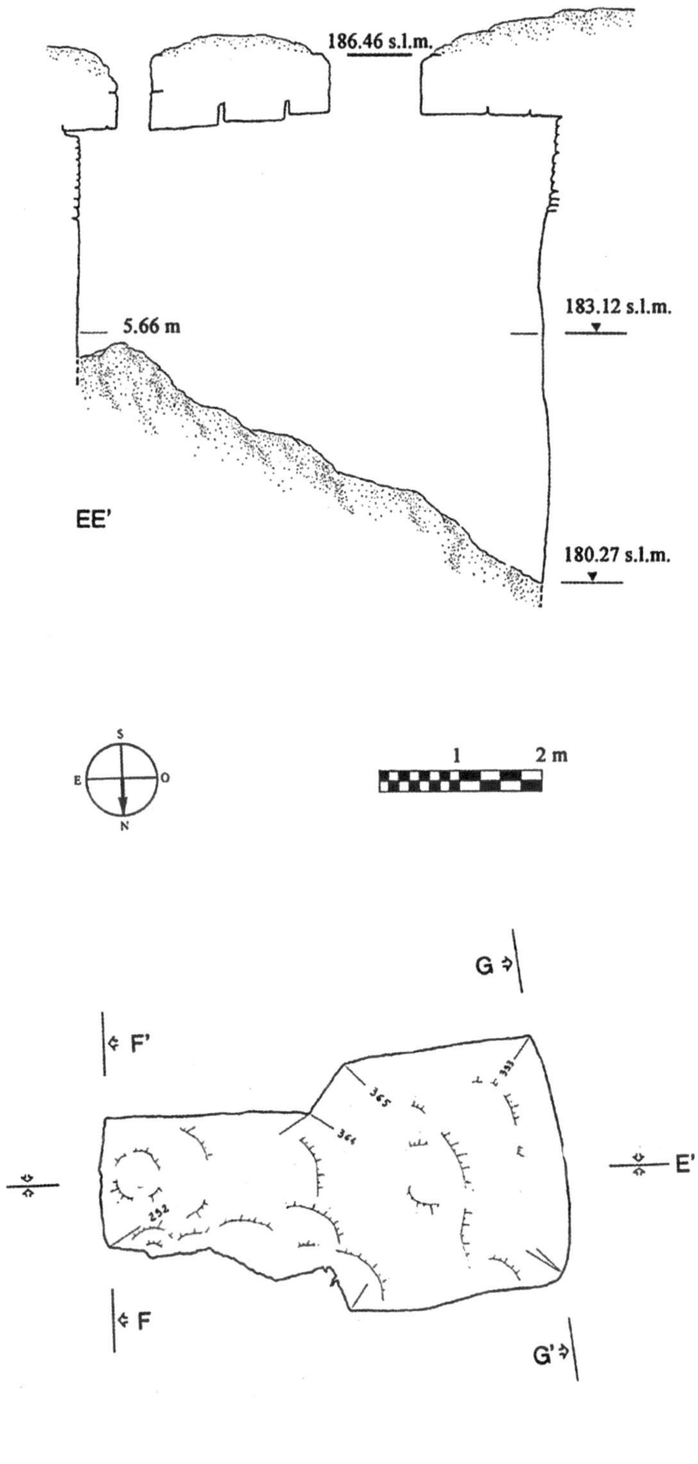

Fig. III.90. *Logge Cistern C1 (CA 01009 TO LI) (Ass.ne S.C.A.M. Archive).*

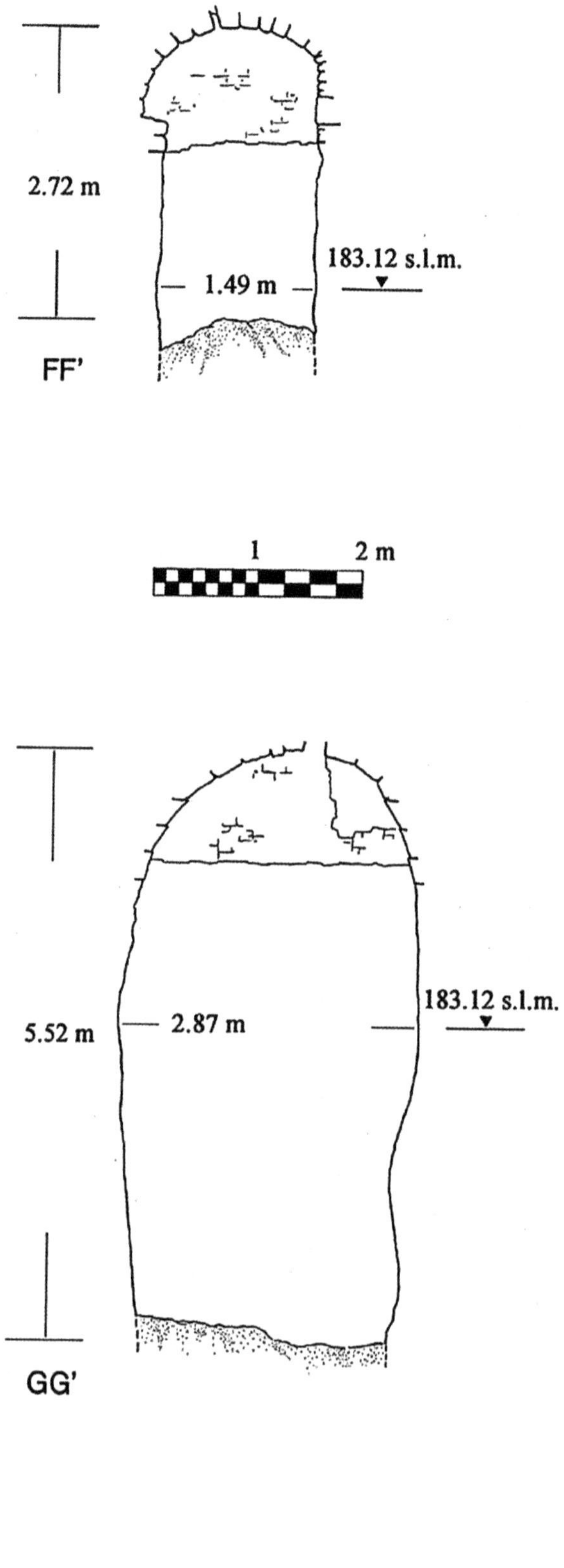

Fig. III.91. *Logge Cistern C1 (CA 01009 TO LI) (Ass.ne S.C.A.M. Archive).*

Fig. III.92. *Logge Cistern C1 (CA 01009 TO LI), stone-vault (photo G. Padovan).*

Fig. III.93. *Logge Cistern C1 (CA 01009 TO LI), wall and vault (photo G. Padovan).*

III.4.11 - *CA 01010 TO LI; Temperino Cistern*

Cadastral number: CA 01010 TO LI
Denomination: Temperino Cistern (fig. III.94)
Region-country: Tuscany, Italy
Province: Livorno
Municipality: Campiglia Marittima
Locality: San Silvestro Archaeological Mining Park
Location: former mining area buildings, reutilised by the Archaeological Mining Park
Ownership: Parchi Val di Cornia S.p.A.
Cartography: C.T.R. 1: 2000
Geological unit: //
Altitude: //
Position: //
Context: situated to the side of the square behind the Archaeological Mining Park Restoration Centre
Operations conducted: exploration
Work carried out by: Associazione Speleologia Cavità Artificiali Milano (S.C.A.M.); November 1998
Warnings: poor air circulation
Typology: 2c - cistern
Description: the cistern consists of a rectangular chamber, accessed through a well situated inside an aedicule. A wall separates the two sections and is parallel to the east blockage. Just over two metres in height, its purpose appears to be that of a small water filtering-settling chamber.
Dating: XIX-XX century.
Notes: the cistern has not been surveyed and a complete, in-depth survey is warranted.
Bibliography: none.
Data ownership: Associazione Speleologia Cavità Artificiali Milano (S.C.A.M.).
Compiled by: Gianluca Padovan (Ass.ne S.C.A.M.).

Fig. III.94. *Temperino Cistern C1 (CA 01010 TO LI), aedicule of the cistern in front of the ancient building of Etruscan Mines Ltd. (photo A. Casini).*

CHAPTER IV

DEFINITION OF AN ARTIFICIAL CAVITY

IV.1 - *Artificial cavities*

An artificial cavity is a man-made structure which is formed by removing soil and rock from both the ground and subsoil, in order to create an underground area for a specific purpose. Such an underground structure has at least two sides, a roof and a floor. The structure may be self-supporting or it may have internal containment structures or pillars. To ensure that the artificial cavity is in line with requirements, it may be partially or fully coated with clay, mortar, wood or masonry. Its internal surfaces may have no covering whatsoever.

Such underground structures can be both excavated below the natural ground level and thus beneath the classic "ground surface" and by boring into the sides of mountain and hills (fig. IV.1). Examples of such structures are road tunnels, certain uncoated Etruscan underground water channels or tombs and catacombs excavated in tuff and trachyte (fig. IV.2). Other examples are provided by impermeable underground aqueducts, sealed with hydraulic mortar.

In other cases, such as in the "ravines" of the Apulo-Lucane Mountains in Italy, the sides of erosion valleys were excavated thus creating dwellings, places of worship and tombs. Underground military structures are often internally reinforced with walls and stone or reinforced concrete pillars to contain the earth's thrusts and the effects of bombs. However, many mines rely on stone, brick, and wood support and containment structures, or supports made of metal in the case of more recent mines. This helps to prevent structural collapse and to limit its effects, even when unconsolidated material is encountered (Padovan 2005 a).

Artificial cavities come in all shapes and sizes. They may consist of a single environment or of several rooms and may develop over one or more communicating levels. They may have one or more entrances, and in particular cases, have external openings or skylights.

IV.2 - *Other structures classified as artificial cavities*

It has been ascertained that limiting research to unequivocally underground environments, with well-defined characteristics, is reductive if not misleading. Numerous different types of environment must be included in the catalogue, despite the fact that such structures sometimes only appear to be underground.

Take a "pit" in the ground for example. The distinction between a "pit" and an artesian well with a depth of 80 metres is essentially in its size: we will not always need to study and take a census of a "pit", which is just a few tens of centimetres deep, while a well will always be investigated. In any case, wells do not normally have a vaulted covering and their bases cannot be compared to the ground surface. The same can be said of particular cisterns, created through simple cylindrical or sub-cylindrical perforation of the ground (Padovan 2002 a, pgs. 57-60).

Open cast mining techniques, used in certain quarries and mines should be considered in their totality and not limited to underground activity. Aqueducts, sewers, underground conduits and countermine tunnels and practicable structures may have been cut and then covered with soil or filling material and are now, in all effects, subterranean. Furthermore, due to changing needs and expansion of the urban network, certain canals, artificial ditches and moats for the protection of city walls are equipped with vaulted roofs and banished underground.

Various bastioned fortifications of the XVII and XVIII centuries still retain articulated environments with casemates, storage rooms, storage tanks and covered passages communicating with adjacent structures (figs. IV.3, IV.4, IV.5 and IV.6). Although these are normally surface buildings, their environments present the features of artificial cavities. Besides, such structures may now be completely covered by soil, as is the case of the fifteenth century Acquasola Bastion in Genoa. In the XIX century, the port town expanded and rather than demolishing the now disused bastion, it was enclosed by a wall and buried. Its remains were found by speleologists in 1989: «In relation to the discovery of the fifteenth century bastion at Acquasola by the Genoa Work Group, two details are significant: the first consists of the fact that the entire fortification is completely "buried" in the centre of the city Its structure, as well as its parts are now underground.

Fig. IV.1. *«La Grotte du Pausilippe, près de Naples» (Badin 1876, pg. 119).*

Fig. IV.2. *«Un escalier Primitif des Catacombes de Rome» (Badin 1876, pg. 67)*

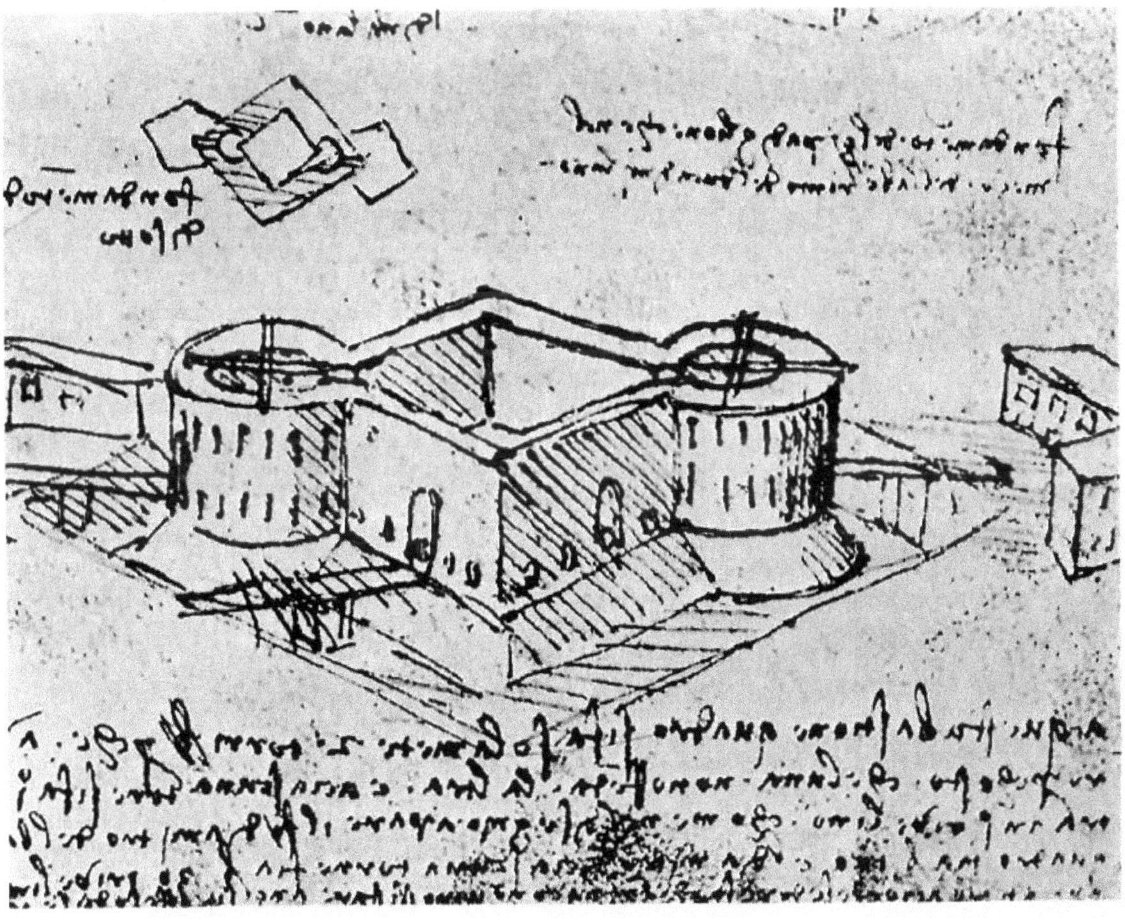

Fig. IV.3. *Study of a quadrangular fortalice with two corner donjons and surrounded by a ditch (Leonardo, Ms. B. fol. 12 recto).*

The second, even more extraordinary aspect, relates to the sensational discovery of large accumulations of human bones within its walls» (Bixio, Sai 1991, pg. 5). Mining demolition passages and countermine defence tunnels can often be found within military buildings: cramped and dark, these are easily explored.

With regard to natural cavities, many caves were subject to extraction activities. All types of extraction must be considered when studying quarries and mines, including cave extraction. The size of the operation is irrelevant. We shall not take the speleological cataloguing system into account, whereby only natural cavities, with more than 50% of their internal surfaces artificially modified, are considered artificial. Even a few traces are sometimes sufficient to provide information and clues and such evidence will always be considered as valid, especially when inserted into the bigger picture.

In addition, all caves presenting evidence of human work can be classified as artificial cavities. Thus burials are also included: the correct choice is clearly dependant on practical sense and on the competence of the investigator. Basic evidence does not necessarily turn a cave an "artificial cavity". On the other hand, the erections of tomb or fortification walls are classified as artificial cavities.

These "vie cave" trench roads have been studied, primarily on account of the fact that they often lead to artificial cavities. The same can be said of military trenches. Completed studies reveal that research and studies have also been carried out in environments which are not strictly subterranean. Returning to underground aqueducts, these often have open-air sections, where the masonry channel is supported by arches or substructions: it is often the case that in researching a man-made structure, its individual aspects must be separately assessed. This means that studies conducted on a basic cavity can lead to extensive research in the broadest sense of the term.

The purpose of a man-made structure will be the determining factor in its classification to a specific typology. It should be taken into account that a structure is not always the result of a sole building project and that it may, upon investigation, turn out to have been created in multiple phases, over a variable timeframe. Generally speaking, an artificial cavity may be subject to interventions which changed both its structure and its original purpose.

Case record analysis and distinction thus proved to be both complex and useless, to such an extent that they indicated that the numerous environments classified as "artificial cavities" could only have the below factors in common:
- lack or total absence of light;
- not always easily practicable.

IV.3 - *Structures to be classified as "artificial cavities"*

The below man-made structures are recognized and classified as artificial cavities without further distinction other that their classification under a precise typology and sub-typology.

1. Man-made structures in the ground and subsoil.
2. Cut and cover structures, like certain types of aqueduct or certain defence structures.
3. Roofless structures which were subsequently covered, e.g. water courses and artificial vaulted cavities, regardless of whether the covering was built at the same time or later than the hydraulic system.
4. Open-air structures, which were subsequently covered, either artificially or following natural events.
5. Particular environments such as "bomb-proof" environments, casemates and modern forts as well as mine and countermine structures built within elevations, whether at the same time as the man-made structure itself or whether subsequently created by excavation or breaking through walls.
6. Natural cavities presenting signs of extraction work.
7. Natural cavities, transformed by man, or rather caves which show signs of expansion or simple adaptation, or defensive, settlement or cultural buildings etc.

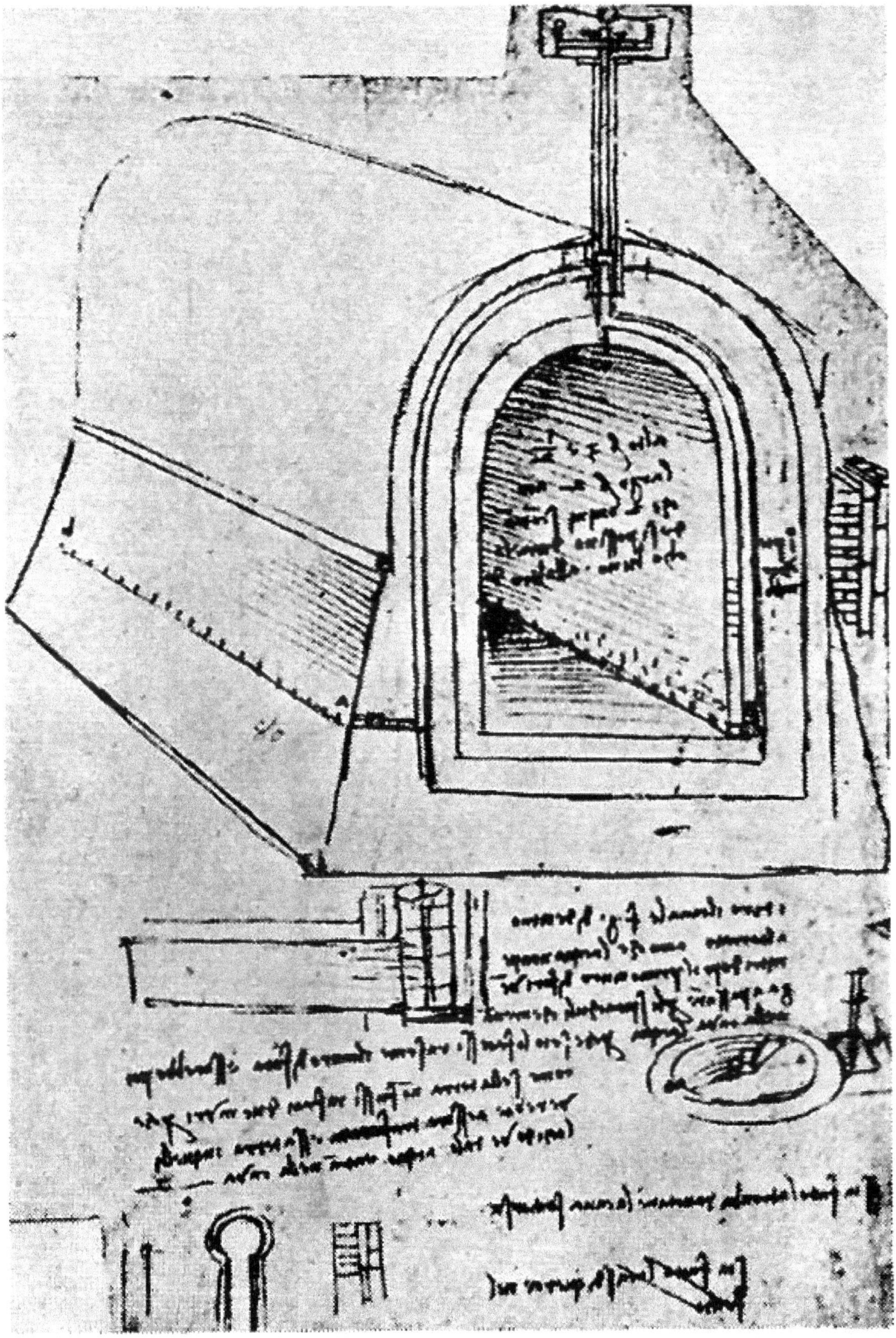

Fig. IV.4. *System used to flood a fortification's internal communication tunnel (Leonardo da Vinci, Codex Atlanticus, folio 359, verse-a).*

Fig. IV.5. *Hypogeum of the Novara Castle: room with a closed well (photo G. Padovan).*

Fig. IV.6. *Tunnel around the Great Cistern beneath Mercato delle Scarpe square at Bergamo (photo G. Padovan).*

CHAPTER V

AN UNDERGROUND WORLD: IDENTIFICATION AND STUDY

V.1 - *Why man creates artificial cavities*

Due to the force of gravity, water infiltrates and flows through the subsoil, creating its own channels and forming galleries, meanders, environments rich in formations and wells of even remarkable proportion. Thus caves, or natural cavities, are formed. On his part, and in response to the continuous and urgent needs imposed by an ever-developing *societas* or community life, Man "yielded" and created countless structures, in a true or illusionary attempt, to take advantage of Earth's almost hidden offerings, even turning to the subsoil.

Since prehistory, shelters under rocks and caves have been used as shelters, temporary dwellings, meeting places and places of worship: «In the history of architecture, caverns are seen as the propelling element in man's efforts in his search for structure, if not as his starting point» (Rossi-Osmida 1974, pgs. 5-19). However, alongside the undoubted advantages, natural cavities brought other factors, such as humidity, dripping water, animals and a location, which did not always meet the needs of its users. It is thought that man originally adapted certain natural cavities to his own needs and that this inspired him to create his own artificial cavities, based on acquired understanding (fig. V.1).

From the adaptation of caves to the excavation of rock dwellings which continued to our own times, built-up areas, of even remarkable size, have developed. The search for the materials required in the creation of tools may have commenced at ground-level, subsequently moving on to the extraction of stone, such as flint, directly from the outcrop, from the surface and from natural cavities. Following the layers of utilisable rock, such as flint for example, Man simply continued his excavations thus creating underground environments. Upon discovering that certain rocks concealed minerals and that these could be treated in order to obtain metal, research and extraction methods were developed all around the world (fig. V.2).

At the beginning of the Metal Age, a major change took place, probably akin to that which took place with the discovery of agriculture in the Neolithic Age (Giardino 1998, pg. 4). It is more than likely that excavation and water channelling techniques for drainage purposes and the search for essential groundwater to supply the growing settlements were developed from the exploitation of mineral deposits.

Forbes believes that the gradual use of water research methods began from both the observation of nature and the experience gained in mining exploration and tunnel excavation (Forbes 1993, pg. 674 and pg. 689). Agriculture remains of fundamental importance. Productivity and demographic expansion are the result of the development, application and constant improvement of soil adaptation techniques. Drower states that, where water in naturally lacking, water irrigation of sown fields goes hand in hand with drainage, that is the removal of excess water from the soil (Drower 1993, pg. 528).

It cannot be excluded that observation of a steam emerging from a cave or from a simple crevice in the ground, prompted Man to dig into the rock whenever a water supply was required. As previously indicated, Man's tradition of living with nature, of observing nature, thus developing specific "sensitivities", taught him to identify, with good approximation, useful excavation points. Ancient Man was, undoubtedly, far more advanced than we, with our electric drills and running tap water, may think today. The worship of the dead is diffused the world over and artificial cavities of all shapes and sizes were created for this very purpose. Equally, the various religious beliefs led to the creation of specific places of worship and devotion, even underground.

Considering artificial cavities in their globality, we must establish what led Man to excavate the ground and how specific survival needs as well as the need to improve life conditions allowed Man to create underground buildings, which are still evolving today.

On a hypothetical level we can understand how living close to nature would lead man to observe animal behaviour and habits and learn or draw inspiration from this. Man had no shortage of application methods. Generally speaking, it is likely that Man's motivations were as follows:
- the need for a safe place;
- the worship of the dead;

Fig. V.1. *«Grotte de Notre-Dame de la Balme (l'entrée)» (Badin 1876, pg. 231).*

Fig. V.2. *«Cave du Diable en Angleterre» (Badin 1876, pg. 217).*

- the exploitation of underground resources;
- the need for suitable environments for food preservation;
- the need for suitable environments for the storage of water;
- generic waste disposal;
- the need to increase water supply.

V.2 - *Underground monuments*

The Earth is like a vault of historical, architectonic and archaeological information. There is also an underground world, the result of economic and social activities, of daily life and culture, which generations of workers, excavators and bricklayers have left in testimony of their presence. Just as he created buildings on the surface, during the course of time, Man pierced the surface of the earth thus creating "spaces". He left behind architecture which is essentially complete, which can be studied, restored and even used.

The technological surge of the Twentieth century caused Man to lose his knowledge of manual excavation and water culture, knowledge which could one day be useful and which would already today improve living conditions in many areas. Such "knowledge" should be pursued, studied and recovered. Not always easily practicable, underground environments require an exploration process which allows for their research, understanding and cataloguing.

V.3 - *Underground investigations*

Artificial cavities can be found in almost any place that there is human activity. They are generally concentrated in urban areas, which retain a larger number of cavities corresponding to the many different needs. If a settlement's future consists of destruction, reconstruction and expansion, it is obvious that such processes will also implicate pre-existing underground structures and may result in their being encompassed in new cavities, re-utilised for a different purpose, buried or simply destroyed.

Given the variety and even the planimetric development of subterranean structures, urban research is undoubtedly of great interest. Especially, as often happens, when placed side by side with surface investigations for the easier understanding of the evolution of a specific site over time. Inversely, the study of the very same development can lead to underground investigation in the attempt to locate specific subterranean environments for completion of the acquisitions framework. It would be better still, if the complexity and chronology of the site itself could be assessed through an archaeological dig. Cities should be regarded as constantly evolving organisms. Our understanding of cities should not be limited to emerging volumes. Underground, kilometres upon kilometres of structures can be explored, whether subterranean structures or structures which became such through rising ground levels and urban growth. Hence, the study of urban settlements must also take into account subterranean realities.

Artificial cavities created to serve expanding cities, such as stone quarries and underground aqueducts for water supply, are to be found in suburban areas. Outwith the strictly urban environment are mines, which may have influenced surrounding town and country planning. New road systems, deforestation, charcoal kilns and industrial buildings can also be found. These sometimes involve a considerable amount of people and can lead to the development of true, structured settlements.

V.3.1. - *One characteristic of man-made underground structures*

If elevated structures are subject to rebuilding, expansion, demolition and drastic reconstruction, underground structures, and those structures which have become subterranean due to the passage of time, are better preserved due to their underground characteristics. An underground structure is more integral, more easily studied and eventually recovered. On the other hand, eventual re-use and consequent changes to the purpose of the structure render the rebuilding of primitive structures more difficult. We find ourselves faced with unfavourable stratigraphic units, which due to their very nature, render the previous phases obsolete. In any event, we will always refer to those environments which we are able to interpret.

V.4 *Favourable and Unfavourable Stratigraphic Units*

Certain phenomena and actions lead to the removal or introduction of material to the environment in which we live. A flood may cause landslides, the removal of material and its deposit elsewhere. The simple flow of a spring etches the ground, deepening the bed and depositing the erosion matter downstream. Man excavates the ground in the construction of buildings or digs canals, raising their banks with embankments. Thus, over time, favourable and unfavourable stratigraphic units are formed. Such stratigraphic units are physically linked, one to the other: a Stratigraphic Unit may cover another: a collapsed wall may cover the collapsed roof of the adjacent building. Or it may come into contact with another: a soil floor comes into contact with the walls of an environment. It may even intersect another stratigraphic unit: the excavation of a well intersects the ground surface. In summary, the physical relationship between Stratigraphic Units is as follows: covers/is covered by; is supported by/supports; intersects/is intersected by; is connected to/is the same as. Such physical relationships determine a temporal sequence relative to the Stratigraphic Units and allow us to understand which came first and which came later. Barker reminds us that the accurate study of a site and the recording of the any phenomena found are required in order to establish how the strata were formed. All types of constructions such as palisades, ditches, bastions as well as any other type of man-made building, leaves their trace. The sequence of such traces provides us with an understanding of the events which have taken place at the site (Barker 1977, pg. 236).

If SU (stratigraphic unit) 1 covers SU 2, this means that SU 1 is more recent than SU 2. If SU 3 intersects SU 4, then SU 4 is older than SU 3. If SU 5 is connected to SU 6, then this means that both SUs were created at the same time. Generally speaking, it can be said that all types of stratification, whether resulting from natural or artificial factors, are the result of erosion/destruction; movement/transport; deposit/accumulation. «Going by the relative chronology, the stratigraphic unit concept aims to reduce the different actions and their spatial relationships to the same degree of abstraction as stratigraphic relationships. This is equivalent to reducing a wall or a sewer to the same simplistic level as a before or after. In order to do this, we must move from the topographic identification of an action to its numerical identification. The wall becomes number 1003 and the sewer number 1027. This leads us to the logical assumption that 1027 intersects 1003 and was therefore of later construction» (Carandini 1996, pg. 75). Artificial cavities can also be defined as "unfavourable stratigraphic units". If a cistern is intersected by a limestone pit, the cistern will be the earlier of the two; similarly, a cistern containing the remains of an underground passage will be subsequent to the passage itself. The very same concept can be applied to the study of mines, bearing in mind that a Mining Unit intersects (removes) another or covers/is covered by, etc. (Casini, Cascone 2000, pg. 95. Padovan 2005 b, pg. 6).

V.5 - *Basic trinomial*

Artificial cases are to be found everywhere: under urban centres, in isolated, almost inaccessible mountain areas and even in lagoon areas. For instance, the city of Venice has many wells and cisterns. In our approach to the study of a site, we should ascertain what the underground exploration options are. The "geological terrain – physical site characteristics – place history" trinomial is a useful criterion for the identification of areas which may hold subterranean architecture (Padovan 2005 c, pgs. 73-74).

The analysis results of each of the trinomial points should produce a thematic map presenting the area's geological features, its morphological features, the identification of springs and fountains, the presence of mineral deposits, the dynamics of the population with localisation of architectonic and/or archaeological elevations and traces of road systems, in chronological order.

V.5.1 - *Geological terrain*

Subsoil composed of resistant rocks such as porphyry and granite, presents certain hardness and is not easily excavated. The main structures to be found in this type of subsoil are extraction works, i.e. quarries and mines. Other types of artificial cavity are to be found where there is softer rock i.e. tuff.

On the other hand, in loose terrain, such as in flood plains, horizontally developed underground structures are to be found. These may also be superficial as under such conditions, the ground water table can be found at a depth of just a few metres. Unconsolidated deposits are relatively easy to excavate, although, given their structural self-supporting inability, they require strong frameworks and coverings.

The identification of the geological and hydrogeological features of a site should be carried out, where possible, during the preliminary research stage and at the same time as the exploration of any existing artificial cavities. Geological knowledge of the underground can assist in the determination of an underground structure's stability, as this can provide information on the level of safety applicable to the place in question.

The ground's thrusts on cavity walls are dependant on the granulometry, cohesive properties and compaction and absorption of the soil itself and on the shape and dimension of the cavity and on its discontinuities (fractures and faults) (Bassi, Berto, Perletti 1996, pg. 20). The comparison of the geological composition of a specific context and the analysis of the can assist our understanding of both the workers and the excavation strategy used.

V.5.2 - *Physical site characteristics*

Elevated positions on the surrounding land have often been chosen as sites for settlements and fortifications. This is because they were better suited for territory and road control. Due to their superior views, they could consequently provide early warning of attacks. If naturally equipped with steep, craggy slopes they also assisted defence. From historical sources and archaeological digs, it would appear that since antiquity, man has opted for settlement locations presenting water sources and which were thus able to meet his needs. But the chosen locations were not always both easy to defend or chosen for settlement purposes nor did they always have natural water supplies or underground aquifers, sufficiently close to the surface to allow water extraction. In such cases, the settlements would originally have been equipped with cisterns for the collection of rainwater, the number and size of which would have increased with urban expansion thus improving storage techniques. Additionally, drinking water could be obtained via the construction of aqueducts.

V.5.3 - *Place history*

As a hypothetical example, let us take a city built in an elevated position, having the below characteristics:
1. easily excavated rock substratum;
2. lack or absence of drinking water sources;
3. intense history, articulated and prolonged over time;
4. surface of several hectares;
5. overlapping architecture.

We can immediately forecast the presence of artificial cavities with specific features and purposes.
1. Here, artificial cavities result from excavation of the rock matrix, carried out through perforation of the unconsolidated soil stratum and the creation of containment structures, the depth of which is dependant on the resistance presented by the unconsolidated stratum encountered.
2. There may be structures for the following purposes:
- a. exploitation of small water sources by means of underground excavation;
- b. rainwater collection and storage;
- c. transport of drinking water from the neighbouring terrain;
- d. disposal structures.
3. Different and diversified installations were used to ensure the longevity of the settlement itself, with the excavation of civil, religious and military structures.
4. The larger the settlement's surface, the greater the chance of finding articulated and diversified underground structures.
5. A site's stratification generally results in building overlap. The external ground surface can sometimes be raised to ensure the survival of the underlying environments.

V.6 - *Understanding the purpose of an artificial cavity*

Like any other man-made building, underground structures are created intentionally through will and the application of both material and intellectual resources. In the face of numerous examples, it can be said that everyone has a will, whether this is expressed freely or under duress. We are not always able to establish the original intention or purpose of an excavation. Over time, the purpose of the underground structure may have changed and the structure may have undergone such transformation that the original purpose has now been concealed or annihilated. If it is possible to deduce the purpose or function from objective considerations, it is also possible that we mat be in the dark about certain aspects. At least upon a first examination.

Conducting operations within an artificial cavity thus means the exploration and acquisition of complete data in compatibility with contingent factors. The following points provide us with solid, fundamental information upon which we can gain direct information about the underground structure:
- systematic prospecting for the identification of access-points,
- geological framework,
- geological, geographic and topographic framework,
- historical, architectonic and archaeological context.

The primary works to be conducted within an artificial cavity are:
- exploration,
- map and section surveys,
- photographic documentation,
- video documentation,
- collection of all data relating to the underground structure.

Full understanding of the information obtained, requires knowledge of the environment in question: observations, comparisons, archive and toponomastic searches are the elements required in order to gain an understanding of the man-made structure and provide a historical reconstruction. Although not always possible or feasible, stratigraphic investigation remains a valid tool. Furthermore, every cavity can constitute an ecological niche. From a bio-speleological point of view, its investigation normally provides information of certain interest on the fauna which transits or lives there. Knowledge or at least a basic knowledge of the restrictions that underground cavities are subject to is also required as is an interest in the relative legislation.

As data collection, graphic rendering and documentation can take place in "difficult" environments, speleological knowledge and awareness of the risks associated to the activity are required. It is clear how an aptitude for unusual environments and training allow operations to be easily and safely conducted, to the benefit of the work to be carried out. In certain cases, advancement difficulties, dangerous situations and the knowledge that safety must always come first, are determining factors in the application of rigorous methods and an investigation strategy, which allow for maximum results with minimum risk. Finally, consideration should be given to the fact that speleological equipment can sometimes come in useful even where the structures are not strictly subterranean in nature.

V.6.1 - *Research and processing*

As with every task, words alone are insufficient and must be substantiated by facts. In summarising the aforementioned concepts, certain fixed points in the investigation of man-made underground structures can be listed:
- Cartographic assessment, aerial and/or satellite photographs.
- Systematic research and identification.
- Accessibility of the structure and capacity for underground permanence.
- Geological, geographic and topographic framework.
- Creation of a planimetric survey.
- Context documentation.
- Analysis of material evidence.
- Photographic and video documentation.
- Historical, architectonic, archaeological, toponomastic, bio-speleological, jurisprudential, research, etc.
- Rendering, processing and synthesis of collected data.
- Publication.

CHAPTER VI

INVESTIGATION TOOLS

VI.1 - *Geology and geophysics*

For a general view of the geological context, a geological map should be consulted. Awareness of the characteristics of the soil and subsoil is fundamental in the correct choice of research and exploration methods. Geological, geomorphological and hydrogeological studies are also useful. A preliminary geological analysis of the area around an archaeological site can cover various aspects of geological science: geology, geomorphology, hydrogeology, surface water hydraulics, geotechnics, geomechanics, regional geology, seismology, mineralogy, petrography, palaeography, etc. Topics of primary interest relative to the typology and assumed or known history of each case will be examined on a case by case basis (Aquino 2005, pgs. 137-138).

Applied Geophysics techniques can be useful in the identification of anomalies in connection to different types of buried anthropic structures. Seismic, electric ad electromagnetic methods are the main methods which may be adopted. These are rapid, non-destructive survey tools for natural and anthropic buried structures. In the archaeological field, such tools «can provide valid information for excavation planning interventions as well as information on larger areas adjacent to the digs and thus allow a fuller understanding of territorial organisation in ancient times» (Carrozzo, Nuzzo, Quarta 2002, pg. 317). For example, during the 2001 archaeological recognition campaign at Populonia (Livorno), geo-electric prospecting excavations were conducted at the acropolis and its surrounding area: «sections were processed by a specific programme able to create instant bio-dimensional resistivity models, to process large volumes of data using mathematical calculations and to assess average resistivity of significant underground volumes» (Cambi 2002, pgs. 101-102). As highlighted by Finzi more than ten years ago, the constantly evolving Georadar provides an alternative to electromagnetic prospecting methods (Finzi 1990, pg. 191).

VI.1.1 - *Cave minerals and formations*

The study of cave minerals and formations has led to the exploration of both natural and artificial cavities, with interesting results. «In the last 10 to 20 years, mainly in Europe, cavers have become more interested in exploring artificial "caves" - that is, old mines, aqueducts, underground passages beneath castles, sewers, and the like. These voids may host formations which morphologically resemble speleothems. But they cannot be considered to be true speleothems since they do not form in natural caves. Generally the formations in artificial "caves" are far less frequent and much smaller than speleothems present in natural caves» (Hill, Forti 1997, pg. 224).

VI.2 - *Mining archaeology*

Geology is required in the study of the ground, subsoil and mineral deposit characteristics. These three elements generally condition a mine's morphology as well as the organisation of both internal and external works. The study of mineral deposits aims to gather information, to provide an understanding of:
a. who worked in the mine;
b. what was extracted;
c. when, how and why the mine was excavated.

The strategic choice of mine excavation can be understood through its mineralogical aspects and the technological level reached within a given area, within a given period. The study thus augurs an interdisciplinary intervention, in that it is possible that certain morphologies, which are incomprehensible to an archaeologist or a speleologist, may be interpreted by a geologist and a mine expert and vice versa. The aims of mining archaeology research are summarised in the following points as first presented at the XV Congress of Lombard Speleology in 1999 through the work of Alessandra Casini and Giovanna Cascone (Casini, Cascone 2000, pgs. 93-122) and subsequently developed (Padovan 2005 d, pgs. 75-100).

The identification of each of these points provides an understanding of the strategic choices made by the miners, of their technical knowledge, of mining topography, of the mine's functional divisions and hence of the work organisation, of the mine's different exploitation phases and insofar as possible, of the chronology of excavation activity. Such "study objectives" can also be used and applied to the study of quarries and related issues.

1. The nature of the mineral deposit and the geomorphological characteristics of the territory.
2. Mineral deposit identification method.
3. Research method.
4. Extraction method.
5. Blasting method.
6. Support structures and progression infrastructures.
7. Ventilation system.
8. Dewatering.
9. Lighting system.
10. Mineral transport system.

VI.2.1 - *Surface prospecting in mining archaeology*

Surface prospecting allows the localisation of mining areas (entrances and waste disposal sites), grinding mills, metal transformation slag disposal areas and production site, roadway and settlement waste disposal areas. Mineral veins are not normally isolated and once one has been located, research and exploration usually extends to the surrounding area. Given that areas rich in mineral deposits may be more prone to atmospheric erosion, where vegetation is dense the veins may be concealed. A vein embedded in soft rock is more likely to be uncovered as erosion will affect the enclosing rock, thus leaving it bare.

Mineral fragments found at the valley bottom or along water courses are good indicators of the presence of mineral deposits. In the past, such indicators were already used in the identification of mineral deposits and resulting mineral exploitation (Biringuccio 1540, I. Gerbella 1947, pgs. 29-32). Areas where forests were once cut down to obtain wood for mining or where "wood charcoal" was gathered for the furnaces where the mineral underwent the "roasting" process or for the transformation furnaces, can sometimes be identified. For a fuller understanding of the panorama, the insertion of data within the territorial context and historic territory map processing, each element must be topographically positioned using appropriate instrumentation. In the absence of reference relating to the chronological distribution of mining works, the relationship between these elements and the settlements or nearby elevations assists in the formulation of an assumption regarding the chronology of each period of activity. Settlements and mining areas are generally closely linked and a settlement may be used to control and protect both the extraction and metallurgical sites and the access roads. Furthermore, accurate tracing of the roads can lead to the identification of the entrances to extraction plants.

VI.2.2 - *Production cycles and historical landscape*

The study of production cycles allows us to deal with human activity issues in relation to natural resources. This ability to exploit mineral resources triggered a series of processes over the years involving the social, political, economic, technological and consequently landscape fields. Knowledge of Mining Archaeology is useful as it can shed light on the excavation technique and methodology used in the creation of the various man-made structures. Where there are mineral deposits, investigation of the historic landscape and settlement dynamics must take extraction results into account. This will assist in obtaining a fuller understanding of the organisation of work, economical and social relationships and the technological revolution stemming from the development of the mining system and subsequent mineral transformation.

Evolution may follow a likely pattern and promote or ease the establishment and development of a large array of underground cultural, civil and military structures. Mine workers may find work in the construction of structures for purposes other than extraction. Other workers simply contribute by sharing their knowledge and experience: choice of excavation site, rock quarrying method, measurement and calculation system, transport of materials, setting-up of covering and containment structures, ventilation, dewatering and water supply, identification of areas at risk of collapse and measures to be taken. Such elements can affect any type of underground excavation.

VI.3 - *Archaeometry*

Archaeometry is an aspect of archaeological investigation which utilises chemico-physical and mathematical disciplines to sort and classify that which can be measured through archaeology. It is a measurement science, which deals with the issues of such a position and particularly with the scientific paradox of measurement relativity: «fortunately, the scientific method is the only steady bridge we can build between reality and the world of ideas» (Gigante 2002, pg. 15). For dating purposes, the following are generally used:

- Carbon 14 dating;
- ultrasensitive mass spectrometry dating;
- thermoluminence dating.

Archaeometry provides information on a man-made structure by means of analyses carried out with natural science instruments. Each attempt at becoming acquainted with the world in fact leads to models and interpretations which are highly dependant on the instruments used. There is thus an apparent inconsistency between what we know and how and what we measure.

Once the man-made structure has been located, geological investigation is of great assistance during the interpretation and dating phase. Absolute and relative dating can be carried out through the study of sedimentary succession (stratigraphic analysis). With a degree of approximation, absolute dating can sometimes provide a specific archaeological age for specific events. Relative dating, on the other hand, attempts to place events in chronological order, by simply establishing that one event took place prior or later than another. Some of the methods used are descried below: dendrochronology, lichenometry, the varve method, palynology and paleoclimatology. Mineral and rock analysis, hydrogeological investigation, geotechnical analysis, geomechanical and structural analysis systems can be useful when researching materials and selecting an excavation site (Aquino 2005, pgs. 139-143).

VI.4 - *Applied topography*

Modern technology can be applied to topographic surveys, which can in turn be used to improve data management in speleological and archaeological investigations. This will provide rapid and reliable graphic renderings, the scale and format of which can be adapted to any situation, thus optimizing analysis and synthesis times. Just think of the possibility of obtaining graphic printouts in absolute co-ordinates, designs in different scales with inserted, to-scale, geo-referenced photographic images and geometric measurements for any stratigraphic analysis.

Linking an underground installation to surface morphology and marking its altitude can be useful in completing information obtained underground, particularly where there are architectonic elements. An underground environment is linked to the surface by means of a single or multiple "umbilical cord", in the form of a horizontal or vertical link, or of a series of rooms to which it is attached. This allows the creation of an overall picture of the cavities studied and of their specific context. External reference points are positioned through the identification of survey stations, upon which the various elements relative to the hypogeum's entrances triangulate and relative and absolute altitudes are established (Basilico 2005, pg. 197).

The science of topography produces data and information, which must be transmitted using language suitable for comprehension and comparison in different contexts. Such language must envisage the use of identical parameters. The co-ordinates of any survey must be uniform, and must be recorded as geographic co-ordinates or as UTM (Universal Transverse Mercator). The accuracy of information relating to geographic positioning is dependant on the type of instruments used and on the geodetic reference network. There is a difference of tens of metres between the co-ordinates obtained from a GPS pathfinder receiver and those obtained from a topographic GPS receiver. This is due to the fact that the former has no land references i.e. it is not connected to a local network and is not therefore very accurate. The latter, linked to a geodetic network, is accurate to the centimetre. Total electronic station survey co-ordinates are accurate to the millimetre, however, if the survey is not linked to a geodetic or system network, the positioning of certain combined points on the terrestrial sphere cannot be properly established (Frignani 2005, pgs. 151-153).

VI.4.1 - *Total electronic station*

On account of their resilience, impermeability and reliability, total electronic stations can be used to survey even underground environments, or at any rate to survey semi-subterranean structures although not without limits and with due precautions. The most important aspect of a total electronic station is integrated control system mapping together with internal computer management of data and the various functions. It can record data recorded internally, externally or on a smart card; the choice of system is dependant on the type of work to be conducted. Given the technological level reached by companies who manufacture precision instruments, it is difficult to ascertain if one instrument is better or worse than another. It would be fairer to say that one instrument may be better suited to a certain type of work than another (Frignani 2005, pgs. 155-156).

The survey of an underground environment requires the use a simple instrument, with few cables and a keyboard, which is easy to use. New generation total electronic stations with internal data recording functions provide the operator with excellent mobility; they come equipped with an alpha-numeric remote control for the following functions:
- data management;
- recording;
- prism reading.

Recording by means of an external recording unit allows a larger keyboard to be used and, more importantly, allows work to be carried out around the instrument, if not at the time of collimation. This factor should not be underestimated as work is often carried out in confined spaces, such as in underground aqueducts or in narrow section haulage roads. Maintaining a distance from the instrument prevents accidental collision with the instrument itself, thus preventing any modification to its position.

The disadvantages of a lack of light can be compensated through the application of basic prism illumination systems. This instrument has its own illumination, allowing the prism to be viewed although the image may not always be sharp. Personal illumination, whether primary or secondary, should not be overlooked. It should be taken into account that a total electronic station can be used until such a time as is permitted by space restrictions, as the instrument cannot be mounted on a tripod in environments below a certain height and width. Horizontal, telescoping mining tripods are available, but even these cannot be used unless certain dimension requirements are met. Where running or stagnant water submerges the ground level by more than 10 cm, ground centring cannot take place.

Where there is no requirement or no possibility of conducting a polygonal survey, the relative survey may still be conducted by positioning and fixing survey stations on the walls and then dealing proceeding with other work using specific underground instrumentation. Upon graphic rendering, work carried out during different phases will be inter-linked, one to the other. The survey of at least three points in common with the previous job must be carried out for each job after the first job and this should not be overlooked. The next step is a triangulation survey, which will ensure that the correct geometries and above all, the correct orientations, are maintained (Frignani 2005, pgs. 155-156).

VI.5 - *Topography and information technology*

Topography is not just a means of representing the terrain or the territory. Thanks to information technology, topography has become an information management instrument. The rapid development of information technology as a science has resulted in the rapid evolution of measurement equipment, transforming it from topographic instrumentation into "design instrumentation".

Software and hardware evolution have increased the direct and indirect functions of measurement instruments, allowing topography to be used for dynamic applications. Such dynamism is at its height when operators are able to move away from standard guidelines and use applications which could be considered as creative. Both traditional topographic surveys and the survey of elevations can be carried out using modern instruments.

Similar results can be obtained by using a digital camera, either in conjunction or in addition to the topographic instrument. During the graphic rendering phase, data is sometimes processed into plans, tables and three-dimensional images. Software for the management of territorial information systems is then used to link information relating to the specific element to the graphic elements. A topographic survey can be the functional backbone of any research in the geographic, archaeological, naturalistic or urban fields (Frignani 2005, pg. 151).

VI.6 - *Photogrammetry and phototopography*

Photogrammetry is a collection of techniques and methods for the measurement of photographic images, through which the size of the photographed object can be determined. The idea of using photographs in topographical surveys was first thought of by Laussedat, a Frenchman, in 1850. The main purpose of photogrammetry is to obtain the exact geometric dimensions of a piece of land (or a man-made structure) from its photographic image. There are two types, aerial and terrestrial: the only difference is in the object itself. Photogrammetry is also used in archaeology. Phototopography is a procedure used in the topographic survey of large surfaces. There are four principal phases: aerial survey, ground preparation (topographic land organisation), rendering (topographic interpretation) and completion (map control and integration).

Photographic images can be broken down according to the following:
- orientation, given by the optical camera axis upon shutter release;
- material, depending on the type of sensitive material used.

The images can be subdivided as follows (Cosci 1988, pgs. 14-15):
- nadir photograph (vertical axis): this is the point of intersection of the camera's vertical optical axis and of the horizontal frame;
- oblique photograph (panoramic and semi-panoramic oblique photograph);
- black and white photographs which record the various colours in shades of grey;
- colour photographs;
- infrared images;
- colour infra-red image (or false-colour);
- thermal infrared images (pseudo-photography).

The nadir photograph is the most widely used in topography. The three-dimensional view of nadir photographs is obtained through the use of a stereoscope. Three-dimensional models of the terrain (DTM) can be obtained using this method. Over the past few years, satellite images, as well as photographs taken from aeroplanes and helicopters, have been used. Stereoscopic or three-dimensional vision is also used in archaeological photo-interpretation.

VI.6.1 - *Photointerpretation*

Photointerpretation relates to the recognition, identification and interpretation of the countless shapes, colours and tones of the earth's surface. This type of interpretation allows the identification of any rocky structures or archaeological wall relics, even below the first layer of soil. The type of film used depends on the type of investigation. Investigation of different sections of the electromagnetic spectrum can take place thanks to different film sensitivity. The availability of aerial coverage taken over different periods allows the identification or clarification of specific anomalies, not always visible from a single image. Underground structures and their entrances can be identified through photointerpretation, e.g. aqueduct service wells or the dromos of certain underground tombs.

With regard to archaeological photointerpretation, various interpretation keys have been proposed (Cosci 1988, pgs. 29-30):
- crop-marks: marks resulting from terrain anomalies and caused by the growth of cereals on top of buried wall structures;
- grass (weed)-marks, marks with similar characteristics to crop-marks and caused by the growth of vegetation;
- shadow-marks, marks left by surface micro-reliefs;
- damp-marks, marks resulting from ground colour anomalies;
- soil-marks, marks due to anomalies caused by the surface recovery of fragments of wall structures or brick material.

VI.7 - *Cartography*

Cartography is the tool which provides direction to the territory and the area to be explored and is vital for marking results and information acquired, even during surface prospecting. It provides physical, geographic, political and toponomastic information. «The cartography used during research, from project planning to field activity, to data processing to the scientific publication and disclosure, is at the same time both a search tool and a knowledge tool» (Cambi, Terrenato 2002, pg. 45).

Where ground supports such as trigonometric points are required, traditional topographic maps are used. Photomaps and photoplans, which can provide toponomastic elements and altimetric references are also useful, or better still are orthophotomaps and orthophotoplans. Paper maps can be converted into computerised maps using computerised systems (raster format and vector format). Computerised cartography allows the acquisition and overlapping of old maps and different scales within the same system. The information contained within specific databases can be linked to the topographic representations. Such data structures are managed by information systems knows as GIS (Geographical Information System). The creation of GIS platforms for historical and archaeological data management, must consider the availability of both raster and vector format maps in the different acquisition scales and at the same time take into account the burden that this entails (Luca, Salzotti, Valenti 2001, pgs. 39-42).

VI.8 - *Research and surface prospecting*

The trinomial mentioned in Chapter V can point, with good approximation, to the existence of artificial cavities. Furthermore, information relating to the cavities and their location can be obtained from historical data, old maps, books, magazines and aerial photographs. Sometimes, a single newspaper article is all that is needed to reveal that beneath apparently natural, subsidence (collapse or subsidence "doline"), lies an underground structure, the roof of which has collapsed. Geological and topographic maps normally record the location of mines which are still well known or were still in use in recent times. Certain toponyms may indicate or simply suggest the presence of hydraulic structures now in disuse or the presence of abandoned mining areas, not documented in archives or maps.

In addition to identifying buried wall structures, aerial photographs can reveal dolines, wells, open cisterns, the outlines of old, buried canals with underground sections, mine and countermine systems, trenches and other fortification structures. Terrain presenting anomalous pools of stagnant water may conceal the outlets of underground drainage systems or ancient, still functional aqueducts. Different soil colouration is evident from the photographs, particularly when shot under certain geographic or climatic conditions, such as in the Mediterranean area during the summer months. The contrast of the green area, saturated with water, will stand out against the dry, surrounding area.

Certain types of plant can be indicative of cavities. For example, wild figs can be indicative of high levels of concealed humidity. Humidity can be trapped by sections of half-buried wall, by the ruins of dwellings now covered by grass or by artificial cavities. In the area around Tarquinia (Viterbo), such plants grow near cisterns, underground passageways and rock tombs. Numerous wild figs can flourish at the outlets of underground passages and hydraulic tunnels. Sometimes, as is the case of the cavity known as "Fig Well" ("Pozzo del Fico") (CA 01039 LA VT), the plant grows out of the well itself and its roots descend around the segments which internally line the structure (Padovan 2002 b, pg. 83). The same peculiarity has also been observed in other parts of Southern Etruria and in certain parts of Northern Italy.

Generally speaking, once it has been identified that an area may conceal the entrance to an artificial cavity a systematic prospective study must be carried out. In flat, non-wooded areas, such a study can be carried out by a team with each person at a distance of no more than 5 m one from the other. In wooded areas, in the Mediterranean brush or in mountainous areas, the distance between each person is reduced down to 1 m.

VI.9 - *Other investigation tools: ancient topography, topnomastics and archive administration*

If topography identifies current sites, ancient topography attempts to locate the sites of previous civilisations. Its primary aims are the «reconstruction of the various phases of landscape anthropisation for each individual area of the classical world and the identification of trends in Man's behaviour in his relationship with the environment in particular, compatible situations» (Uggeri 2000 a, pg. 15). Ancient topography encapsulates territorial research carried out using following different layouts and refers «to those studies, whose object is the reconstruction of the topographical structure, and ultimately, of the past evolution of man's relationship with the environment, be this in the remote or more recent past» (Alfieri 2000, pg. 13). Furthermore, certain settlement choices can be understood through the reconstruction of an area's physiological situation and the input of historical data (Dall'Aglio 2000, pg. 177).

During the study of an area, toponyms provide essential support in the spatial identification of what has been modified or cancelled by time and anthropisation and also provide clues on the culture which contributed to this process. These shall be interpreted and assessed both literally and metaphorically (Del Lungo 2005, pg. 261). For instance, in the Italian territory, there are toponyms which point to the existence of not only caves but of disused hydraulic passageways, cisterns and various structures with underground areas, which were abandoned centuries ago. In Liguria, for example, «the localities of *Argentera, Argentina, Auriola, Oriola*, etc. are significant in that from an etymological point of view, they do not normally fall under "metalliferous". On the other hand, some of these localities are undoubtedly linked to the exploitation of mixed sulphides with traces of silver and gold, substances which could today be disregarded but which in the XII and XIII centuries supplied a high number of *silverware stores* and may have given origin to additional toponyms recalling the presence of gold, whether real or assumed» (Pipino 2003, pg. 40).

Within the province of Latina, the Capo d'Acqua toponym could indicate the point of origin of an ancient or mediaeval aqueduct «its specific semantic efficacy and interventions relative to the capture of spring water, the keeping it use almost until this very century. The first known reference appears to be that contained within the papal seal through which Pope Sylvester II grants Count Daiferio seigniority of Terracina and its territory on 26 December 1000. The confines of said territory are known as *incipientes a Capo de Aqua,* as part of the territory itself lies on the Pontine Plain where the springs of the *Forum Appii (*Borgo Faiti*)* are located, between Sezze and Priverno» (Del Lungo 2001, pg. 4).

Toponomastics can provide ethno-linguistic information which archaeology is unable to provide and the existence of specific populations within the territory can be determined by the language of the toponym (Uggeri 2000 b, pg. 120). Historical research on underground structures is normally carried out on the basis of indirect evidence: from books, local history, periodicals and other similar text. This type of investigation generally reveals a lot of information however the limitations of such sources should always be established. Literature can contain many errors and voluntary or intentional omissions. Or be a source of unreliable information. Modern authors sometimes report information from earlier literature in critical fashion and a series of unfounded or imprecise facts is thus transferred from one text to the other without any form of verification ever being conducted (Pesaro 2005, pg. 219).

On the other hand, an archive document ensures the quality of information, and generally renders such information immune to those misunderstandings and distortions which can compromise the value of other indirect sources. The potential of working from first-hand information need not be underlined: targeted archive investigations have often overturned the interpretation of a man-made structure or have contributed to accurately explain the structure's history. Of the many possible examples, the Roman aqueduct at Asolo (Treviso), where the investigation of direct sources resulted in the identification of original parts from the barely recognisable, modern restorations (Riera 1991, pgs. 191-197).

In Tuscany, documentary sources for the reconstruction of mediaeval settlements have seen a decisive quantitative and qualitative increase «to such an extent that the selection of potentially useful documentary typologies has now replaced the perusal of every possible piece of evidence as the fundamental operation» (Ginatempo, Giorgi 1996, pg. 1). Inscriptions constitute another direct source. The consultation of epigraphic sources can provide interesting clues on the geography of the ancient territory as well as on the location of underground structures.

VI.10. - *Planimetric survey*

A survey is not the simple graphic reproduction of a space, but is a means of understanding the hypogeum and the information system used. Additionally, strange though it may seem, we have a tendency to "look" without actually "seeing". Generally speaking, our lifestyle is not conducive to analytical vision. To put oneself in a position to directly survey means to direct one's attention on the object in order to pinpoint details, which are not always identifiable.

The time available, the accessibility of the location, the equipment used, team spirit and the experience of survey team members, the risks involved and whether returning to the site is a possibility are some of the factors which should be considered and which can affect the positive or negative outcome of our investigations. In the study of hypogea, *conditio sine qua non* is the first-hand assessment of the hypogeum. The point upon which the accuracy of a survey should be based is consequential to the fixed objective and to contingent reality. The approach must take several factors into account:
- accessibility characteristics of the environment to be surveyed;
- what is to be surveyed;
- how the survey is to be carried out;
- what degree of accuracy is required or can be adopted;
- what degree of accuracy is obtained during the course of the survey.

With regard to the accuracy of the survey, this will be determined by the interaction between the data collection method, on identification of the primary points upon which subsequent study phases are to be based and on the outcome of the following five factors:
- environment usability;
- psychological endurance to the environment;
- training;
- equipment used;
- data rendering ability.

An environment with evident structural subsidence poses various risks and permanence within such an environment for any length of time can be dangerous. Under such circumstances, the survey should be carried out as quickly as possible with rapid-survey instruments and should only take into account those elements which are deemed essential. Understanding those details providing the purpose of the investigated location can increase focus on the identification and surveying of elements, which will help confirm its nature. Finally, should it be possible to obtain precise, accurate data without direct exploration of a cavity, we would be unable to provide a complete assessment (Basilico 2005, pgs. 187-188).

To those undertaking the survey, one thing is immediately apparent: there exists no foolproof method, utilisable under every circumstance. The main driving force in the choice of an appropriate study method should be based on the nature of the location and on experience. Conducting a survey outwith a unitary and inflexible analysis system, towards an immutable approach can be of use. An ergonomic usability perspective, definable as efficacy, efficiency and completion should be pursued. Thereafter, flexible study plans ascribable to three different survey levels are preferred:
- running analysis level;
- detailed analysis level;
- in-depth analysis level.

By adhering to the task in hand, we shall have the opportunity of carrying out running surveys through to architectonic surveys. The people involved should at least have a basic knowledge of the subject matter and the study of an artificial cavity cannot dispense with graphic-descriptive rendering ability, dependant on the lexicon associated to the idea to be expressed.

The representation of an artificial cavity envisages consequential survey and data rendering phases that can be assimilated to that required in the description of an architectonic man-made structure. In the survey of architectonic buildings, a plan, the first phase of which envisages the design and measuring of maps, sections and tables as well as of those aspects which are deemed to be of primary importance should be followed. This is followed by graphic rendering. For this very reason there may be situations where maps, sections, tables and details must be prepared with a univocal analytical-temporal method, between the positioning of two survey stations. On the other hand, planned phases of multiple analyses can be carried out by adopting a temporal-analytical plurality, through the initial plotting of landmarks useful for the planimetric definition, which is followed by the other types of analysis.

VI.10.1 - *Running, detailed and in-depth analyses*

A running survey is carried out when safety is an issue or where there are insufficient factors to allow permanence within the hypogeum. Similarly, where the analysis is of little interest or where available instruments are essential, the approach will fall under this typology. Environmental factors often prevent the use of sophisticated equipment: how to deal with such situations and achieve acceptable results will be explained further on.

Detailed analysis envisages an approach directed at a higher-level survey. With more available time and being in an environment where the risk is contained, a more accurate study, which takes other factors into account, can be carried out on the underground cavity. In-depth analysis utilises specialised study methods. Laboratory tests may be required, especially with regard to the dating of certain elements. This method allows increased control of the data acquisition process for optimal definition. Any corrections are dynamically linked to the collected data, thus creating an operating path with dual analysis/checking/correction access (Basilico 2005, pgs. 188-196). The resulting checklist allows the identification and correction of errors. The steps are summarised below.
- preliminary survey for the identification of subunits and for the preparation of a detailed survey;
- plotting of the reference altitudes using appropriate instrumentation;
- morphologic and topographic survey;
- merging of primary unit subunits;
- graphic rendering;
- topography and integration checks;
- identification of construction phases;
- material analysis.

VI.11 - *Video and photography*

Underground photography, whether this takes place in natural or artificial environments, is a very popular practice and in an essential work tool. To photograph means to document the object of study. The choice of camera is dictated not only by its cost but by the environment to be photographed. There are "large format" cameras (24x30, 13x18, 9x12, 6x9; values expressed in centimetres), "medium format" cameras (6x9, 6x6, 6x4.5) and "small format" cameras (24x36 mm). There are "APS panoramic" cameras, "mini cameras" and "digital cameras". Large format cameras use photographic plates: they are bulky and expensive cameras and are not suitable for underground photography. Medium format cameras are also delicate and expensive. APS cameras are perfect for panoramas but deform the image in environments where space is limited; mini cameras are used for snap-shots. Digital cameras are light-weight, immediate viewing of the photograph and deletion if necessary, however their performance is proportional to their cost. The most expensive and sophisticated digital cameras require little light for close-ups and are suitable for many, although not all, underground environments; in any case, digital cameras are delicate and do not react well to dust and humidity (Esposito 2005, pgs. 223-224).

The best small format cameras are "reflex" cameras. In this, the XXI century, which has seen digital cameras rapidly supersede all other types of camera, a discourse on mechanical and electronic cameras could seem anachronistic. Perhaps it is. However, for the purposes of the job in hand, they are still the best. Especially under difficult conditions. In extreme conditions, such as in flooded or submerged environments or where there is heavy internal dripping, underwater cameras are the only cameras which can be used. Special waterproof camera cases can be used for cameras that are not for underwater use.

VI.11.1 - *Surveys and digital photography*

Computerised digital photogrammetry does not necessarily require the purchase of a digital camera: photographs can be taken with traditional cameras and the images obtained can then be scanned. The advantage of a digital camera is primarily linked to avoidance of the intermediate steps, which are obligatory when using film (developing, printing or framing of slides and scanning) and to the possibility of taking numerous shots without worrying about the cost of developing.

Good, low-cost digital cameras are available on today's market, but they are not all suitable for photographs which are to be rectified and geo-referenced. There are also semi-professional, 6 million pixel digital cameras at affordable prices, which can be used with inter-changeable lenses, just like reflex cameras. The requirements are:
- 2 million pixel images;
- a zoom lens which suitable for use with focals which do not excessively distort the images (ideally focals of between 50 mm and 65 mm or higher as these will flatten the image).

In archaeology, photography is normally used to record images for archive-creation purposes, images which document the status of an excavation, the archaeological finds, etc. The interesting aspect is that images (following appropriate treatment and processing) can be rectified and geo-referenced in order to obtain a geometric basis for the graphic rendering. In synthetic terms, we can move directly from the photograph to graphic rendering without carrying out a detailed survey; in particular, the graphic rendering system of monuments or objects, such as, ceramic fragments, can take place (Frignani 2005, pgs. 166-167).

In the latter application, design tools and measurement gauges will be replaced by a digital camera, software for the rectification of films and their geo-referencing and by image digitalization software.

VI.11.2 - *Image rectification and geo-referencing*

Rectification and geo-referencing software sometimes include automatic graphic rendering functions (vectorisation); AutoCad is the software of choice in that it is powerful and has flexible image management, which allows graphic rendering, similar to that which can be achieved by hand. The operating procedure is as follows:

1. Image acquisition. Close observation of the object to be photographed can be useful; if the object is very large, image mosaicing, that is, the placing of sequential images one next to the other and subsequently one above or below the other, is recommended. To start with, a complete image of the subject is required. Shots of the details, such as doors, windows, columns, etc. are then taken. Shots of the details should be taken at a perpendicular angle.

2. Image processing. Once downloaded, the images are treated with specific rectification and geo-referencing software. Rectification is a mathematical operation which is carried out by defining the horizontal and vertical lines of the image; the image is then geo-referenced by defining certain characteristic points within the image, to act as "targets" or simply as corners, the dimensions of which have been ascertained. Adaptation of the x, y relationship is

thus imposed on the results: these measures were taken while photographing and represent the length and height of the object to be geo-referenced. At this stage the photograph is rectified and geometrically compensated. The use of photo retouching software in not recommended as although this rectifies the image, proportions are not retained.

VI.12 - *Video films*

Metaphorically speaking, video gives us the opportunity of bringing the cavity into the office and is an irreplaceable documentation and disclosure system. The choice of video camera is important and should meet both our needs and that of the environment to be filmed. The same applies to electricity and lighting systems and also to the filming techniques used.

The cameraman should get used to "editing while shooting", that is filming in a specific order, as if following a screenplay. Cinematic post-production is generally not an option to the cameraman on account of the high costs involved; an editing control unit complete with video recorders, titler, effects etc., requires a large investment. However, good digital editing results can be achieved with modern information technology. It should not be forgotten that the positive outcome of any job, is dependant on correct editing (Wilke 2005, pgs. 233-239).

VI.13 - *Advancement: speleology and underwater speleology*

Both natural and artificial cavities have several common denominators:
- lack of light, in the case of cavities of modest proportion and/or cavities directly connected to the surface;
- total absence of light, in the case of larger and deeper cavities;
- humidity;
- possible presence of water, whether stagnant or flowing water.

With few exceptions, environments of modest proportions are not difficult to explore and can be investigated without specific expertise. In the presence of the below three factors, the situation changes immensely and speleological experience becomes a requirement:
- considerable spatial development;
- environments of limited dimension;
- vertical shafts.

Exploration and documentation times increase in articulated environments and even more so in environments of limited dimension. For vertical shaft exploration, a series of anchors must first be set up. Speleology requires training, skill and physical as well as psychological resistance. A speleologist is psychologically prepared for moving around in total darkness, for running the associated risks and for collecting of data, under even extreme conditions. He has mastered both exploration techniques and speleological equipment, which permit his underground descent, allowing him to remain there for numerous hours in order fulfil exploratory and other speleological requirements, such as: research.

VI.13.1 - *Speleology and artificial cavities*

As a general guideline, an artificial cavity will not require the same level of physical input that is required for cave exploration. Comparing the two for the sole purpose of establishing measurement parameters would be pointless. Especially when we consider their diverse origins. However, it has been found that operations carried out in artificial cavities can result in increased psychological strain, not only attributable to the type of work carried out, but primarily on account of the static and sanitary conditions, which are rarely optimal. It can therefore be stated that attitude, preparation and equipment are always required, irrespective of the circumstance.

In this respect, we would like to provide just a few pointers regarding the equipment and the risks involved. Mastery of correct speleological methodology is down to course-work and specific manuals. Unfortunately, various fatal and non-fatal accidents have taken place involving people, with little knowledge and inadequate equipment, who chose to become make-shift speleologists.

There are two basic rules:
- before venturing underground, at least two separate lighting systems are required. This ensures that should one system cease to function you will not be left in the dark and thus run the risk of an accident;

- speleological equipment should always be used when faced with a vertical shaft, even if this is just a few meters deep.

If the characteristics of an artificial cavity are not comparable to those of a cave, this does not mean that its exploration should be taken lightly.

VI.13.2 - *Underwater speleology*

Underwater speleology combines its own specific speleological and underwater methodologies and relevant techniques. Without prejudice to the fact that we should already be experienced divers, it is important to first attend a standard speleological course and then an underwater speleology course before venturing into submerged cavities.

A significant number of immersions must be carried out in different types of situations: salt water, fresh water, hot and cold water and wreck immersions. Excellent psycho-physical balance is also required. In the speleological field (and not only in this field) to take risks means putting one's life on the line, as well as the lives of those who come to the rescue, or whose task it is to recover the body (Bertulessi, Padovan 2005, pgs. 151-153).

Underwater advancement can take place even at considerable depth, in confined spaces, often characterised by loamy deposits, which cloud the water upon movement thus reducing or obscuring visibility to such an extent that personal instrumentation cannot be read. This type of activity normally falls under solitary diving. In such a situation, the choice of diving alone, without a companion is due to the fact that in deep and/or cramped environments, where there may be limited visibility, a companion would be of no assistance whatsoever in the event of an "unexpected difficulty" and would often be counter-productive. Analysis of fatal accidents involving underwater speleologists reveals that in 80% of cases, those who go to the rescue of a companion also die. But this relates to "extreme" or very specific explorations.

In artificial caves, the "golden" solitary diving rule is not normally adhered to during the survey of chamber cisterns or of large spaces where there are limited risks. But even if the complexity and risk levels involved in deep cave immersions are not comparable to those carried out in artificial cavities, these should not, under any circumstance, be approached with superficiality.

Equipment and techniques are far more demanding than those required in the speleology of aerial environments, another type of speleology which does not fall within the scope of this manual. Nevertheless, it seems appropriate to point out the importance of taking specific courses and to apply oneself to continuous training.

VI.14 - *Risks*

There are multiple, although infrequent, risks associated to activity in artificial cavities. These are mainly associated to the type of cavity and its specific morphology. The risk typologies of underground structures are entirely dependant on the purpose allocated to the structure over the years. Take for example a cistern created for the collection of water (therefore a low-risk environment): this could over time, due to new urban requirements, become part of the sewage collection system (high-risk environment). All this imposes that careful assessment, even historic, of the places to be explored and documented be conducted at all times.

As a general guideline, there may be:

- falling masonry, of modest proportion, from the structures;
- collapse, or structural subsidence;
- toxic substances, pollutants or explosive materials;
- animals;
- inadequate equipment;
- inadequate compliance to safety measures.

VI.14.1 - *Structural collapse*

As a general guideline, each environment is destined to settle naturally due to the passing of time or due to secondary factors. Abandoned quarries and mines in particular, present areas of collapse. Less frequent in old mines, where only manual tools were used for extraction (no explosives were therefore used); areas of collapse are more frequent in mines of recent construction, where different demolition and mining techniques were used.

Underground passages and tunnels, whose arches were created with typical wooden frames, could have rotten or collapsed structures. Even "filled" spaces could be unstable. Submerged wells, which are difficult to identify and quick-sand may be encountered in flooded sections. The static condition of the puteal or very lining of cisterns and disused wells is sometimes precarious. Prior to descent, an external metallic tube frame must be set up. Ropes can then be attached to it, thus preventing unnecessary stress on the structure.

VI.14.2 - *Explosive materials and war residues*

Abandoned explosives which are not to be touched under any circumstance can sometimes be found in quarries and mines. The passing of time and humidity may have rendered these unstable and thus highly dangerous. It should furthermore be taken into account that there may still be unexploded loads within the remains of demolition chambers. Live devices can still be found along the fronts of the two World Wars, both outside and inside fortification structures with underground sections. It should not be forgotten that chemical irritants, such as chlorine, carbonyl chloride, hydrogen cyanide, yprite ("mustard gas"), etc., were used during the First World War. The casing of a chemical grenade is generally thinner than that of an ordinary grenade: due to deterioration this may give way and release the irritant without exploding. When found, residues of war should never be removed and the relative bodies should be immediately notified.

VI.14.3 - *Gas*

Gas is a common occurrence in artificial cavities. This may have developed from the putrefaction of dead animals or from various types of refuse concealed within disused underground structures. Additionally, gases, naturally found in the surrounding soil, may be present. Such gases find easy passage through the artificial drainage channels and form gas pockets, which should not be ventured into without due precaution. This is often the case in coal-mines where firedamp pockets are commonplace. Gaseous presence within underground structures can be attributed to the following factors (Gibertini 2005, pgs. 265-266):

1. Permeation from areas of thermal, volcanic, pseudovulcanic or mineral deposit activity. The presence of sulphurous or nitrogenous compounds results from thermal or volcanic activity. Light hydrocarbon mixes (substances derived from the alteration of deep sedimentary organic layers, e.g. petroleum, but which have a lower molecular weight than gaseous substances such as methane, ethane, propane, butane etc.) may be found in areas of mineral deposits (coalmines, petroliferous areas, etc.), or where the soil surrounding underground structures is reactive, that is to say, where underground minerals readily react to permeating substances, such as sulphur minerals.

2. Gas stagnation resulting from biological activity (putrefaction). In the case of gas deriving from biological, putrefaction activity, thought should be given to the oxygen consumption relating to the metabolic processes of numerous micro-organisms and to the inevitable establishment of anaerobic conditions (lack of air) and consequent formation of gas in reduced form (the chemical composition of gas does not include oxygen, in that the formation of gas occurs in situations where there is little, if any, oxygen). In this regard, there have been various accidents, some of which were fatal, which can be attributed to the presence of now rotting wooden boarding or roof reinforcement.

3. Pollution from external activities and the release of toxins. When working in man-made areas, the most common and most dangerous situation is that of finding civil or industrial discharge effluents, which are sometimes not neither officially registered nor authorised.

4. Infiltration resulting from the leakage of nearby or invasive pipes. In addition to the previous situation, one may find himself faced with the more or less conscious utilisation or interception of old disused sewage conduits, of any other type of underground structure or with recent sewage structures or gas pipes. Sometimes the positioning of such structures in nearby areas or in areas near the cavity is sufficient to ensure that in the event of sewage loss or gas leaks, the cavity itself takes on a drainage function and collects any leakages with consequent atmospheric changes.

In general, it should be remembered that underground structures may have been utilised as cesspools, dispersion tanks, authorised or unauthorised sewers and dumping grounds for waste and toxic waste (usually in certain abandoned mines or quarries), sewage and solvents. Various substances, such as rotting wood, can cause the formation of gas or reduce the amount of oxygen in the air. It is recommended that flameproof lamps or waterproof lamps, such as underwater lamps, be used for illumination purposes. In any case, acetylene should be avoided. Most important of all is that air analysis equipment be on hand at all times. (Gilbertini 2005, pgs. 267-270).

VI.14.4 - *Animals*

The common traumatology factors generically linked to physical activity in "particular" environments aside, some mention of infectious or toxic risks, resulting from direct contact with germs or animals carrying infection or vectors of other germs, should be made.

In relation to the evidence of risk, we will leave aside matters relating to bites from animals such as snakes, stray dogs, foxes and from the reptiles found in tropical areas, from spiders, scorpions, insects or from other animals, as all such unpleasant eventualities are part of the general risk linked to unusual, untamed areas. Precautions should, however, be taken against rabies and common snakebites. The areas to be found below inhabited centres are to be considered the unhealthiest (and rightly so). In addition to what has been deduced, the main drawback is linked to the possible presence of rats. There are two common types in Europe: the black rat (*Rattus rattus*) and the sewer rat (*Rattus norvegicus*), and both can be carriers of several diseases. An appropriate mask with replaceable filters for the various situations should be worn in the presence of mould, fungi or dust. Hands must be carefully washed and wounds must be thoroughly disinfected and medically treated after each operation while clothing and equipment must be washed and disinfected (Bregnani 2005, pgs. 277-281).

«The primary risk is linked to contact with pathogenous micro-organisms. The risk of intoxication by germ-produced toxins can be generically treated together with the risk of infection. This risk can be sub-divided into two categories, according to the infection method:
- infections from contact, ingestion or inhalation of water, air or contaminated ground, shall be referred to as "direct contact";
- infection from carrier (or "vector") animal bites (primarily anthropoids), shall be referred to as "vector infection".
The first risk category includes histoplasmosis, transmitted by the spores of a fungus which can become airborne thus contaminating air and water. The fungus is often found in caves and artificial cavities, especially those inhabited by bats or birds» (Bregani 1999, pgs. 96-100. Bregani, Tien, Ceraldi, Delfitto, Figini 2000, pgs. 396-401).

«Very frequent, at least as concerns the exploration of artificial cavities, is the risk of infiltration from the sewage collection system or from cesspools in those areas without a sewage disposal system, or from animal farms. Not to be underestimated is leptospirosis, a serious illness of difficult diagnosis, which is transmitted from soil and water, contaminated with infected rat urine. In respect of the second category, the focus is drawn to fleas, nits, ticks, flies and mosquitoes as the number of infections transmitted by arthropods (fleas, phlebotomine sand flies, horseflies, rat fleas, etc.), particularly in tropical countries, is extremely high. In relation to foreign activity, it is advisable to make sure that there are no epidemics in the country in question, particularly when travelling outside Europe. The below internet site provides information in this regard: http://www.port.venice.it/sanimav/epi.htm» (Bregani 2005, pgs. 277-282).

VI.14.5 - *Equipment*

The greatest risk associated with the use of speleological and underwater speleology equipment is essentially linked to two factors:
- use of the above equipment without prior familiarity;
- inappropriate use of general equipment.
It is strongly recommended that operations within environments requiring the use of other types of equipment, such as drysuits, masks with organic vapour filters, etc. be avoided.

All materials are subject to wear and tear. The incorrect assumption that the speleological equipment used in artificial cavities is subject to less wear and tear and therefore "lasts longer", certainly does not make the equipment "eternal". Hydrocarbons, quarry silt or even worse, mine silt, acid water and so forth, can weaken and erode equipment (particularly carabiniers, ropes and lanyards) far more rapidly than in karst environments. For all our efforts to gain adequate anchorage and the correct equipment, all too often the factual realities are far from acceptable and all too often cords are thrown against sharp edges. This results in extremely rapid deterioration as well as rupture of the inner strands and laceration of the outer mantle.

Steel carabiniers and identification tags last longer than aluminium ones and are therefore preferable for speleological use. This means that ropes, harnesses, lanyards, carabiniers etc., should be replaced more frequently and should always be washed and checked prior to use.

VI.14.6 - *Underwater speleology*

Underwater speleological operations in artificial cavities are less complex and present less risk than cave operations. There are no great depths or kilometres of development. The only exceptions are certain underground mines, which may extend on multiple levels and which became submerged when extraction works ceased and the drainage system was deactivated.

Work on a mining complex in the Municipality of Olgiate Molgora (LC), which was excavated at the beginning of the XX century, has recently been completed. These are the Pelucchi, Cepera and Valicelli mines. Their explorations were carried out by the Almè (Bergamo) Diving Group: «For the purposes of the exploration and the associated survey work, to date, many metres of line have been used in order to provide lifelines to the various tunnels, in fact, more than three kilometres of lifeline has been placed. Furthermore, more than 700 hours were spent diving and 2,000,000 litres of air and 150,000 litres of pure oxygen were used during the decompression phases» (Bertulessi, Rota 2005, pg. 52). In submerged areas, speleologists explored to a depth of -64 m below the water's surface.

As previously mentioned, it should be taken into account that the waters may be polluted. There should be no reason to repeat that water should always be analysed in advance. Operations have been abandoned on more than one occasion due to the floating carcasses of small animals, including mice and rats on the water. In any case, it is recommended that drysuits be worn at all times. The golden rule should always be followed: never remove the mouthpiece from the mouth, especially in environments located on the other side of a siphon (Bertulessi, Padovan 2005, pgs. 251-252). Samorè's article "Analysis of fatal underwater speleology accidents and their prevention" reports: «Two divers trapped by mud blockage. Two inhale sulphur dioxide from lignite deposits in abandoned cave-mine, just beyond the siphon; the third becomes, now aware of the sulphur dioxide, replaces the others' mouthpieces before going in search of help. But all in vain» (Samorè 1979, pgs. 63-64).

V.15 - *Biospeleology*

In recent decades, entomologists and more specifically, biospeleologists, have discovered the existence of numerous species of living animals in natural caves. Research was recently extended to include artificial cavities, which were previously only marginally considered. In every ecosystem, small or large as it may be, the integration of its biocenosis and its own biotype is influenced by and is strictly related to abiotic and biotic factors, which are characteristic of the environment itself. The fist information that a biospeleologist takes into consideration is in fact information relating to environmental factors, on both a qualitative and quantitative level (Zanon 1996, pg. 148). Each element contributes to our understanding of the environment to be investigated, allowing for more complete data acquisition, which upon first examination may appear superfluous.

Given their morphological heterogeneity and their vast range of distribution, artificial cavities cover a large ecological area. For further examination of the various environmental features and the consequent biocenosis, reference should be made to the typology breakdown (Zanon 2005, pg. 277).

Underground extraction works are representative of the typology which almost entirely covers traditional biospeleology. These are often found in, or next to, calcareous complexes and their morphology normally presents long tunnels, wells and inclines. Mines sometimes intersect meanders, water tables and natural vertical shafts. The primary abiotic factors (darkness, humidity and temperature) can reach the same levels as karst complexes. A large part of the species known in the biospeleology field can be found in these environments.

In the complex typology which covers hydraulic structures, the most important structures are those destined for the transport of water, especially if now in disuse. The flow or oscillation of water rushes along the walls, not giving fauna the chance to rest there. Whether in use or in disuse, it should not be forgotten that hydraulic structures can retain pools of perfectly still water and humid walls, ideal environments for small fauna. Water capture structures are of course important, although not so much from an environmental point of view as for the opportunity they provide for aquifer investigation, particularly in areas which are geologically free of natural caves. Storage and disposal structures often have dim lighting, high humidity and a temperature which is not especially stable. These factors are characteristic of the "shelter biotype" of various groups of fauna. Aquatic fauna largely fall under this typology, particularly specialised types of crustaceans. There are also numerous diptera and consequently spiders. Amphibians which live in locations with calm or stagnant waters should also be mentioned (Zanon 2003, pgs. 78-82).

Places of worship or funerary structures fall under a biotype, the environmental characteristics of which derive from the context in which they exist and from various collateral factors. A hermitage, built within a natural cave somewhat differs to a crypt located in an urban church vault. A rocky place of worship, created by adaptation of the entrance to a mountain cave, is ecologically different to a catacomb, excavated in tuff and situated on a plain. The morphology of a structure and its lighting, whether this is dim or completely lacking, further influence dwelling opportunities. All previously mentioned factors concur: depth, presence of water, organic material, etc. Troglobitic fauna may be found where there is karst substratum. On the whole, the fauna consists primarily of species which only use such places for shelter.

From an ecological point of view, the most promising structures falling under the vast typology of structures for civil use are those created through the adaptation of cave environments and which therefore present environmental characteristics found only in areas near natural caves: little illumination, relatively stable temperature and humidity. Disused tunnels and underground structures are in general only interesting from an ecological point of view. They do not always offer satisfactory humidity, however, occasionally they are excellent "shelter biotypes" for various groups of fauna.

Of the most interesting structures for military use are those which exploit natural caves and those excavated in calcareous rock. Under the environmental profile, abiotic factors are sometimes apparent and can reach values similar to those found in caves. The biocenosis of such biotypes is highly dependant on the substratum, on the morphology and on the way in which the environments were insulated (for humidity). Walls were often erected between the environment and the rock face thus creating an air space and preventing direct contact with the interstitial area. The state of neglect of such structures and exogenous trophic contributions somewhat impact on the qualitative and quantitative aspects of the fauna. These sometimes also include the troglobite species and various types of troglophile. The most interesting aspect is provided by the trenches, which "winding" through the subterranean environments, provide us with exploration opportunities without any need for excavation. Interesting results can also be obtained from the underground sections of military structures situated on foothills or plains. The location and environmental discontinuity provided by the underground chambers of the Porta Giovia Castle in Milan, has permitted the development of various anophtalmous entities, some of which are linked to the caverniculous environment as opposed to the underground environment. For example, a colony of *Mesoniscus alpicola*, never before observed on the plains, has been identified as have some *Paraleptoneta spinimana* (Zanon 1996, pgs. 150-164), never before seen north of Tuscany (Zanon 2005, pgs. 292-294) (figs. VI.1 and VI. 2).

VI.16 - *Legislation*

Numerous laws relative to subsoil ownership and specifically mine ownership have been enacted in Italy. Each country will have its own laws, and time should be taken to find out what these are, at least in their general guidelines. In the XIX century, Abignente states the following in relation to the quarries and mines of the Roman era: «There is no doubt that a singular right was established for the excavation of marble from 320 A.D. as in that year, Constantine the Great ordered the Rationalis of Africa, thus allowing anyone there the right to excavate marble: *etiam didistraendi facultatem* and in 363 A.D., Giuliano extended this right to the Oriental Diocese, having taken into account the usefulness and decorum which would result from the extension of the use of marble to the building trade» (Abignente 1888, pg. 57).

The "subsoil" is a meta-judicial concept: in order to define it, reference must be made to judicial disciplines as well as other subjects. «Subsoil is to be understood as the layer of soil under the surface soil; it consists of the lowest layer of soil, which acts as a reserve and, which rests upon rock (substratum). Ground, on the other hand, relates to surface soil, which is in direct contact with the surface and upon which agricultural and construction works are built. From a judicial point of view, the notion of ground is solely descriptive. Legislation does not recognize any definition of subsoil other than that which applies to the ground: judicially, ground and subsoil are one and the same. Similarly, it would prove difficult to think of a right (e.g. right of ownership), which had for object the ground rather than the subsoil. Even if, as we will see, certain *utilitates* (the term *utilitates* is often used in reference to any benefit, even if not financial, which improves fund utilisation), deriving from subsoil extraction – such as minerals, hydraulic or energetic resources – may be the object of interest, which has connotations other than ground utilisation and could therefore constitute equivalent assets in the judicial sense, which in turn can become the object of separate rights, entirely distinct from ground ownership» (Nesti 2005, pg. 319).

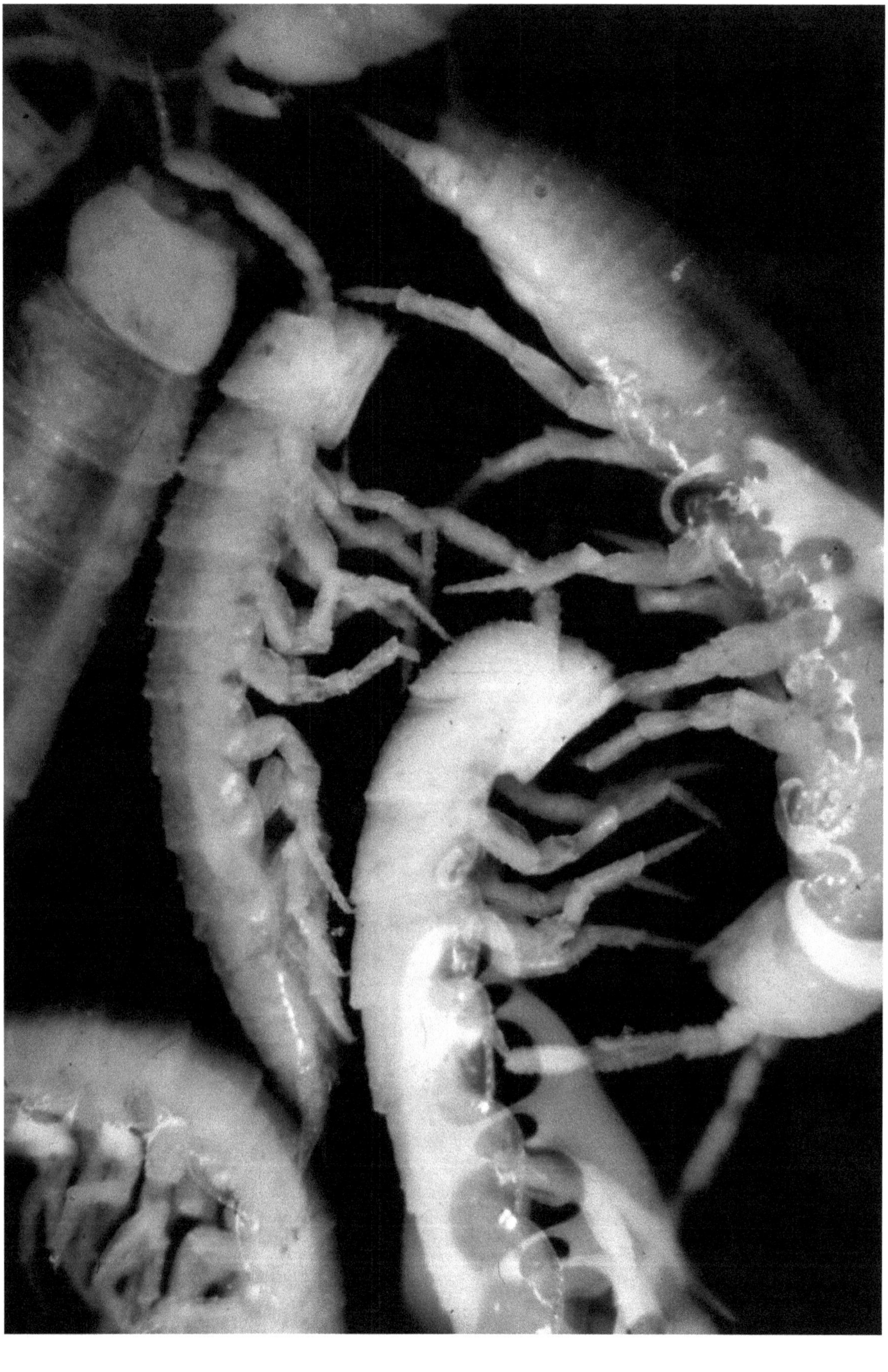

Fig. VI.1. ISOPODA TRICHONISCIDAE: *Mesoniscus alpicola, 8 mm. Species found in the underground level of the Castle Porta Giovia in Milan (photo D. Zanon).*

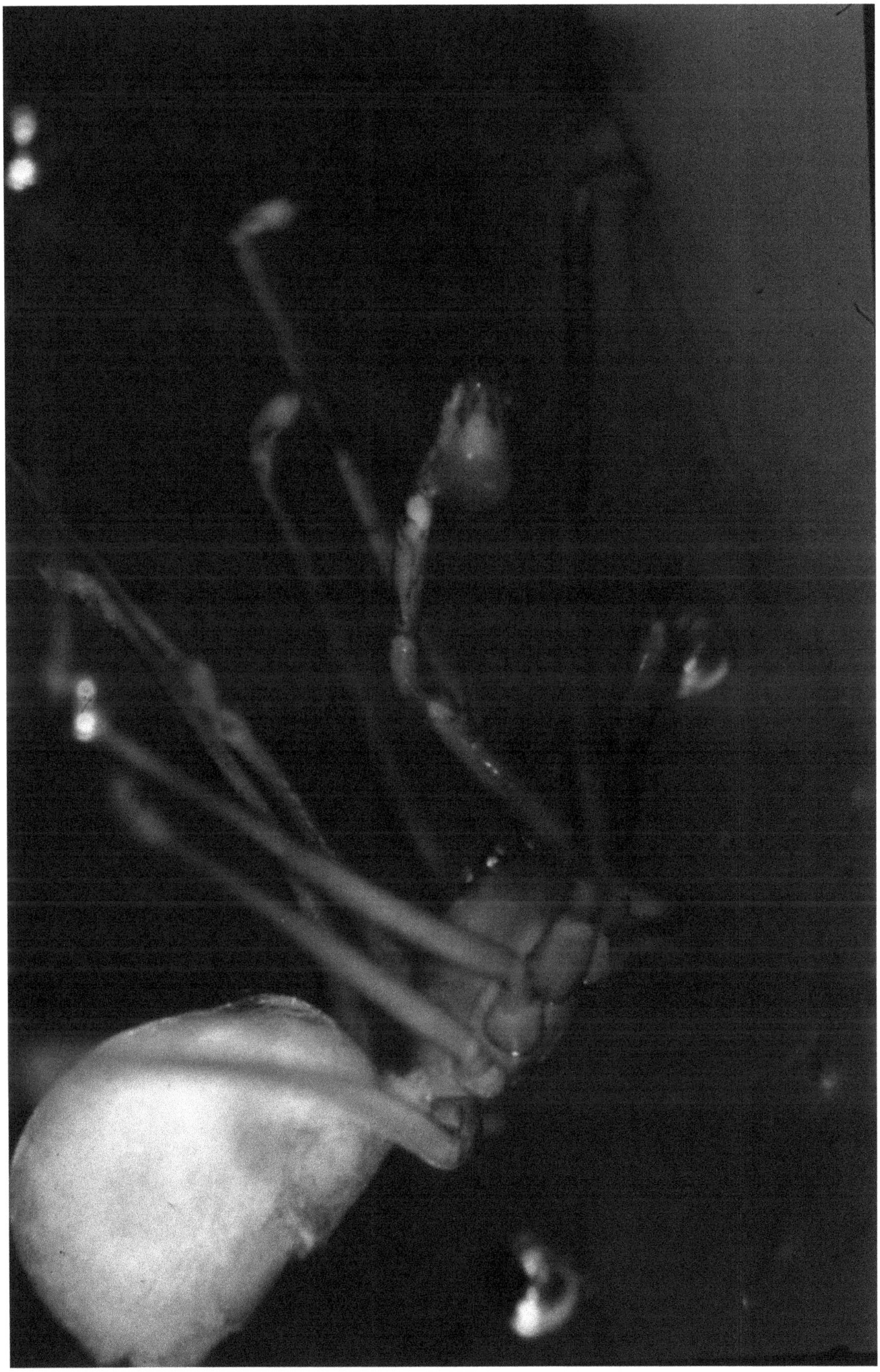

Fig. VI.2. ARANEAE LEPTONETIDAE: *Paraleotoneta spinimana, 5 mm. Species found in the underground level of the Castle Porta Giovia in Milan (photo D. Zanon).*

VI.17- *Construction site regulations: workplace safety*

Italian workplace safety regulations follow the guidelines stipulated by Legislative Decree No. 626/1994. This decree, stemming from European Community directives, fully and specifically deals with the methodologies required to ensure optimal hygiene and safety within working environments. Risk assessment is one important aspect introduced by this regulation: the employer must identify the risks and must have a document outlining how such risks should be dealt with.

In the workplace, danger does not equate to a risk; the first is the source of a possible health risk while the second is linked to the probability of such a danger posing a health risk to the employee. This decree relates to the public and private sector fields and there are a series of provisions relative to workplace characteristics, to the use and operating methods of work tools and to protection against the harmful agents a worker may find himself dealing with. In relation to the latter, training/information is envisaged for employees.

The law identifies the protagonists of relations who determine tasks and responsibility within the workplace. Under health and safety criteria, each of these figures has specific rights and duties, in relation to the execution of works. We thus have:

- EMPLOYER, person in charge of a firm or production unit;
- WORKER, a person who carries out work and who is employed by an employer;
- MANAGER OF PREVENTION AND PROTECTION SERVICES, suitably qualified person appointed by the employer;
- QUALIFIED DOCTOR, professional specialized in workplace medicine whose duty it is to monitor employee health;
- WORKER SAFETY REPRESENTATIVE, person appointed by the workers and who deals with health and safety issues.

The transposition of work safety criteria for plant safety is expressed within Legislative Decree 494/96, as amended by Legislative Decree 528/99. These laws derive from the adoption of European Social Directive 92/57EC, "Plant Directive", which focuses on the health and safety prescriptions to be adopted by temporary and mobile plants. The law is essentially aimed at construction workers whose work is of such a dimensional, temporal and organisational nature that specific control of the various work phases, in accordance with specific planning is required. The minimum requirements for application of this law are detailed within the law itself. Such requirements result primarily from the existence of subcontracts or multiple firms and in the case of individual firms, to the estimated number of operators (men-days), to the type of work carried out and to the presence of specific risks.

In similar fashion to decree 626, this decree also envisages the below key figures: the client, subject who commissions the structure, the project supervisor, figure also present in public structures, the planning phase safety co-ordinator and the work execution safety co-ordinator, two professional figures appointed by the client or by the project supervisor, company employers and self-employed workers. The co-ordinators deal with the drafting of the Safety and Co-ordination Plan (SCP) during the planning phase and ensure that this is followed by companies during the execution phase.

These figures have the power to amend the SCP where this is found to be inadequate for plan requirements or where unexpected issues arise; if the companies do not follow co-ordinator guidelines, they can face criminal penalties and administrative fines and be removed from the plant. The client and the co-ordinators may also incur criminal penalties and administrative fines if they do not comply with legal provisions. Companies are required to prepare an Operational Safety Plan (OSP), a document similar to a risk assessment, which must be created ad hoc for each plant; regulation sets out minimum OSP requirements. We further remind you that decree 494/96 refers to temporary and mobile plants: particular activities such as mines, quarries and marine works are excluded and are subject to other regulation.

CHAPTER VII

CLASSIFICATION OF ARTIFICIAL CAVITIES BY TYPOLOGY

VII.1 - *Typology 1: extraction works*

Outcome of mining science operations for the research and exploitation of minerals and useful rock.
Mining science identifies and exploits useful mineral deposits both on the earth's surface and underground. Various types of science merge to fulfil this objective. Mining science has also been known as "mining art" as special aptitude, as well as specific scientific knowledge, is required of the technician (Gerbella 1947, I, pg. 1). There are two types of extraction works: quarries and mines. The first term refers to the extraction of consolidated and unconsolidated rock, while the second refers to useful mineral extraction.

At the current point of research and as far as the development of extraction techniques is concerned, it would appear that mining systems saw little or no development between the Neolithic extraction of flint and throughout the mediaeval period. There was more of a development in the materials used for mining and transport equipment than in the functional work structure. Over the course of just a few centuries, there was the rapid introduction of efficient and constantly evolving tools: gunpowder, electricity, steam tunnelling machines, nitro-glycerine, helical-wire (for quarries), dynamite and spark-ignition engines. In particular, the gradual introduction of explosive materials, primarily for mining use, resulted in the modification of advancement systems in the XVII century. In any case, it should be taken into consideration that mines may also have been used in the past and not exclusively in Europe, although there is no evidence of this and if such evidence exists, it has yet to emerge from the archives.

Vergani states: «It has previously been confirmed that the use of explosives in metalliferous mines constitutes, from a qualitative point of view, the most important civil use of gunpowder in the XVII and XVIII centuries. However, before entering into the subject, certain unfounded myths must first be expelled: the first in regard to the use of gunpowder in one of the mines in Rammelsberg, Harz during the XII century and the second in regard to its use in the Transylvania goldmines around 1395-96. In the first case, there has been clear confusion with the old fire-setting technique which was highly diffused in Harz since mining activity began in that area. With regard to the second example, this is in fact a misunderstanding, originating from a French eighteenth century text, the subject of which, was not in fact *mines* in the sense of "mines" but *mines* in the sense of "underground tunnels excavated for military purposes"; this misinformation, coloured by a little imagination, then passed on to Romanian technical literature where repeated examples have and continue to appear In reality and as already demonstrated, gunpowder was first used in metalliferous mines in around 1574 by Giovanni Battista Martinengo, in the Schio Mountains. But more importantly, the first concise, net and undeniable description of the specific aspects of this new technique, the bore hole (*bohren und schiessen, boring and shooting, drilling and shooting*), appears in the relative documentation: twenty years after Filippo de Zorzi, mining official of the Republic of Venice, Martinengo writes; "boring a small hole in the mountainous rock using black powder artillery meant forcing and splitting the mountain and thus revealing that, which was concealed within". Almost sixty years passed before equally vivid and direct evidence of the bore hole and its use in mines emerged when Caspar Morgenstern travelled from Harz to Freiberg in Saxony in 1643 to demonstrate its use» (Vergani 2003, pgs. 865-878).

In relation to the quarry situation, he states: «Although ancient technical literature attests that the use of black powder in quarries dates back to well before this time, research into the matter has not revealed any evidence of its use prior to the XVII century. The first known reference goes back to 1621, when a newspaper in Bautzen, Saxony recounts in vivid detail (the size of the chamber and the weight of the charge the latter being equal to 11-12 pounds of explosive) the local quarry's use of explosive mines; they immediately explain, however, that the experiment was not worthwhile on account of the high cost of black powder (Wind H. W., Die Entwicklung des Zündens von Schwarzpulververladungen von den Anfängen bis zur Erfindung der brisanten Sprengstoffe, in Bergbau - Zeitschrift für Bergbau und Energiewirtschaft, 46, 1995, pg. 459)» (Vergani 2003, pgs. 865-878).

The latter decades of the XX century saw an even more rapid progression with the introduction of modern automatic machinery: the compressed air hammer drill became a museum piece. Although applicable to the majority of cases, this was not always the case. It should be taken into consideration that in some of the mines, which were still operational in the second half of the XX century, manual blasting and transport tools were used (and are still used today) primarily (or exclusively), due to lack of available finance. In Europe, the majority of mines are now closed and raw materials are imported from other continents.

Mining can take place both on the surface and underground and both systems may be used simultaneously. Opencast mines which evolve underground are common; in more recent times, the type of mining adopted is partly dictated by its impact on the environment. The nature of the deposit to be extracted, its dislocation, the structure of the mining plants and the extraction system used collectively determine the mining method. The various types of "opencast mining" depend on their positioning: Stratum mining aside, other types of mining are "valley mining" and "mountain mining" (foothills, drift mining, mountain-top mining), which can be subdivided into "mining of unconsolidated materials" and "mining of consolidated materials". The latter, is in turn subdivided according to whether a regular or irregular product shape is required (Gerbella 1947, II. Frare 1996).

Depending on the type of material to be extracted and on the actual extraction process, there are thus various extraction methods. It should also be taken into account that one or more types of extraction may be adopted within the same plant (fig. VII.1). These extraction methods refer primarily to mines in use until the mid XX century. Generally speaking these are:

single or multiple bench mining: used in the extraction of loose or slightly consolidated material such as gravel and sand; also used in clay quarries;
step mining: generally applies to clusters, which are visible or of little depth;
amphitheatre extraction: normally used in the extraction of stone, the terraces in this type of mining are set out in the shape of an amphitheatre;
single terrace or single front mining: used in the presence of visible (or barely covered) strata, sub-horizontal or sub-parallel to the area's topographic structure;
multiple step or bench mining: for consolidated rock; the tread dimension of each step is relative to the type of material, the methods used, safety, profitability and to the current restoration project (Frare 1996, pg. 34);
pit extraction: for loose, unconsolidated materials or consolidated stone material;
step extraction using underground transport: where the orography of the area permits, material can be removed from an extraction pit by means of tunnels at a level with the steps or by tipping the material into a chute and transporting it via a tunnel to the surface. The chute can also be used in the drainage of meteoric waters;
funnel extraction: used in the extraction of consolidated rocks; at the base of a funnel-shaped terrace ditch is a chute which connects to an underlying tunnel; excavated material in the funnel is removed via the chute and is either channelled to the surface or lifted within an extraction shaft;
shaft extraction: normally used for the extraction of ornamental stone;
depression extraction: used in areas characterised by subsidence and vertical erosion;
horizontal slice extraction: normally used in plains, for the extraction of unconsolidated material;
top slicing: for the downward extraction of materials;
panel extraction: for the extraction of small blocks or segments, subdivided into thin horizontal slices.

U*nderground mining*: in basic terms, this type of extraction normally takes place within a cavity that fulfils the following criteria: entrance, circulation and mining plant. Access is gained by means of vertical shafts, inclined wells, descending tunnels and drift tunnels. From these branch the main haulage roads and from these the secondary haulage roads, which lead to the mining plants (fig. VII.2). There are many methods of underground mining and their structure is often complex, particularly in mines from the industrial era. These may take the form of "room and pillar mining", "stope and retreat" or "stope and fill mining" (Gerbella 1947, II, pg. 80). In the XX century, various methods and techniques were replaced by systems in rapid and constant evolution.

Room and pillar mining: this simple method was the most used in the past although there are examples of its use even in the XX century. Room and pillar mining permits the extraction of all the available mineral from the vein, without the risk of collapse. This can be subdivided as follows:
open stope mining with isolated rooms: the removal of useful material together with non useful material creates underground rooms, the roofs of which require no support;
underhand stoping without supports: when descending to follow the mineral vein, vertical or traversal support or containment pillars are left in the areas of little or no mineralization;
multiple pillar extraction using only pillars: when removing useful mineral, non-useful mineral is left behind, in the form of pillars, for roof support;
downward slice room and pillar mining: similar to the latter method, mineral is also extracted from the floor; extraction can thus takes place at lower levels;
mining with numerous artificial pillars: when the tunnel cannot be completely filled, additional rock and mortar pillars are created to prevent subsidence;

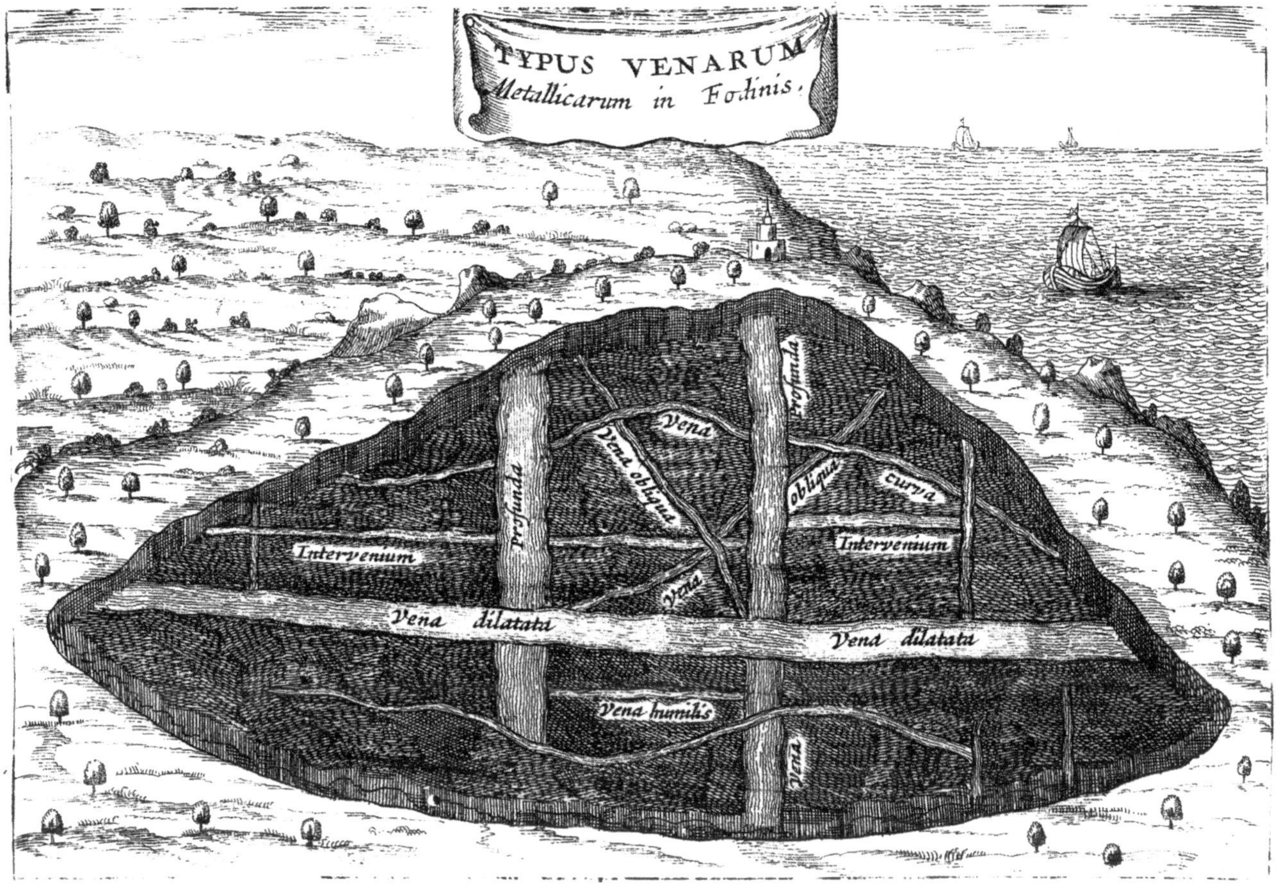

Fig. VII.1. *The underground vein structure according to Mundus Subterranean (Kircheri 1678, X, pg. 198).*

Fig. VII.2. *The First Book of Biringuccio's "De luoghi de le miniere" by Biringuccio shows a close-up of a person pushing a wagon. The mine's entrance can be seen at his shoulders (Biringuccio 1540).*

mining with multiple pillars and crown pillars: used in multiple pillar mining, where the roof is prone to collapse; sections and sometimes slabs of mineral, are left between one pillar and another to prevent the tunnels becoming too narrow;
multiple barrier extraction method with depots: used where the roof is prone to collapse in order to obtain parallel tunnels, which are separated by mineral barriers;
multiple barrier extraction method with rooms and barriers: involves the creation of mining plants at the location of mineral deposits; these are separated by roof support barriers;
multiple barrier extraction method with depots: a series of rooms are made within the deposits and these are filled with useful mineral, excess material is removed upwards by means of hoppers and chutes.

Stope and fill mining: this method is suitable for almost all types of mineral deposits and involves the backfilling of gaps formed following the extraction of useful minerals. The entire deposit (or most of it) can be removed, with the removal of oxidisable or combustible substances, which could give rise to spontaneous heat or fire. Sterile material is not taken to the surface and phenomena of surface subsidence are thus limited or avoided. When compared to block caving, this allows improved underground stability and decreased risk of accidental falling rock. In addition to the sterile material which is left on site, an adequate system for completion of the fill works is required: fill material can be introduced via tunnels, shafts or special chambers. Special machinery is also used to carry out the various works. For instance, hydraulic fill is carried out by introducing the fill material together with pressurised water via specific piping. Sometimes the stope is only partially filled to avoid having to transport fill material form the surface. There are numerous methods of stope and fill mining, the choice being determined by the type of mineral deposit and by its location. Even so, various methods of stope and fill mining were used at least until the first half of the XX century.

Stope and retreat: this method involves caving the enclosing rock, resulting in the extraction of a higher percentage of mineral compared to multiple pillar extraction; this also reduces costs as no filling is required. The disadvantage is that this method often causes subsidence. There are many caving extraction methods and each method has its own variants. Tunnels (vein mines) are excavated at the site of the mineral deposit, on either one or more levels (in which case they are equipped with connecting shafts); excavation continues as far as it can go, thus a type of multiple barrier extraction is obtained. Alternatively, the tunnels on one level can be linked by means of perpendicular tunnels, thus obtaining a type of multiple pillar mining. In both cases both the stopes and pillars cover larger areas. Upon completion of this initial phase, the stopes and pillars are removed (stope and retreat) and the roof cave-in is adequately controlled. There are several types of stope and retreat mining.

In addition to that already mentioned, other methods are adopted for the exploitation of salt deposits (with the introduction of water), sulphur, coal, petroleum, natural gases and endogenous force deposits. Of the many aspects to be taken into account in the study of underground mining, are the study of mineralized rocks and deposits and identification of the support structures utilised (figs. VII.3 and VII.4). This has two functions:
- underground tunnel thrust resistance, contributing to renewing the rock's equilibrium;
- to protect excavation works from falling rocks and keep both men and machinery safe.

«So you are surprised that they go by this name: "*Cave*"? ["Quarries"] What you are thinking of are the souterrains, the "*cave*" of the Vatican. My dear Mr Fleurissoir: "*cave*" is a Latin word which simply means: "be careful"» (Gide 1995).

VII.1.1 - *Mine (Cava)*

Complex of sites, tunnels, inclines etc., for the extraction and transport of mined material.
A mine is a complex, consisting of mineral resources of industrial interest and of the works and equipment necessary for its exploitation. The deposit is classified as "metalliferous" or "non metalliferous" depending on whether metals or non-metals are extracted. Minerals are naturally-occurring, solid compounds, formed through inorganic processes; one exception is mercury, which, despite its liquid form, is still classified as a mineral. Minerals are also characterised by physical homogenous properties, each mineral having a particular chemical composition and a unique arrangement of atoms (Mottana, Crespi, Liborio 1993, pg. 8). For those minerals which are primarily used in road, building and hydraulic construction, the complex is generally known as a quarry.

With regard to the structure, underground works are accessed by means of drift tunnels or vertical shafts (fig. VII.5), or inclines and inclined wells. The main haulage tunnels, constituting the main structure of the mine, branch off from such tunnels or wells. Secondary tunnels, in turn, branch off from the haulage tunnels and extraction plants, subject to

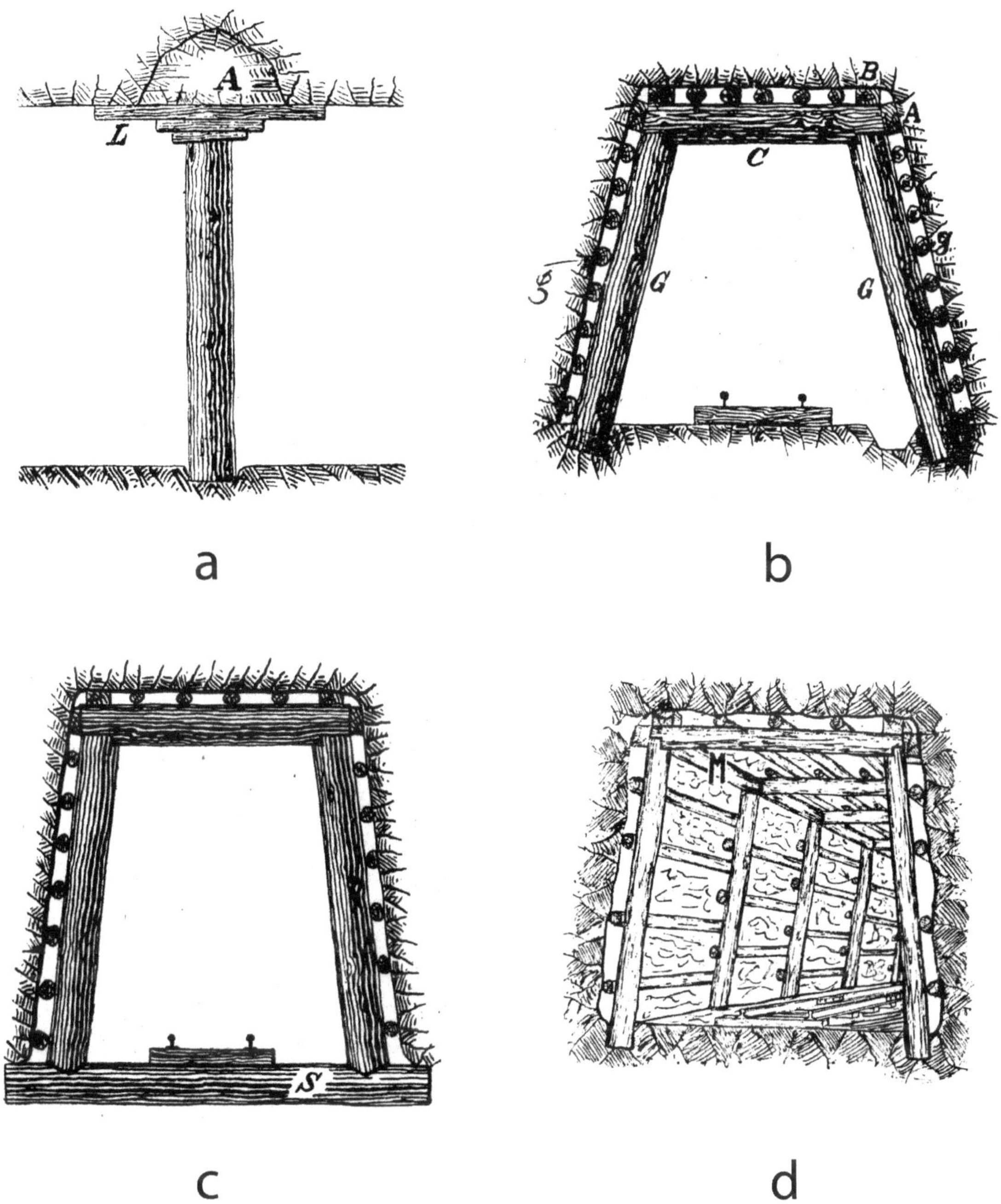

Fig. VII.3. *The most common type of mining frame, at least until the first decades of the XX century, was the timber frame. The type of timber was selected according to the purpose of the frame. For example, oak varieties are especially suited for use as beams (drawings and captions, from Gerbella 1947, I, pgs. 194-199).*

a. Support beam for an unstable roof (A: section of rock subject prone to becoming detached; L: timber planks placed between the rock and the head of the beam).

b. Standard timber frame (g: lining; A – B: clamping wedge; C: cap; G: legs, also known as "supports").

c. Timber frame with base sole (S: timber sole).

d. Tunnel reinforced with timber prop (M: props, also known as stulls).

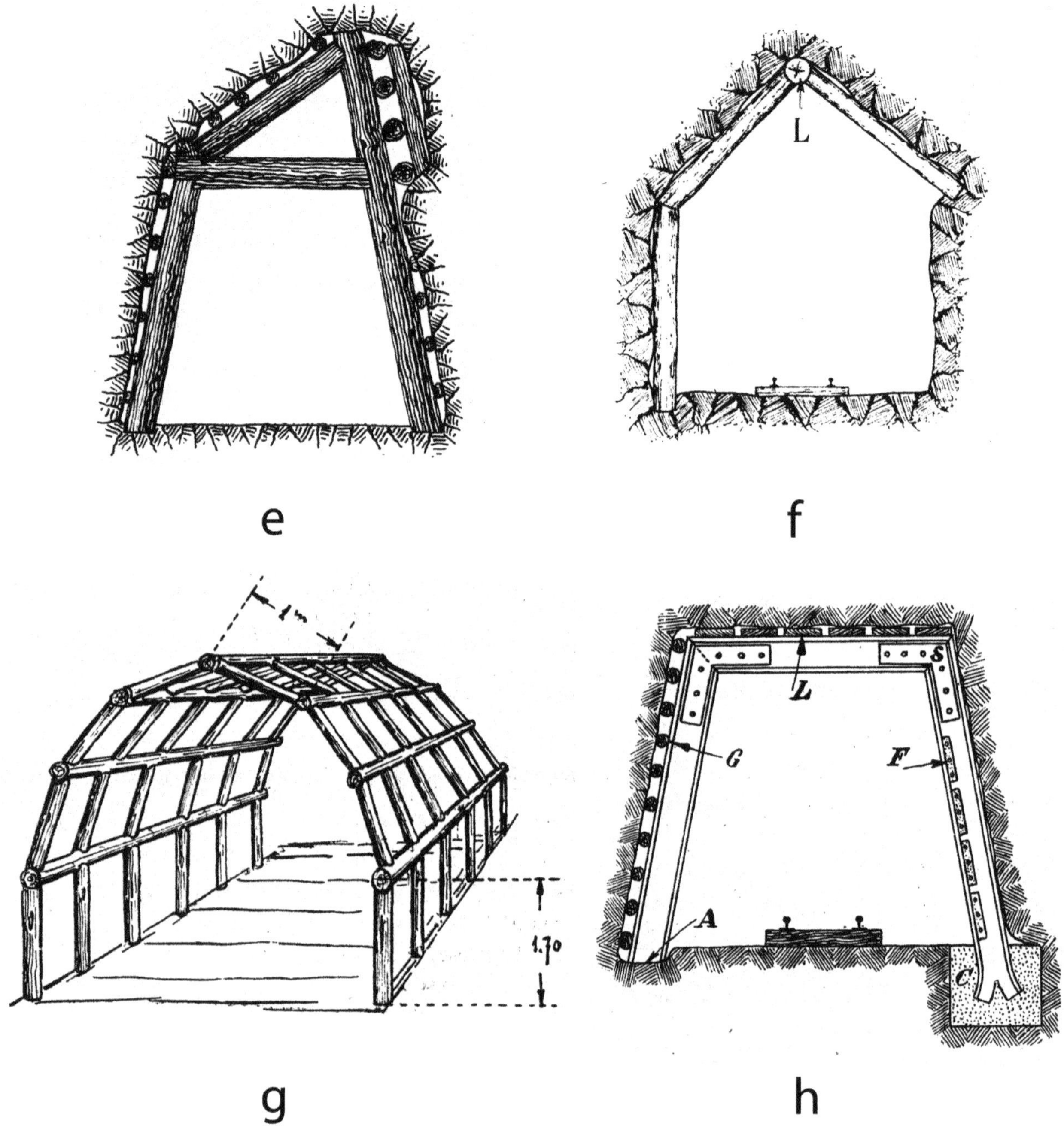

Fig. VII.4. *Other types of mining frames (drawings and captions drawn from Gerbella 1947, I, pgs. 200-214).*
e. Timber frames in a section of tunnel which has caved in.
f. Incomplete roof support frame with longitudinal L connecting member.
g. Underground chamber timber frame.
h. Metal or trapezoidal T-bar framework connecting the legs to the cap by means of bolts and brackets (A: strut; C: curved end strut set in concrete; S: metal L-shaped bracket; G-H: round timber and timber plank lining; F: reinforced concrete perforated brick lining).

Fig. VII.5. *Mine advancement by means of a long ladder:* «Qui in fodinis operantur, ardentibus continuo candelis utuntur, es distributione temporis, ut duodecim horis alterna operarum requie à laboribus abstineat. Venas fractas, comminutasque dorsis impositas, per scalas quasdam, ut in Figura patet, mira arte constructas efferunt. Sunt hae scalae tribus ex bovino corio contortis funibus factae, quibus trasversi baculi graduum specie inferuntur, ita ut duo invicem occurrentes, commodè sibi cedere queant. Scalarum quaelibet decem oryiarum longitudinem habet, ejus extremis alia subnexa: Ad scalarum latera nonnullibi scamna ex ligno confecta adfecendentibus quieturis exstructa funt. Operarum primus cereum ardentem praefert, ad lucem tum sibi, tum subsequentibus se sociis necessariam praestandam; & uti profunditas adeo horrida est, ut vel intuentibus primo formidinem incutiat, ita quoque miserandi fossores, vel fractis scalis coriaceis, vel venarum, quibus onerantur, mole praeponderante, laboribusque fracti, dejectique aerumnosam vitam formidanda morte finiunt» *(Kircheri 1678, X, pgs. 228-229).*

continuous development, eventually cover the entire deposit area (Gerbella 1948, II, pg. 64). Haulage tunnels are used for wheel-barrows, hand-pushed wagons and in more recent times for track-driven (Decauville railway) mining wagons. As previously mentioned, the set-up of the mining plants and the manner in which useful mineral is worked, determine the various extraction methods.

In summary, horizontal and/or sub-horizontal structures include:
the entrance tunnel or main haulage tunnel: providing access to underground mining plants;
the secondary haulage road: connecting the primary haulage road leading to the mining plant;
vein tunnel or passage: the tunnel is crated by removing the mineral and following the mineral vein;
bank tunnel: extraction takes place by excavating from within the mineral deposits;
assay tunnel: excavated within the mineral deposit itself to establish its strength;
traverse-bank tunnel: the rock is excavated until the deposit is reached;
shaft: a downward driven inclined tunnel connecting two separate levels for transit or ventilation purposes; can also be used for access to the mineral deposit or the outside;
winze: where excavations have reached a certain depth, this allows rapid access to the lower extraction level;
raise: an upward driven inclined tunnel.

Vertical shafts include:
external shaft: generally connects with the outside;
master shaft: external shaft which links the mine's dewatering, ventilation and other services. (fig. VII.6);
internal or secondary shaft: does not connect with the surface but connects two or more tunnels for the transport of minerals, people or materials;
extraction shaft: used for extraction purposes; can sometimes be used for ventilation (*ventilation shaft*) or the pumping of water; in some cases, the shaft is not vertical and is known as an *inclined shaft*;
upcast shaft: for the external deflux of air;
circulation shaft: for the movement of mining workers;
rock chute: vertical-cut shaft or shaft with an inclination of more than 45°; used for movement or ventilation;
rock chute: used for mineral or sterile loads, its lower section, the hopper, was devised for wagon loads.

Quarrying is the fragmentation of rock and progression is the advancement of extraction works through quarrying. Various tools are used, in particular: pickaxes, rock picks, mallets, wedges, clubs, chisels, bradawls, etc. Learning how to recognise the marks left by the various tools can be of great use, as it is in the study of quarries or any man-made underground construction.

Old work-tools have often been found during archaeological mining digs. Numerous and varied archaeological finds have been uncovered in old Spanish mines: stone clubs and mallets dating back to the Bronze Age in the *El Piconcillo, Los Pobos* and *Arroyo de los Almadenejos* mines; iron picks and rock picks dating from between the II and I century B.C. in the argentiferous lead mines of *La Loba*; bronze cochlea and pump in the Roman Age mine at *Sotiel Coronada* (Domergue 1991, pl. XV-XXII).

The drill, also known as "drill rod" arrived with the introduction of explosives. This drill consists of a steel rod with one or more terminal cutting edges, which was hand hammered to bore through the rock in the preparation of bore holes; it was subsequently attached to a boring machine. Modern times have seen the steady use of mechanical equipment such as drills, cutting machines, rotary mining tools etc.

A mine encapsulates man's "labour" and intelligence, his need for raw materials and also his greed; things that come at a high price for those who worked (and still work) there: the miner. In order to highlight the dangers involved, which increased with the introduction of explosives, Simonin named Chapter VIII of his book on XIX coal mines: «Le champ de bataille». And states: «Les quatre éléments des anciens, le feu, l'air, la terre, l'eau, sont conjurés contre lui. Le feu le menace dans les coups de mine, les incendies du charbon, les explosions du grisou; l'air, en se raréfiant ou mêlant à des substances méphitiques, détonantes; la terre, dans les éboulements; l'eau, dans les inondations. Le houilleur oppose à tous ces ennemis, souvent invisibles, ce calme stoïque, ce courage à toute épreuve, cette science pratique qui font les vaillants et habiles mineurs» (Simonin 1867, pgs. 156-157). It was the battlefield where people fought against destitution.

Fig. VII.6. *Lower mining plant water-lifting machine.* «Axis Statutus A. Rota dentata B. Dentes C. Axis stratus D. Tympanum, quod constat ex fusis E. Alterum tympanum F. Catena ductaria G. Pilae H» *(Agricola 1621, pgs. 151-152).*

VII.1.2 - *Quarry*

Complex of sites, tunnels, inclines etc., for the extraction and transport of rocks, the object of extraction.
The term "quarry" refers to the excavation of material useful for civil construction purposes and thus, by extension to the place of work, which may be either opencast or underground. There are quarries for unconsolidated materials (gravel, sand, pozzolan etc.), magmatic rock (granite, diorite, porphyry, basalt etc.), sedimentary rock (conglomerates, sandstone, limestone, tuff etc.) and for metamorphic rock (gneiss, marble, schist, skarn etc.). Central and Southern Italy sees the extraction of pozzolan materials, esteemed since antiquity for its physical and mechanical characteristics. The extraction method used, known as "room and pillar" or "multiple pillar" was used until the XX century (Lombardi, Polcari 1984, pgs. 26-27. Cherubini, Sgobba 1997, pg. 55). The tunnels follow an essentially regular geometric system and the excavations, at right angles one to the other, eventually join to form a "grid-like pattern" of isolated pillars. This not only ensures the stability of the entire quarry but also the permanence of agricultural and pastoral surface activities (Cherubini, Geminario 1991, pg. 102).

Shafts which communicate with the surface allow aeration and evacuation. For instance, if the expansion of the Neapolitan urban network now extends above the ancient quarries, it is equally true that due to the lack and cost of public transport, construction materials were also extracted in cities, if not from directly beneath building sites. Various cavities, thus created, are subsequently turned into cisterns for use by the above building (some being supplied via links to underground aqueducts), or are used for storage or dumping purposes. In the XX century, some were re-utilised as air-raid shelters. The situation is similar in other Italian cities.

Funnel-shaped quarry (inverted): in Sicily, this type of quarry was used for the extraction of calcarenite blocks.

Latomia: in antiquity, this was the term for stone quarries. The Syracuse *latomie*, used by the Siracusians as prisons ("Latomia dei Cappuccini") during the war between Athens and Sparta in the V century B.C. and quoted by Thucydides, are well-known: captured Athenians and their allies were imprisoned here (Thucydides, VII, 86, 2).

Muchata: in and around Palermo, this dialectal term of Arabic origin, refers to "multiple pillar" extraction, which can take place over two levels (Todaro 1988, pg. 52).

Opencast quarry: the opencast extraction method is used and is subdivided according to the method permitted by the type of extracted rock and by the rockbed.

Underground quarry: underground mining method with variations depending on the type of material extracted.

Not all stone is destined for the same use and not all stone is worked in the same way. The choice of stone for construction had to take the purpose into account, i.e. (Cagnana 2000, pg. 34):
- hardness, in the sense of cutting resistance;
- tenacity, in the sense of impact strength;
- compression force resistance;
- resistance to tensile forces which are prone to yielding and breaking-down;
- cutting phase divisibility following mineral beds and/or microfault plans;
- polishability, for the obtainal of flat surfaces;
- conductibility and thermal stress resistance.

The ability to recognise the extraction systems and materials used can come in very useful when studying quarries, even if this only involves deciphering and understanding the tool-marks left on the rock face. Partially worked blocks are sometimes found on-site, as in the Inferniglio quarry on the outskirts of Corneto (now known as Tarquinia). Here, alongside evidence of recent mining activity, a series of stone bricks were found. The bricks were of the same size as the square blocks utilised in the construction of the defensive walls of the Etruscan town, Civita di Tarquinia (Padovan 2002 b, pg. 124).

Quarries cause irrefutable damage to the environment. Not to mention the numerous examples of quarries, which have intercepted or destroyed natural caves and artificial cavities or examples where these have led to the partial or total destruction of settlements or works of archaeological, historic and architectonic interest. The quarry, which was opened in Verrua Savoia (Turin) in the mid 1950s almost entirely destroyed the XVII century bastions of Verrua Fortress. Part of the underground system and evidence of previous settlements dating back to prehistoric times were also destroyed (Padovan D., Padovan G., Bordignon, Ottino 1997, pgs. 193-195).

Such issues are not only linked to historical and naturalistic impoverishment but also result from the stability of abandoned quarry sites, which can present evident risks due to underground and surface collapse as evidenced during the course of studies conducted at the Vezzano sul Crostolo (Reggio Emilia) chalk quarry. The particular geological and structural situation, the resulting karst morphology and the underground extraction areas, render these areas dangerous: walls have been known to suddenly collapse and rock slabs have unexpectedly sunk (Frignani F., Sassi, Frignani F., Casini, Padovan 1997, pg. 270).

Although the quarries concerned do not fall under the usual definition of the term, this typology also includes natural cavities, which were used for ice extraction purposes. They cannot be classified under "Typology 2 b" as the ice (or snow) forms naturally within such cavities and is not simply placed there for storage purposes.

The "Ghiacciaia del Moncodeno" (Moncodeno Icehouse) can be found at Massiccio delle Grigne (Lecco), just past the Passo del Cainallo and near the path leading to the Rifugio Borgani. Utilised for the extraction of ice-blocks until recent times, the remains of the wooden ladder used by the excavators could still be seen in 1982. In 1671, Niccolò Stenone carried out its topography and provided details of this in a letter to Cosimo IV, Grand Duke of Tuscany, where he proposed some interesting considerations on the subject of the internal formation of ice and its relation to air circulation (Buzio, Casini, Padovan 2000, pg. 142).

VII.2 - *Typology 2: hydraulic structures*

Structures for the capture, transport and storage of water and the drainage of waste water.
Water is essential to life. This is an obvious concept, which must be considered in the study and understanding of settlements and more generally in the history of Man, in all his manifestations. There are many water utilisation methods, depending on whether water takes the form of water vapour, liquid or is in its solid form i.e. ice. Such methods are determined by:
- the nature of the location;
- the type of settlement;
- applied knowledge (in function or in subordination of the different variations);
- availability of funds;
- subsequent progression, adaptation or involution of the method applied.

On the basis of the available funds and technology for the supply of water, different types of structures were created both on the surface and underground. Taking the differences between the various types of structure into account and without being tied to examples and concepts not destined for use, hydraulic works were divided into four sub-groups:
- water capture and transport;
- vertical capture;
- storage;
- discharge.

Archaeological research has highlighted the importance of water, both in terms of the establishments and development of settlements and in terms of the role of lakebeds, springs and water courses. In relation to the role of water in the holy places uncovered in Basilicata: «The first indigenous place of worship is in the area bordering the Metapontine *chora* and has Greek-style structures. This is the Garaguso sanctuary, close to a spring and extending along an old shepherds' trail, which connects the River Basento to the River Salandrella. A channel and a cistern again attest the use of water during sacred rites» (Russo 1999, pg. 105).

VII.3 - *Typology 2 a: water supply and transport*

Operations and systems for the movement of water masses in the transport of drinking and non-drinking water as well as in the creation of hydraulic engineering works.
A furrow in the ground allows water to be collected from a spring, a torrent or a river. A hollowed tree-trunk, split lengthways, fulfils the same function. A simple excavation can therefore be the first step (or one of the first) in the development of hydraulic works for the capture and transport of water. When made deeper and covered with stone slabs, the furrow ideally becomes an underground canal and a hollowed-out tree-trunk acts as a conduit and when buried, becomes an underground conduit. Such concepts convey how a specific action and its multiple applications may have contributed to the creation of hydraulic techniques for the transport of water.

With the onset of inhabited and agricultural centres, came the need to no longer exclusively rely on water from natural sources, with its variable quantity, excesses and shortages. A relatively simple but functional system was the creation of large brick basins or basins excavated in the rocky ground, which are gradually filled with rainwater during the year. Storage, including underground storage systems was created in anticipation of dry spells. The water could be easily lifted and distributed: «The five primary issues associated with modern-day water supply that is its collection, lifting, transport, storage and distribution were already studied and resolved thousands of years ago» (Motta 1981, pg. 11). Water generally was and continues to be used for various purposes:
- for drinking water;
- for agricultural use;
- for industrial use;
- for the operation of sewer system;
- for defence purposes;
- for roadways.

It should not be forgotten that a continuous supply of water, such as that from the capture of perennial water streams involved and still involves a second factor: drainage. This task is fulfilled by sewers, which collect water discarded by consumers (Frega 1984, pg. 49).

VII.3.1 - *Aqueduct*

Complex of hydraulic works for the capture of water and its transport from the point of origin to the consumers.
The term 'aqueduct' refers to a system, whether simple or complex, which allows the transfer of water from the point of supply to the point of use. The first distinction is between drinking and non-drinking water. Channelled drinking water acquires social and political connotations as well as financial connotations. When the community exceeds a certain population threshold, the continual intake of a significant amount of water becomes necessary (Pisani Sartorio 1986, pg. 28). At least in the past, exceeding a certain population threshold could be directly linked to the possibility of increasing water intake.

The main works required in the creation of an aqueduct are the following:
water capture works: for water capture in places where water is naturally available;
supply conduit (or supply channel): this is needed to transport water from the point of capture to that of use, where "conduit" refers to generally cylindrical piping and "channel" refers to the closed channel or *specus*, with free surface water flow; an earthenware, *eternit* (made of asbestos and Portland cement fibres) or other type of conduit is sometimes positioned in the *specus*; water is channelled through this to ensure quality and prevent pollution;
reservoir or storage works: used in the storage of water when consumption is inferior to supply and in the supply of water when the situation is reversed;
distribution network (network conduits): complex of small conduits or pipes for the transport of water to the supply points;
private installations: network of small channels or more commonly piping, linked to the distribution network and which directly supplies private consumers.

There are also:
mechanical lifting system: to compensate for any natural-occurring altitude variations so that adequate amounts of water may flow within the conduits;
potabilisation plant: provides water with the chemical and bacteriological qualities required for human consumption; part of modern systems, these were introduced in the XIX-XX century.

Water which may be captured can have various sources and can derive from: springs, lakebeds, rivers, underground, artificial basins; an appropriate capture system will be chosen according to the water-type.

Spring-water capture: this requires hydro-geological research of the spring's source and the land through which it flows. In stratified rock, water capture is carried out through the removal of the layer of soil and detritus, which covers the rock. Where there is a single source (or where there are numerous nearby sources), an impermeable capture chamber is created to contain the water. Where there are numerous water veins to be linked and these are distanced one from the other, a water inlet is created for each vein. Where the sources are distributed lengthways across fissures in the rock, collection will also take place along the length of the rock, possibly in the form of a tunnel running along the rock face. For talus springs, if the water flows from the rock, the above procedure is followed. If, on the other hand, the water flows along the detritus mass, collection can take place through the construction of impermeable barrages on the bedrock or by draining the water mass and channelling it into a water collector. In alluvial soil, springs generally emerge along the outcrop of impermeable clay strata (or clay and sand strata) and water sometimes flows in an upwards direction into pools. In this case, water is collected by means of a work which encapsulates the pool, the walls of which are placed at a depth sufficient to prevent surface water infiltration. If the springs are located at the base of hills or mountains, these often emerge through masses of detritus, which is generally removed to allow water capture.

Lacustrine water capture: water is captured at a depth and far from the banks by means of pipes laid along the bottom of the basin; in antiquity this was carried out via the creation of appropriate basins along the bank itself. In certain cases, water is captured via tunnels, which are cut into the side of the valley.

Capture of fluvial water: where the water-level is sufficiently elevated, water is captured using "bank drains" (brick works placed along the river's banks). Where the water-level is too low, this can be raised using dams or alternatively, by the excavation of filtering tunnels under the river-bed.

A*rtificial basin water capture*: water is normally captured using dams, at a suitable depth which prevents silt from being removed from the bottom of the basis.

Underground water capture: prior to capture, it should be ascertained if the water is:
a. groundwater originating from meteoric or flowing water;
b. deep subterranean water, separated from surface water by strata of impermeable soil.

The capture of groundwater can be carried out by digging a trench or a ditch in the surrounding land. Alternatively, practicable wells are excavated, at the bottom of which, a series of drainage galleries or pipes horizontally connect with the groundwater; the water is then collected into the well. The water can be mechanically lifted from the well and piped; alternatively a lightly downward-sloping underground aqueduct tunnel or culvert can be built. Capture of deep underground water takes place through hand excavated wells and more recently through sinking or drilled wells. Sometimes, when drilled, a deep stratum allows its waters to rise as far as the surface. A well thus formed is known as an artesian well. Should the water not rise to the surfaces, mechanical lifting systems or underground conduits can again be used.

From our knowledge so far, we can see that aqueducts with drinking water capture, transport and distribution systems were already utilised in archaic and classic times (figs. VII.7 and VII.8). Many predated the fall of the Roman empire and many were built thereafter. The five factors associated with current water supply that is collection, transport, lifting, storage and distribution had already been resolved some two thousand five hundred years ago. Wells (or rarely inclines), which served the below functions, may therefore be uncovered:
- wells which reach the depth required for the creation of an underground channel;
- wells used for the removal of excavated material and for ventilation purposes;
- wells used to lift the liquid to the surface upon completion of works;
- wells utilised in aqueduct maintenance.

The building of aqueducts primarily exploited the law of gravity as well as many variants of this law. At certain times in the past and for specific purposes, aqueducts were created just as we might imagine: kilometres of underground conduits, which are almost always impermeable, as well as spectacular paths over archways and substructions (fig. VII.9). The introduction of the motor pump, of iron and cast iron pipes transformed the natural, gravity flow water supply system into a pressurised water supply system (Capacci 1918, pg. 30). However, an endless number of similar works were created with wooden, earthenware and stone pipes or from basic, unlined passages excavated directly in the rock.

There are a significant number or underground passages for the transport of water in central Italy, normally excavated in tuff. There is no shortage of limestone examples, such as those in Tarquinia (Viterbo). Each individual case must be analysed to identify whether the work is an aqueduct as such (figs. VII.10 and VII.11). On the basis of his research, Del Pelo Pardi tells us that such underground passages should be subdivided into two main types: The first type is excavated in lithoid tuff and is used for drainage purposes. The second type is excavated in granular tuff and is used in the supply of water (Del Pelo Pardi 1943). Judson and Kahane from the British School in Rome, make reference to the extensive underground systems uncovered in Southern Etruria and Northern Latium. More recently, they analysed the Veii and *Caere* areas and the western and southern slopes of the Alban Hills. They retain that such hydraulic systems should not be placed in direct relation to urban development (Judson, Kahane 1963, pgs. 75-93).

Fig. VII.7. Aqua Claudia *Specus - Cross Section (CA 02100 LA RO), altitude 294.4 m a.s.l. This Roman aqueduct section is situated in the gorge of the Aniene River at Vicovaro (Rome). It was researched by Thomas Ashby, an English archaeologist, at the beginning of the XX century (Ashby 1991, pgs. 219-302) and is currently being studied by speleologist from the Italian Foundation of Artificial Cavities (photo R. Basilico). Frontino informs us of the following*: «Claudiae ductus habet longitudinem passuum quadraginta sex milium <quadringentorum sex>: ex eo rivo subterraneo passuum triginta sex milium ducentorum triginta, opere supra terram passuum decem milium <centum> septuaginta sex: ex eo opere arcuato in superiori parte pluribus locis passuum trium milium septuaginta sex, et prope urbem a septimo miliario substructione rivorum per passus sexcentos novem, opere arcuato passuum sex milium quadringentorum nonaginta et unius» *(Frontino, XIV).*

Fig. VII.8. *Aqueduct from Roman Empire named "Buso della Casara" (Padua) (photo G. Padovan).*

Fig. VII.9. *Aqueduct section at the Vetulonia (Grosseto) Hellenistic settlement, in the Poggiarello Renzetti locality (photo G. Padovan).*

Fig. VII.10. *Docciola Fountain in Volterra (Pisa). It draws its water from two tunnels carved into the rock. The rock consists of lower and middle Pliocene marine sedimentary deposits «The first documented reference of a Docciola fountain dates to 1224: a chapter of the Volterra municipality statutes for that year, provided that a road from the Docciola front to the River Era, cutting through the city's lower ditch, be built (ASCV, G nera 4, cap. CCLIX,* De via mictenda)» *(Furiesi 1998, pg. 89) (photo G. Padovan).*

Fig. VII.11. *San Felice Fountain in Volterra (Pisa); the entrance to the main channel of the hypogeal aqueduct can be seen at the back (photo G. Padovan). «The archaeological remains of numerous Etruscan and Roman structures were found nearby. From these it has been found that the spring has been a popular destination since antiquity. The remains of Villanovian huts were also found directly opposite the fountain» (Furiesi 1998, pg. 92). The commemorative epigraph, on the wall of the fountain's mediaeval external façade, is dedicated to the* Fons Saracinorum *fountain, which once stood here.*

Another important contribution is that of Ravelli and Howarth, who essentially affirm that «The *cuniculi* found in the volcanic rock in Latium, similar to the qanats of Iran and to the khettara of Marocco, were excavated by Etruscan and Latin civilisations primarily for the collection of drinking water and, secondarily, for soil drainage or for the derivation of water (probably to be used also for irrigation)» (Ravelli, Howarth 1988).

In his *De Architectura,* the treatise writer, Marco Vitruvio Pollione talks very clearly of the water, its collection and pipe distribution (Vitruvio, VIII, VI). He states that water is more easily collected where there are surface sources and that where none are present, underground springs must be located and channelled. Once the water has been found or selected, the planning phase ends and the underground and/or surface structure is plotted. Vitruvio also mentions the use of water level measurement equipment: «*Cuius ratio est prima perlibratio. Libratur autem dioptris aut libris aquariis aut chorobate, sed diligentius efficitur per chorobaten, quod dioptrae libraeque fallunt*» (The water level should first be established using dioptra, levels and chorobates. The latter being the most accurate instrument, while the others may prove to be misleading) (Vitruvio, VIII, V, 1).

Water was captured by means of wells, inclines or tunnels leading underground or through hillsides to the groundwater aquifer or pressurised layer, by means of reservoirs which enclosed the pools or through the capture water from rivers, streams and natural and artificial basins. Decantation basins (*piscinae limariae)* were usually positioned at the beginning of the conduit (or channel) and the channel (*specus*), a masonry wall within a trench was cut into the rock (Botturi, Parecini 1991, pgs. 20-21), on top of masonry substructions or, where the ground was particularly uneven, on arches, to prevent repeated water loss. With regard to the plotting of the underground excavation, this generally commenced from the base of wells (more rarely from the passages, inclines, steps or ventilation shafts), where at the appropriate water flow depth, two tunnels were excavated in opposite directions. Each tunnel branch would meet the tunnel from the adjacent well; linking all the wells allowed the level and direction to be maintained and thus the channel was formed.

The calculation of the slope was important. A channel with no slope or which sloped in the wrong direction would have led to water stagnation. An excessive slope would have caused water erosion and ultimately destroyed the channel. Where relatively short, uneven sections were to be covered certain expedients were used, such as the creation of drops at the base of which, impermeable compact stone slabs, immune to water erosion, were placed.

The Greeks and the Romans were aware of the pressurised water channel (forced) technique, but construction resources only permitted their use for short tracts and specific situations. The work usually extended via a channel (underground or surface *specus*)*,* which led to a discharge tank at the confines of the depression (valley) to be overcome. From there a downpipe surpassed the inferior limit by means of arches or substructions and then moved upstream to the charge basin (oscillation tank) where it would continue its journey via a second *specus.* Around 400 A.C, in Olynthus, city on the Chalcidice peninsula, Heron has an aqueduct built with pressurised water channels and earthenware piping «From the spring's source, approximately 12 kilometres from the city, Heron plotted an aqueduct across a plain and from there to the Northern high-ground of Olynthus in order to supply water to the lower and central areas by means of a tunnel, underlying the main road» (Tölle-Kastenbein 1990, pg. 90).

The base and the piers of hydraulic works were generally lined with hydraulic mortar (*opus signinum*, concrete). There are also many examples of unlined channels, where the compactness of the rock matrix permitted this. Sometimes the tunnels and passages were partially or fully lined with segments or masonry (bricks or flat roof tiles), for example, at their point of contact with natural cavities or unconsolidated material or following rebuilding works due to structural collapse. Aqueducts were subject to inspection and required constant maintenance. In his treatise on ancient Roman aqueducts, Frontino discusses aqueduct maintenance (Frontino, 119-123).

In antiquity, calculation of the water flow rate, that is the volume of water passing through a specific section in time units, was not unknown. At the end of the channel and after having passed one or more silt basins, the water flowed to the *castellum*, a constant level tank, within the walls of which, *calices*, bronze gauge pipes had been inserted, at constant head. These delivered water to the various beneficiaries, which then passed through lead or earthenware channels (*fistulae*). Frontino provides a detailed list of the pipe sizes then in use (Frontino, 24-63). The aqueducts and certainly the larger ones were equipped with a *castella* for water distribution and also with end tanks.

Generally speaking, aqueducts, and Roman aqueducts in particular, complied with these regulations resolving the various problems along their path, both above and below the ground. A perfect example of this is the Gier aqueduct, which with its overall 86 km extension reached the city of Lugdunum in France (Burdy 1996). On the other hand, the Virgin aqueduct in Rome ran almost entirely underground. Frontino tells us that its length was of 14105 steps, of which 12865 in an underground canal, 1240 on the surface, 540 on support walls in various locations and 700 steps on arches (Frontino, 10). The Roman Setta aqueduct is entirely underground. It captures the waters of the river Setta and transports these from Val di Setta to below *Bononia*, the modern-day Bologna, with an extension of 19735 m (Giorgetti 1985, pg. 47).

An interesting example of a modern aqueduct built according to ancient criteria is provided by the Campiglia Marittima aqueduct in Livorno, now in disuse (figs. VII.12.a and VII.12.b). Created circa the 1920s, the aqueduct captured modest springs by means of small chambers carved into the rock and channelled water to the masonry surface tank via cement-asbestos pipes. The supply conduit still branches off from this today and extends underground on substructions and on two sets of brick arches (no longer intact), until it reaches a second tank. A motor pump then transported the liquid to the charging tank, next to the Campiglia Castle lookout tower, for free-flow distribution to the town below. It should be taken into account that the only differences between this aqueduct and the ancient aqueducts is the use of asbestos-cement, the water potabilisation plant and the motor pump, which replace the bucket elevators and bucket augers.

Water was needed for the operation of both public and private fountains, laundries, thermal installations and mills, as documented in Ancient Ostia (Ricciardi 1996, II, pgs. 14-185). The Fortuna Primigenia sanctuary at Praeneste, now known as Palestrina (Rome), was supplied by an aqueduct, which most probably also supplied the fountains in the man-made recesses on each side of the external steps, below the Terrace of the Exadrae (Coarelli 1987, pg. 46).

Between the XII and the XIV centuries, Siena saw a demographic increase and was in the middle of political and economic expansion; at this time, the governor's main concerns would have been the restructure and upgrading of the underground aqueduct: «our attempt to prioritise and increase the Fontebranda water load, which supplies the leather, hide and above wool manufacturing plants, is no coincidence» (Balestracci 1987, pg 29). As evidenced by Frontino on the subject of Alsietinan water (Frontino, 11), not all aqueducts carried drinking water. Built in 2 B.C., the aqueduct carried water from Lake Alsietino to the *naumachia* at the foot of the Janiculum Hill (Panimolle 1984, pgs. 175-176).

VII.3.2 - *Artificial underground canal*

Canal, either underground or through a relief.
An underground canal is a work normally cut into the rock or at any rate, directly into the soil. For example, in France, in the latter half of the XVII century, an underground canal was built to connect the Atlantic Ocean to the Mediterranean Sea. The canal is 165 m long and has an average section of 6.6 x 8.1 m. The Saint Quentin Canal in Picardy has the longest navigable underground tract in France.

Canals for hydroelectric and industrial use were built both by underground excavation and by the cut and cover method (artificial vaulted canals). One interesting example, built according to the designs of architect Gaetano Moretti at the turn of the XX century, can be found at the Trezzo sull'Adda hydroelectric centre in Milan. The pronounced and rocky western slope where the curve of the River Adda is intersected by a tunnel which carries water from the hydroelectric basin and releases it back into the river a little further downstream.

Forced channel: used for its driving force, this type of canal can sometimes be cut into the rock or consist of pipes placed inside the tunnel.

Water diversion: this type of work is used to divert a watercourse. For example, in Italy sometime around the VI century B.C., the Etruscans built the Ponte Sodo near Veii (Rome). This tunnel was cut into the rock to channel and divert the watercourse. There is a similar work, known as Ponte Vivo, in Cerveteri (Rome). Cut in tuff ("cappellaccio"), this tunnel is 19 m long, 4 m wide and 5 m at its highest point. Its base is full of detritus, and it is here that the waters of a stream are channelled, its course having first been artificially diverted (Rizzo 1988, pg. 107). Petra (Jordan) has an artificial sandstone tunnel, which was excavated by the Nabateans to divert a stream.

Levada: type of irrigation canal found on the island of Madeira, predominantly excavated from the XV century onwards. Over 2,000 km of canals, 40 of which underground, to a total of approximately 200 "levadas" are still in use today (Bodini 2002, pg. 72).

Fig. VII.12.a. *Arched section of the Campiglia Marittima aqueduct, built in the 1920s (photo G. Padovan).*

Fig. VII.12.b. *Subterranean part of the Campiglia Marittima aqueduct, built in the 1920s (photo G. Padovan).*

VII.3.3 - *Artificial vaulted canal*

Artificial vaulted canal – created either during construction of the canal or subsequently.
The floor and banks of an artificial canal constructed by surface excavation can be either natural or masonry-lined. There are many types of canal, due to different needs and purposes. Over time a canal may be provided with a vaulted covering, although there are many examples of these being created at the same time as the canal is built. An artificial vaulted canal may consist of large surface or underground water conduits.

Canals built around castles or bastioned fortresses and around walled cities, guaranteed a good defence as they prevented the immediate approach of siege machines to the defensive perimeter and made mine excavation difficult. Initially, canals were often open-air channels with no covering and they were only subsequently provided with vaulted coverings. This may have been for hygienic or viability factors or simply because the canal was no longer required and was to be reutilised as a sewage conduit. If many cities have artificial canals for primarily defensive purposes and to a lesser extent for the disposal of organic waste and viability, with the expansion of the urban network these can often be "banished" underground. Having lost their defensive function, such canals restrict urban traffic and pose a danger to injection vehicles as they are not subject to regular maintenance and maintained mud-free.

Prime examples are the Navigli of Milan, a water and defensive network, which covered and surrounded the city. Most of the aqueducts were covered with brick vaults between the XIX and XX centuries. Of the ancient urban canal system only a few remain in use: the Naviglio Grande, the Naviglio Pavese and the Darsena of Porta Ticinese (once known as Darsena di Sant'Eustorgio). Yet many kilometres of underground canals remain, although the presence of rats and stagnant exhalation gas discourage their exploration (Padovan 2002 c, pg. 408).

Discharge channel (or discharge tunnel): this work helps keep certain works dry, such as certain fortification moats for example. At Fort Demonte in Valle Stura (Cuneo), an underground, XVIII century brick installation serves as a drain for rainwater and ice water, which would otherwise stagnate in the ditch surrounding Bastion of Saint Ignatius. 23.1 m of this is still practicable and leads to a circular chamber with three small drainage conduits. At approximately half-way is the floodgate housing, which can be actioned by an overlying countermine system (Padovan 2003, pgs. 33-37).

VII.3.4 - *Connecting shaft*

Vertical structure for the passage of water.
Despite the attempts of our ancestors to avoid the flow of large masses of water through masonry wells or wells cut into the rock or through highly inclined channels of either material, certain hydraulic works for these very purposes have been found. One example can be seen at the San Cosimato rock on the River Aniene (Rome): the connecting channel between two aqueducts, the *aqua Claudia* and *aqua Marcia*, which supplied ancient Rome (Basilico, Casartelli, Frignani, Lampugnani, Ninni, Padovan 2006, pgs. 150-151).

The continuous cave-ins, to which the crag was and still is subject, resulted in continuous maintenance works being necessary. Sections of both aqueducts have in fact been internally built-up, thus creating by-passes (Ashby 1991, pgs. 123-126 and pgs. 222-229). A few meters past the Claudio aqueduct by-pass is a trapezoidal section well, of which the first part is made of brick while the underlying part was cut into the rock: this opens directly onto the underlying and primitive conduit of the *aqua Marcia* aqueduct. Dating of the bricks and formations, which partially cover the rock section, will provide the structure's chronological date. In any case, the vertical system was used to divert water from the *aqua Claudia* aqueduct into the *aqua Marcia* aqueduct in the event of downstream interruption of the former or upstream interruption of the latter and probably also during maintenance works. Similar systems (having a different function) can be found, for instance, in mills and other works which use hydraulic energy to power tilt-hammers.

VII.3.5 - *Drainage channel*

Group of structures, in this case underground works, which drain and improve land subject to regular flooding and prone to water stagnation, for both production and sanitation purposes.
In central Italy, various underground structures were created for the drainage of land prone to water stagnation. Not always easy to comprehend, such installations consisted of one or more wells for the collection and removal of water. They were connected by underground passages, whose job it was to transport the liquid elsewhere. In Lalibela in

Ethiopia, underground tunnels, cut into the rock, connected the ditches surrounding the monolithic churches to ensure the flow of meteoric water.

Drainage channels have been uncovered inside the artificial platform upon which San Lorenzo, in Mexico, is built (Miller 1988, pg. 19). Established in the X century B.C. by the Olmec population, the city was later destroyed. There may be other aqueducts and sewage systems among the hydraulic works.

VII.3.6 - *Filtering gallery*

Hydraulic work for the collection and supply of infiltration water.
Masonry, tunnel-shaped structure located on one or more flood plain depressions where underground filaments of water merge. The side of the structure is equipped with weepholes, for the capture of even modest amounts of groundwater. It can be used for drainage purposes or for the supply of water where water is scarce.

Under-river drainage tunnel: this is a specific work generally utilised for the capture of fluvial water, from beneath the riverbed. One specific example is the "Traversante del Trebbia", built in the Mirafiori di Rivergaro (Piacenza) in 1865. This probably served to ensure a supply of irrigation water to the surrounding fields during the summer months. Extending in rectilinear fashion for approximately three hundred metres, it crosses the River Trebbia. Its 82 drainage mouths, consisting of openings of various sizes positioned along the wall upstream of the tunnel, all at the same level (Chiesi 2001, pgs. 15-28).

VII.3.7 - *Underground emissarium*

Canal or channel which drains water into a drainage basin, channels river-water or connects two basins.
The territory's control and hydraulic works include underground man-made emissarii for the natural basins. In central Italy and particularly in the Colli Albani area in Latium there are a high number of underground passages plus documented evidence of at least two effluents, probably created by the Etruscan and Latin civilisations. The polygenic volcano at Colli Albani is part of the "magmatic Roman province". Its last eruptive cycle led to the formation of characteristic conical buildings with only the slight hint of a slope, often containing small, active lakes, such as Lake Nemi and Lake Albano or dry craters such as Prata Porci, Pantano Secco, Valle Marciana, Giuturna and Ariccia (Società Geologica Italiana 1993, pgs. 94-98).

The emissarium which serves Lake Nemi was manually cut into the rock on two opposing sides. It is 1650 m in length, and once flowed into the Arriccia crater. (Ucelli 1942. Castellani, Dragoni 1991, pg. 54). The emissarium of Lake Albano was explored and surveyed in 1955 by the "Circolo Speleologico Romano" (Roman Speleological Society) and again in 1970 by the Rome-based "Gruppo Speleologico URRI" (URRI Speleological Group). The structure, manually cut into the rock, extends for approximately 1400 m, has a maximum height of 2 m and an average slope of 2.25%. Its central section contains numerous calcareous formations and there is a trapezoidal section near its outlet in the locality of Le Mole (Dolci 1958, pgs. 17-19. Chimenti, Consolini 1958, pg. 20. Cardinale, Castellani, Vignati 1978, pgs. 20-24). «The underground passage has two wells, the first (3 m deep) at approximately 80 m from its outlet and the second (34 m) at approximately 400 m from the same downstream outlet» (Castellani, Dragoni 1991, pg. 48. Castellani 1999).

Giovan Battista Piranesi (1720-1778), architect and engraver, left a series of views and illustrations of the emissarium of Lake Albano, entitled: «Descrizione e Disegno dell'emissario del Lago Albano» (Description and Design of the Emissarium of Lake Albano) (Ficacci 2000, pgs. 434-440) (figs. VII.13 and VII.14).

Umbria has the emissarium of Lake Trasimeno, which was excavated or simply unblocked and returned to functionality by Braccio Fortebraccio da Montone in 1420. 1057 m in length, with a short open-air section (Castellani, Dragoni 1981, pg. 38), its purpose was to regulate basin waters. It later underwent maintenance and is no longer accessible. «In 1895, a new emissarium was planned for Lake Trasimeno and this was completed in 1898. Parallel to the mediaeval emissarium, its total length is of 7314 m, 896 m of which is tunnelled. A masonry dam, between the inlet and the underground channel acted as water level regulator» (Castellani, Dragoni 1981, pgs. 38-39).

From 41 to 52 A.D., the first of several underground structures for the drainage of Lake Fucino in Abruzzo was created, without lasting result. In 1854, the banker Alessandro Torlonia, initiated drainage works on the basin through the creation of a tunnel. Today, a second more recent tunnel is in use and part of the Claudio-Torlonia emissarium can now be visited.

Fig. VII.13. *Internal view of the mouth of Lake Albano's Emissarium mouth. Engraving by Giovanni Battista Piranesi.*

Fig. VII.14. *Internal view of the mouth of Lake Albano's Emissarium. Engraving by Giovanni Battista Piranesi.*

The precise purpose of the artificial emissariums created in antiquity is debatable. Some believe that these were used for the drainage of lake basins, while others believe that they served to regulate or simply limit water level fluctuations during rainfall: as the basins had no natural effluents, flooding within the troughs would not have been infrequent. Given the range of hydraulic works known today, it cannot be excluded that some may have been drained intentionally; however, such cases only go to confirm that ancient populations were far more aware of the territory's ecological structure than we are today. It should be taken into account that a basin was primarily an immediate source of available water, however, through fishing, it also provided a source of food, and this was not to be underestimated. For these very reasons, even smaller basins may have been retained and not drained and would have been easily regulated by a spillway. In addition, this was an excellent source of irrigation water for cultivations situated beyond the confines of the water: passages and tunnels may thus have been designed with irrigation in mind and not solely for the regulation of water levels. On the other hand, the rise or persistence of malaria or the urgent need to cultivate land may have led to the basins being drained. An extended basin was difficult to regulate and even more difficult to drain, at least when using underground tunnels despite the fact that this may have been attempted by means of the Fucino in Roman times. The modern emissary (or rather the discharge or drainage tunnel) had a drastic environmental impact and the Fucino plain suffered from the lack of water; this inconvenience was foreseeable and would not have been unknown to ancient civilisations. In order to avoid assessment errors, each and every emissary must be considered on its own right.

VII.3.8 - *Vaulted, natural watercourse*

Natural watercourse with subsequent addition of masonry banks and vaulted covering.
Over time and due to the intervention of man, a watercourse, be it a river or a stream, may change in appearance and move underground. This is especially true where urban settlements develop along its banks. Part of its natural bed may be replaced by an artificial canal or its banks may be reinforced with brickwork and covered to meet various requirements. In time it will become in all effects, subterranean.

VII.4 - *Typology 2 c: vertical capture shafts*

Outcome of actions and vertical installations for the purposes of groundwater aquifer capture.
The term 'well' generally refers to an artificial tunnel with vertical axis. By extension, the term well also applies to the vertical tunnel of a natural cavity. Vegezio states that a city should always have perennial sources of water within its walls. Where there are no perennial water-sources, wells must be excavated and water must be lifted using ropes (Vegezio, IV, X).

A well's purpose depends on the geological terrain within which it has been excavated, on the type of architecture used in its lining and above all to that which it connects. At first glance, the purpose of every well would seem to be that of collecting groundwater. However, after due exploration, it is often "revealed" that the well leads to an underground aqueduct or is in fact a cistern or a man-made structure which requires further investigation before its purpose can be ascertained. In use since antiquity, wells were manually excavated until at least the beginning of this century, despite the introduction of boring machines. Having explained that the ground may naturally release exhalation gases, Vitruvio recommends that a lit oil lamp be lowered into the well: should the flame go out, two lateral wells should be dug to release the gas from the soil. Once water has been reached, he recommends that the well be covered to avoid blocking the vein (Vitruvio, VIII, VI, 13). If the purpose of the excavation it to reach a groundwater aquifer for drinking water or irrigation purposes, ordinary wells and artesian wells will be created.

By extension, the section surrounding the mouth is also known as well, or more appropriately as edge or parapet, or as puteal or well-curb. The external part of the well consists of a pedestal, on which the puteal rests. Sometimes made of stone and elegantly shaped, the well could be closed by a lid (or hatch). Wells sometimes had a lid support or an architrave, to which the water pulley, with its rope or chain and bucket were attached. Metallic arch-shaped elements had the same support function. All such elements could be found at the uppermost access of both wells and cisterns. The section, which widens just below the pedestal, at the beginning of the actual well, is known in Italian as "gola". There are sometimes beams or arches to support the puteal and the vault near the well's opening. However, two stone rafters support the arches in the Sorbello Well in Perugia (Stopponi 1991, pgs. 237-240).

Wells may be lined with stone, cobbles, blocks and bricks or with pieces of special curved earthenware, held together with lead clamps or strips. Forbes tells us of Mycenaean and Cretan wells, the bricks of which were replaced with earthenware pipes while in temporary dwellings the ancient Romans used timber frames or barrels (Forbes 1993, pg. 674). In 1938, several wells lined with curved sheets of tuff and equipped with footholds, were uncovered in the

Quirinale area of Rome (Pisani Sartorio 1984, pg. 41). A quadrangular mediaeval well with interlocking timber lining and approximately 7 m deep was uncovered in Happisburg, Norfolk (Forbes 1993, pg. 674).

If excavation takes place in unconsolidated soil, a lining is required, as exemplified by the wells in Milan (Castoldi 1996, pgs. 113-122) or those of Ancient Ostia, where earthenware "dolio" or "orcio" vases were sometimes placed beneath the puteal (Ricciardi 1996, I, pgs. 25-88). However, even rock-cut wells may be lined. A XV century well, engineered by Vercellino can be seen in a house in Trezzo sull'Adda (Milan): the well was cut into well consolidated conglomerate ("Ceppo d'Adda"*)* and is fully-brick lined. Perfectly cylindrical, it is just over 40 m deep. Towards its base, at the site of small vacuums containing sand and on a lesser scale, clay, the lining has come away. Wells are sometimes lined with lime.

Footholds are a characteristic feature. These are indents in the wall, which eased both descent and ascent during the well's construction. Footholds were normally carefully cut into the rocky walls and positioned at regular distances along adjacent or opposite sides of the well. Less frequently, these were irregular and placed without apparent order. Footholds can also be found in specific types of cladding, such as in Greek and Roman cladding (Pisani Sartorio 1986, pgs. 37-40). They are also to be found in large structures, which do now allow opposing movement and may be linked to larger indents, some of which may point upwards and were especially made to support beams used as ladders. In some Milan wells of uncertain chronological date, there are regular spaces in the brick layers, which may have been footholds. These may also have held timber frames, during the placing of the cladding.

The drainage of certain wells, such as the Marzabotto and Populonia wells (Minto 1943, pg. 26), has revealed a terminal indentation in the rock. Presumably, its purpose was to collect sediment. Several wells with filtering systems have been uncovered in Milan. These are thought to be Roman, however, analysis of the construction method used has proved insufficient. As such, precise dating has not been possible. «A block of wood, with two cylindrical lead pipes has been found at the very bottom of each of the three wells at Via Speronari, Via Unione and San Giovanni in Conca. This is thought to have been a type of filter, fitted to the bottom of the well, which in this case would have had a timber section the same size as the structure's external diameter, instead of the usual ring; the section would have isolated the well from the water table and the pipes would have allowed for the passage of water, which now free of the water table's sand residuals, would have run clear at the bottom of the well» (Castoldi 1996, pg. 116).

Although normally circular, the shape of the section may be elliptic, square, polygonal or mixed. In one of the semi-subterranean environments of the Cloisters of Saint Eustorgius in Milan, the first section of the well's shaft is square, while the second section is circular. St Patrick's well in Orvieto is also worthy of mention: built at the beginning of the XVI century and designed by Antonio da Sangallo the Younger, it is of particular interest in that a double-helix staircase, with internal windows as a source of light, lines the walls of the 62 m well. Thus, different shapes and sizes can be adopted within the same work from the start and does not only result from subsequent restoration.

The depth, on the other hand, depends on the aquifer's altitude, rarely exceeding 60 m. There are of course, exceptions, where depths of up to 100 m have been reached, such as the XVII century well in the Verrua Stronghold (Turin), which captured under-river water from the River Po (until its improvident obliteration on the part of Cementi Victoria at the end of the 1950s). The Montecrivello Well (Vercelli) is another example. Excavated in moraine terrain and 85.48 m deep, its internal diameter varies between 1.3 m and 1.2 m, the well is fully brick-lined and was held not water at the time of exploration (September 2005).

Water was obtained by means of cylinder or other structure, equipped with a rope and turned by a crank. Alternatively, the rope was lowered by means of a wheel mechanism (or pulley), which was hooked onto an, often elegantly shaped, overlying structure. Another system was that of hinging a long rod, with a bucket on one end and a weight on the other. This simple and discontinuous method of obtaining water ("shaduf") is still used today in some areas of North Africa and the Far East; ancient depictions in the form of a cylindrical seal from the accadic period (circa the third millennium B.C.) and from several Theban tombs, circa 1500 and 1300 B.C.) (Drower 1993, pgs. 528-533. Forbes 1993, pg. 686). Despite the fact that relatively basic systems could be used, ropes and chains were still directly attached to the puteal. Water could be lifted to the surface using bucket wheels, bucket chains, pumps and pistons. The exact excavation method for building wells in antiquity is unknown, however some indication is provided by mining engineering treatises. Alternatively, some indication of how they were built can be found by tracing the well's history, as some wells were manually excavated until the beginning of the XX century. Drower tells us that the ancient system could not have been very different to that used by the Bedouin nomads in Arabia.

As Renate Tölle-Kastenbein attests, when man chose his dwelling place and took the decision to build a permanent settlement, he created wells. However, wells were also required for crop irrigation purposes (Tölle-Kastenbein 1990, pg. 32). The *aqua Appia,* built in 312 B.C. was the first Roman aqueduct. Frontino explains that prior to this, the Romans made do with water from the Tiber, from wells and from springs (Frontino, 4).

Stepped well: the structure's particular design, allowed deep spring water and gravitational water to be reached by means of a vaulted stairway. Sardinia has various examples of stepped wells from the Nuragic period (Contu 1985, pgs. 124-128), some of which may have been used for religious purposes (please refer to paragraph VII.7.3).

From our knowledge of both the territory and the soil, man undoubtedly began to make tunnels in the ground during his search for water. It could also be hypothesized that after inhumations and private dwellings, wells and cisterns were the most commonly built architectonic works. Normally found near towns and villages, inside houses, in courtyards, near cisterns and icehouses, or linked to public structures, in temples and similar buildings, or in squares, wells are to be found almost everywhere. There are also no shortages of wells in fields, where they were used for irrigation purposes or to provide drinking water for the herds; or in the desert, along the caravan routes or along heavy traffic roads.

VII.4.1 - *Artesian well*

Vertical well for the capture of a pressurised underground groundwater aquifer; the well shaft can be circular, square, polygonal, elliptic etc.
Where water is contained in the permeable strata below impermeable strata, natural pressure sometimes causes the water to rise and flow to the piezometric surface, known as an artesian well. The name originates from "artésien", or rather from "de l'Artois", the region in France where the artesian well has been in use for a very long time.

VII.4.2 - *Gauge well*

Vertical perforation in the ground used for measuring water levels.
A gauge well is an ordinary well, which, thanks to the depth markings created inside it, was used for measuring water levels and predicting floods. The most famous example of such a well is the Egyptian Nilometer.

«*Niloscopium, celeberrimum in Aegipto, uti olim ita nunc celeberrimum est Hidrometrium. Est autem fabrica rotunda in ripa Nili adinstar putei, in cujus medio columna marmorea spectatur 20 parallelis circulis, numeris suis columnae incisis; ex adsensu vero & descensu aquae, tempore inundationis Nili, Aegyptii cognoscunt anni futuram constitutionem vel ad sterilitatem, vel ad ubertatem annonae inclinantem*» (Kircheri 1678, pg. 310).

VII.4.3 - *Ordinary radial well*

Vertical shaft in the ground with one or more passages (branches) at or near the bottom.
If the well is not particularly deep or has been excavated in soil where water is scarce, one or more branches can be created to increase the well's water capture capacity. There are radial wells with one or more branches at or near their base, which link to a water table or drain even small aquifers.

VII.4.4 - *Ordinary well*

Vertical well for the capture of an underlying groundwater aquifer (filtering or groundwater well); the well shaft can be circular, square, polygonal, elliptic etc.
When an ordinary well comes into contact with a water table, water never rises above the natural ground level unless the well is near the site of groundwater discharge. According to Vitruvio, by lying down and keeping your chin to the ground you will be able to identify underground springs by the thin and fleeting bursts of vapour: this is the point where digging should begin (Vitruvio, VIII, I,1). The well at Pavarolo Castle (Turin), situated inside a building, on the internal part of the surviving wing of the mediaeval building, is one such example. Created in fossiliferous sandstone during the Miocene epoch, the well is 64.48 m deep, of which 6.26 m is submerged; its access measures 1.38 m in diameter while from -56.63 m, the diameter remains an almost constant 3.08 m Its brick masonry facing, under which the rock is visible, is of 14.6 m; it was probably excavated over two separate periods, the first excavation taking it to no more than 20 m in depth (Basilico, Bianchi, Ninni, Padovan 2003, pgs. 284-289).

In the XIV century, Bonvesin da la Riva refers to the quality of well-water in his chapter entitled «Elogio di Milano per la sua posizione», and states: «The city has neither cisterns nor conduits bringing water from afar but does have natural springs, which are potable and thus suitable for human consumption. The water is clear, wholesome, easily accessible and never scarce, even during dry weather and they are in such abundance that even basic houses almost always have a natural spring, which is known as a well» (Bonvesin da la Riva, I, IV).

VII.5 - *Typology 2 b: storage*

Structure built for water collection and storage purposes, also used for collecting and storing oil, wine, ice and snow.
The need for water storage, above all drinking water, has resulted in a wide range of storage works. Due to the advent of modern aqueducts for drinking water, these have largely fallen into disuse or are now used for irrigation purposes. Cisterns, for the storage of aqueduct water are not included in this typology. Water is collected in its solid form and deposited in special areas (icehouse and snowstores) for food preservation purposes. This typology also includes oil storage cisterns and wine treatment and storage cisterns.

VII.5.1 - *Cistern*

N*ormally a semi-subterranean or underground structure, for the collection and storage of rainwater from a collecting surface.*
A cistern can be described as a large container. These are normally to be found underground, although there are many semi-subterranean and surface examples. The many types of cistern are to be found in all soil types and are used for the storage of rainwater, collected from roofs or special collecting surfaces.

Cisterns were used when there were no other means of obtaining water. The system was simple and efficient so long as periodic maintenance took place. Vegezio ordered that all public buildings and where possible, private buildings, build cisterns for the collection of rainwater from the roofs (Vegezio, IV, X). Its task being that of collecting and storing water, a watertight 'container' was preferable. Cisterns come in many shapes and sizes, dependant on multiple factors, e.g. the material to be used, available funding, available technology, their purpose (potability being a major concern) and last but not least, the geological soil and the environment where they are to be built. A list of the most common types of underground cistern is detailed below.

Pit cistern: simple ditch-shaped excavation in either soil or rock, for the collection of meteoric water; financially economic. The so-called 'giants' kettles' (more or less cylindrical cavities formed by the action of sand and coarse gravel dragged by subglacial stream currents, etc.) were used to collect and store water. One example can be seen at the Belvedere locality of Chiavenna (Sondrio), together with the remains of the mediaeval castle.

Open-air cistern: similar to the pit cistern but larger with articulated collection and distribution systems. Laureano mentions «open-air cisterns» when describing the water-collection systems used in Qana (Yemen), consisting of tanks and filtering and decantation devices (Laureano 20001, pg. 88). However, Nicolletti states: «The open-air cisterns, known as *maj'il* are worth mentioning; some of these undoubtedly fall under Yemen's most fascinating architecture. Fed by wadi or rainwater filtered by agricultural land or channelled by the roofs of buildings, these man-made structures consist of an underground tank, with masonry walls which have been sealed with strong mortar. The base of the tank sometimes opens onto smaller basins, which collect sedimentary deposits» (Nicoletti 1985, pg. 276).

Single chamber cistern: this is the most common and undoubtedly the most well known, providing a wide range of architectonic solutions. In its most basic form, a cistern can be cylindrical, tronococonical, bottle-shaped, demijohn or dome-shaped *(tholos)* or irregular in shape, with multiple variants. One type is known as the "bagnarola cistern". Rectangular in shape, its lower sides rounded, records of this type of cistern were documented at the *Tharros* settlement in Sardinia (Acquaro, Francisi, Mezzolani 2002, pg. 61). The same shape can be found in the cisterns of the ancient city of *Cosa* (Grosseto). The most common type of cistern are those with a, more or less regular, parallelepiped shape although it may well be that these appear more common since many such examples survive.

Multi-chamber cistern: this type of cistern is less common and generally consists of two or more cisterns, which have been joined together. Sometimes this type of cistern was created in environments, which were only subsequently used for water storage and whose original purpose is unknown.

Dual-chamber cistern: this consists of two concentric chambers, which can be round or square, the internal chamber being the storage chamber and the external chamber being used for filtering purposes. The two chambers are joined via the transfer outlet. One example of such chambers is provided by the Veracchi-Crispolti Palace in Perugia (Associazione Subacquea "Orsa Minore" 1981, pgs. 91-104).

Cistern with passages: normally consists of a network of connecting passages; in its more complex form, it is not dissimilar to a room and pillar mine. These are often, as claimed by Riera, structures for the capture of water (Riera 1994, pgs. 313-321).

Filtering cistern: of the various types of cistern, this ensured a good degree of potability. The most common type is known as the "Venetian-style cistern". This consists of a conical excavation, which is at least 3 m deep, the walls and base of which, are lined with a layer of compressed sand and clay. A cylindrical well, which communicates directly with the lower section of the conical excavation rises centrally from the cistern's base. The gap between the well and the wall is filled with washed silica sand. Collected rainwater is filtered by the sand and channelled into the well, from where it is then removed (Riera 1993, pg. 29). During dry spells, the cistern was often filled with water collected in barrels or other containers.

Going back to the basic types of cistern used in rock settlements, pear or bell-shaped cisterns, cut directly into the rock, are often found inside dwellings in the Sassi di Matera settlement and in minor settlements situated along the edges of the ravine. Laureano reports of a particular type known as *roof cistern*, to be found on the Murge plateau, used to provide drinking water to livestock (Laureano 1993, pgs. 140): water is collected directly, by micro-infiltration, into a rectangular chamber, cut directly in the rock. Covered by a double-pitched roof, a central opening is utilised for withdrawal of the water; the water then flows into a chute and is channelled from the roof into the water troughs. Laureano also refers to the "condensation chamber", supplied by the condensation which forms on the cover of perforations. On the subject of water supply systems used by the Yemenite population, he states: «A famous inscription, by the Shaeban people asking God to bring "the autumn and winter rains" was like asking for a miracle, given the already arid, climatic conditions of the time. However, replacing the word "rains" with "dew" truly reveals the populations' worries and concerns at the time. The mahfid are also included in this category: these were devices for the collection of water, which used double masonry cladding as condensation chambers» (Laureano 1995, pg. 96).

The Conversano cisterns in Puglia are another type of collection and storage chambers (Palmisano, Fanizzi 1992, pgs. 35-53). Underground water storage chambers were built both in or on the banks of lakes, often seasonal lakes. These were restored and utilised almost to the present day. These structures are *tholos* or bell-shaped, are lined with rocks or stone segments and are of between 4 m and 12 m deep. Their horizontal sections can be either circular or elliptic. The base of the structures is no longer visible due to the accumulation of silt or soil deposits. Palmisano and Fanizzi detail the section of a structure, investigated in 1887, where its bowl-shaped base could clearly be seen, which was not lined and which had a small central well (schematic section of a Lake Iavorra well by Simon S. in *Norba e Ad Veneris*, report transmitted to the Accademia dei Lincei, 1887; covered by Palmisano, Fanizzi 1992, pg. 49). These ensured a water reserve when the lakes dried up and at any rate, reduced the basin-level to a pool of mud. Additionally, in the Gargano area, structures for the capture and storage of water are associated to fosilliferous valley rifts, dolines and natural depressions, known locally as pool, collection basin or well.

Laureano made some interesting comments relating to the various water storage systems used by the Maya in Yucutan: «In order to obtain drinking water reserves, bell-shaped cisterns called *chultun* were cut in the rock. In the classical period, from IV century A.D., large cities were developed around natural depressions called *aguada*, into which, water collected by the dams and cisterns along the slopes collected. The topmost parts of the *aguada* were paved with flat stones, the connections of which were rendered waterproof by red and brown clay. There were wells and *chultun* at the bottom of these, which stored water when the *aguada* was dry. The system is altogether similar to the cistern system used in the karst area of Apulia in the south of Italy» (Laureano 2001, pgs. 225-228 and pg. 359).

Another water storage system, also used in northern Yemen, was to build a dam to block the course of a *wadi:* the basin thus formed is alluvial in nature and is reliant on the wadi's inconsistent flow, which alternates between dry periods and floods depending on rainfall (Nicoletti 1985, pg. 276). The most imposing was the Ma'rib dam which, blocking the course of the Wadi Adhan, was 15 m height and extended for approximately 600 m; it had three sluice gates, which regulated the flow of water for irrigation of the underlying crops. The Koran tells of the collapse of the dam, which took place in the mid VI century (Peirone, XXXIV, 16).

Further distinction can be made where a cistern has been found in a quarry (in which case it will be classified and registered as a quarry for water storage). The "Cisternone Vittorio Emanuele II" in Cagliari is a Punic quarry, which was re-utilised during Roman times as a reservoir. In the nearby amphitheatre, meteoric rain was collected by special canals cut in the rock and by an underground channel equipped with an underground settling basin, rendered waterproof through the use of *opus signium* (lime mortar with an aggregate of coarse pieces of broken terracotta) (Floris 1988, pgs. 22-29 and pg. 120). There are many lining solutions: There are structures with stone, brick and stone segment cladding, which are rendered waterproof with clay or hydraulic mortar. In more recent examples, or following re-utilisation cement or concrete are used (fig. VI.15). There is no shortage of unlined structures, cut into impermeable rock and presenting no fissures. The vaulted roofs may consist of overhanging, semi-circular, drop, pointed or conch vaults or of arches supported by pillars. The convent of San Cosimato (Rome) has a rectangular cistern, which was cut directly into the rock. Two pillars were uncovered here. Chambers may also have particular architectonic shapes or elements.

Until the 1960s, in the Alberobello, Cisternino, Martina Franca and Locorotondo areas of Puglia, the ground was excavated to a depth of 50-70 cm, to obtain reddish and pasty clay called "vuolo", used as a waterproof lining for the cisterns used to collect rainwater. Cut into the rock, the cisterns were demijohn, flask or pear-shaped. Every one or two years, these were de-scaled, cleaned and sometimes disinfected with lime before re-applying a 5-10 cm layer of clay. On the subject of water circulation in caves, Leonardo da Vinci mentions the impermeable properties of clay: «the density of clay could well obviate and prevent underlying water penetration as demonstrated by saline water cisterns, which have outwith their walls and gravel, a layer of the fine clay from which vases are made. Saline water shall never penetrate this material thus the water within the cisterns *(caver)* shall always be fresh» (da Vinci, Cod. Leicester, F. 3 - r.).

Rainwater, collected for drinking purposes was generally decanted and filtered. One system may have been the addition of a small compartment to the cistern, which was then subdivided into two smaller compartments: the first is essentially a settling pond - from here water overflows into the next compartment; the second serves as a filter and contains layers of gravel, sand and wood charcoal through which the liquid must pass before reaching the storage chamber via one or more pipes. During investigations, it is not always easy to establish whether a cistern has these elements.

Many storage chambers still have internal waterspouts or water supply inlets, although it cannot always be established whether these originate from decantation and filtering systems unless excavation of the surrounding areas can be carried out. Furthermore, the decantation-filtering system can take place by using various construction solutions as exemplified by double-chamber and Venetian cisterns.

The XIV century well and the XV century rectangular cistern of the Sant'Agnese Monastery in Perugia both have a collection basin ("cutino" or "catino"); the planimetric survey of the latter clearly shows that the inclined floor slopes towards the circular depression (Associazione Subacquea "Orsa Minore" 1981, pgs. 43-51 and pgs. 105-111). The same can be said of the well overlooking the Civita di Tarquinia Gate, drained by Romanelli ("Pozzo E"), who reports in his field report that a 20 cm by 10 cm depression is visible on the well floor (Romanelli 1948, pg. 227).

The epidemic, possibly of plague, which also strikes Athens in 430 B.C., should be regarded as a culmination of events. In addition to describing its symptoms and effects, Thucydides tells us about water supply in Piraeus, carried out without *krenai* (streams, fountains) and about how the 'epidemic' almost always affects the populated centres to which war refugees flock, thus rendering the sanitation situation precarious (Thucydides, II, 47, 1-54, 5). Like wells, cisterns are also structures which accompany Man in his journey through life. There are many examples of oil and wine storage chambers.

Cisterns, such as the one investigated in Pizzighettone (Cremona), can be true 'mines of information'. Equipped with fortifications in different historical periods, the village takes on the aspect of a bastioned stronghold, astride the River Adda and from 1648 to 1656, Marchese di Caracena (the Governor of Milan) has the entire structure reinforced (Gambarelli 1995, pgs. 11-13). An archaeological operation carried out inside the town walls in 2002, reveals a cistern, once used as for the disposal of organic and inorganic waste. The organic matter was recovered and dated back to the latter half of the XV century and the first few decades of the XVI century. «The presence of 'fine' crockery and the total absence of fired, enamel ceramics, indicates that such tableware came from rooms directly above the underground space» (Perego 2005, pg. 29). Other finds include metal, work scrap, tripod distancers and «1186 bones, 60% of which were recognisable on at least a general level» (Di Martino 2005, pg. 95). Partially recognisable vegetable matter was also studied. «Such finds provide information on both the vegetarian diet of the communities that used the waste disposal pits and on the site environment at the time they were deposited» (Rottoli 2005, pg. 99).

Fig. VII. 15. *Cistern "del Lantro" at Bergamo (photo R. Basilico).*

A water capture and storage tunnel made of brick and concrete was found during the excavation of a sector of the Gothic Line, near Bologna: «the tunnel was approximately one kilometre long; there were 17 arched openings inside, in the east wall, approximately 50 metres one from the other and these could be accessed by a corridor, which was approximately 10 m long; there were also enormous reinforced concrete cisterns, with circular walls, each cistern being approximately 10 m in diameter by 100 m in length» (Benuzzi, Relli, Fortuzzi 2005, pgs. 224-226). The structure, probably built in the first decades of the XX century, was used as a shelter towards the end of the Second World War. Numerous relics and personal effects of German soldiers, including documents, playing cards and newspapers were found.

VII.5.2 - *Icehouse*

Normally a semi-subterranean or underground structure, used to hold ice for the preservation of food.
Room for the storage of ice either collected during the winter or dug and transported from natural cavities or mountain glaciers. There were various methods of building or obtaining icehouses:
external icehouse;
semi-subterranean icehouse;
underground man-made icehouse;
underground icehouse, excavated in the ground or cut into a rocky wall;
icehouse obtained from the exploitation of a natural cavity.

The most common types of icehouse are those constructed with thick underground, semi-subterranean or even surface walls. Thermal insulation is improved by empty or filled gaps and the floor and the roof of such structures must also be thermally insulated. For additional insulation purposes, those structures which were only partially underground could sometimes be covered with soil, until a mound was formed. These generally circular-shaped rooms are rendered impermeable to external infiltration, have their own ventilation which can eliminate or limit the formation of condensation on the walls and a system for the disposal of icewater. Ice deposits were made via the access corridor or via special external, inclined channels which led directly into the chamber. The ice could be stored and sold during the summer months or could be used on-site for the preservation of food such as meat, fish, butter, etc.

As recently as the beginning of the XX century some fields at the former Olivetano Monastery of S. Maria a Baggio (now a district of Milan), better known as the Cascina Monastery, were still purposefully flooded during the winter so that water would stagnate and freeze over. The fine layers of ice were then cut, laid one on top of the other and once again frozen at which point the ice was collected and taken to the large underground icehouse. The ice, preserved between layers of sawdust and rice meal was sold in Milan during the summer months (Rognoni 1983, pg. 25). The Comabbio (Varese) "giazera" is a circular, brick structure, with a diameter of approximately ten metres and a similar depth. It emerges only slightly above the ground. It is accessed via a corridor ante-chamber with three doors. Most probably built at the end of the XVIV century, it was used by local fishermen for the preservation of fish (Caramella 1999, pgs. 64-65).

The Piantelli icehouse in the municipality of Cairo Montenotte (Savona) is a semi-subterranean icehouse built at the end of the XIX century. It has four monumental preservation chambers with an overall storage capacity of over 6,000 m^3 and a section of railway tunnel which was used for loading; this is now in disuse and one chamber is completely flooded. The ice produced at San Giuseppe di Cairo was transported and used in the large Genovese hospitals and in riverside fish markets (Verrini 2002, pg. 44).

VII.5.3 - *Snowstore*

Semi-subterranean or underground structure, used to store snow for the preservation of food.
Cave, cellar or special chamber used in the past to hold snow for the chilling of food and drink and the preservation of perishable food. Icehouses are often referred to as snowstores. Instead of removing layers of ice, snow was placed inside the room and compacted.

Inside the Fort at Fuentes (Lecco), in the terreplein of the bastion near the Governor's Palace, is an irregular cylindrical well with a masonry access; overhanging parts of its vaulted roof have survived to this day. The well's facing is made of local stone and the lower part of the well rests on the rock; the facing is rendered impermeable by mortar. With an internal depth of 4.4 m and a maximum diameter of 3.3 m, it falls under the classification of snowstore of icehouse (Padovan 1997, pg. 297).

VII.6 -*Typology 2 d: disposal*

Group of structures and systems for the disposal of organic and/or liquid and/or solid matter by transport and/or dispersion and/or storage.
The waste disposal system became necessary when Man, developed self-awareness and organised his community. With the advent of urban complexes, white water and black water disposal systems became vital. This typology therefore deals with the disposal of organic waste, waste water from mills and meteoric water as well as of natural watercourses which have become real open-air sewers. For example, with urban development, sewer systems were built in Florence, even if «the idea at the time was that the numerous streams, ditches and canals which ran along the perimeter or internally intersected the city would act as open-air sewage ducts and discharge the sewage into the river Arno or its tributaries» (Ottati 1988, pg. 42).

In urban centres, disposal which is not in line with expansion, caused (and still causes) serious public health issues with the manifestation of infections and epidemics. Similar issues arose during heavy rainfall, when the systems were unable to adequately drain the matter, resulting in flooding with evident consequences.

VII.6.1 - *Absorbing well*

Ground depression with brick or concrete dry stone fill and through-holes for the filtering and diffusion of waste water into the underlying subsoil. Can be of any shape and/or size.
Generally, a cylindrical stone masonry, brick or concrete rimless depression with a stone base for the collection of sewage from a clarification system. «A layer of broken stones is also laid in a ring, outside the opening-free wall section, which crosses the impermeable soil (0.5 m thick). In proximity of the weepholes and under the layer of stone rubble, the stones are generally larger than the remaining overlying material. The well's capacity is entirely dependant on the characteristics of the soil and on the groundwater level; an indication of the well's depth can thus be obtained, while empirical criterion can be applied to determine its diameter» (Frega 1984, pgs. 410-412). This system is not suitable for compacted clay soil or porous limestone rock with underlying usable water table.

VII.6.2 - *Cesspit*

Pit for the temporary storage of waste material.
Sealed, impermeable hole in the ground for the temporary accumulation of waste material from drains, which is periodically removed (Frega 1984, pgs. 404-405).

Complex hydraulic systems for the channelling and disposal of meteoric water have been uncovered at the ancient city of Heraclea (Matera). Furthermore, the street axis between insulae III and IV «had a narrow, slightly inclined, paved sewer, covered with stone slabs, which channelled the sewage into a cesspit» (Bianco 1999, pg. 80).

VII.6.3 - *Clarification (or biological well)*

In sewage works this refers to a type of septic tank.
The work can consist of two chambers: the upper clarification chamber and the lower, sedimentation chamber. Consisting of a well or chamber cistern, the septic tank is one of the most commonly used systems in the treatment of domestic sewage within rural areas. Partial sedimentation takes place inside the chamber. A spillway or syphon carries the sewage to permeable underground ditches filled with broken stones where the sewage is filtered and drains through the soil where it is undergoes anodic oxidation. The sediment remains in the ditch, where it is anaerobically decomposed and subsequently removed.

VII.6.4 - *Drainage shaft*

Structure which assists the flow of water through land with little permeability; used for both the collection and drainage of water.
In hydraulics, drainage wells are built in land with little permeability to facilitate the flow of water to underlying permeable soil. Their method of construction is no different to that of ordinary wells unless the well is to be used only for water dispersion, in which case small holes in the cladding allow the liquid to be slowly released. There are also structures for the storage of excess water, the spillways of which simultaneously store and drain water.

Drainage trench: this term refers to a trench or well-shaped ditch made in permeable land to drain areas where surface water tends to stagnate; it can be lined with dry stone fill or can have through-holes or be simply filled with faggots or stone.

VII.6.5 - *Septic tank*

Part of the domestic sewage disposal system in towns and cities without sewers.
Also known as a *biological tank* or *biological purification tank,* the septic tank is usually located underground and generally consists of two completely enclosed, watertight chambers. Solid waste is channelled into the chambers where anaerobic fermentation takes place.

VII.6.6 - *Sewage system*

Series of channels and other structures for the removal of meteoric and waste waters from a given area.
A sewer is an underground canal for the collection and disposal of wastewater. The terms "cloaca" and sewer refer to the underground channel which collects and transports rain water and liquid waste elsewhere. Sewers can be classified as either static or dynamic sewers.

Static sewer: collects, purifies and disposes of wastewater via biological or septic tanks or via cesspits.
Dynamic sewer: collects and continually dispels wastewater via a network of channels, generally after having purified the water in special installations.

A distinction between the sewage systems can be made according to the type of water intake:
- white waters essentially consist of meteoric water, which generally contains insignificant levels of impurity;
- black water or lavatory water, is waste water from urban centres, so-called on account of its content (human and animal faeces).

A second distinction is made according to the system itself:
single sewer (single channel or mixed sewer or a Roman system): if white and black waters are transported in the same channels;
separate sewer system (with separate channels): where white and black waters are transported in separate channels;
mixed separator sewer: where a certain amount of meteoric water (usually groundwater which is full of impurities) is permitted within the black water network.

There are various types of sewer, depending on whether the full separator system is used and on the method utilised to channel the sewage waters into their respective canals. The black effluent consists of waste water from urban centres, namely from houses, public buildings and industrial sites, from irrigation and from the cleaning of public areas. Waste water carries solid human waste and other organic and inorganic residue, which must be removed as quickly as possible for two reasons: such waste is putrescible and may contain pathogenous germs The volume of a city's waste water (urban effluent) is strictly linked to the volume of water used and is generally equivalent to 4/5 of this. The proportion of industrial water, which may contain specific pollutants (from abbatoirs, tanneries and chemical industries) depends on local conditions. The white effluent consists of meteoric water, the volume of which can be calculated from the region's pluviometry. This assists in the estimation of maximum volumes, required for the implementation of a collection and disposal system. The network of waste water removal channels must be impermeable to prevent harmful water from infiltrating the urban subsoil. The network must also be distanced from city to prevent sewage gases from polluting the air. Watercourses and artificial cavities, which were subsequently covered, may be used as sewers (fig. VI.16).

Drainage channels are the outer ramifications of the sewage system and are located under gutters and lavatories. They carry waste water to *primary canals*, generally situated under roads; these in turn carry the water into *sewage collectors*, which open onto the *sewage effluent canal.*

Drainage effluent: in sewage systems this is the general collector and usually flows underground, although there are open-air canals in uninhabited areas. The effluent directly links urban centres to the rendering plant or, in the case of natural disposal, to the place where water is discharged. The layout of the sewage system is determined on the basis of underground itself, the presence of any watercourses, the road network and the location of drains.

The most common solutions are summarised below, by characteristic type (Colombo 1933, pgs. 332-346).

Fig. VII.16. *The Dragon Canal is an artificial underground deviation of the River Garza in Brescia. The channel has been utilised for the transport of waste water since the XVII century (photo A. Busi).*

Perpendicular type: collectors from each individual drainage area run perpendicular to the watercourse and each reach said watercourse independently. Now considered antiquated, this is currently used for the discharge of white waters using the separator system either where a watercourse cuts through inhabited areas, or alternatively at the seaside or near a lake.

Interception type: carries final discharge from the main collectors away from the urban centre to a container or purification plant by means of an *interception canal.*

Longitudinal type: this consists of collectors which are either parallel to the recipient's axis or inclined (river or beach). These flow into the effluents and divide the urban centre into stepped areas at different levels.

Fan type: the urban areas are divided and the various collectors carry water to one single collection point.

Terraced type: within urban areas there are specific collectors for upper and lower parts of town; this subdivision is necessary when the former must remove cloacal water and possibly rainwater. The collectors may be linked together by subsidiary canals.

Radial type: the urban centre is divided into areas, each with its own networks and radial collectors, which are may be used for other purposes; this is the most common type of collector used for flatland cities.

In Mohenjo-Daro, in Pakistan, the drainage pipes and gutters of houses dating to the second millennium B.C. are linked to secondary collectors which are, in turn, linked to primary collectors. Studies carried out on ancient cities with sewage systems reveal that the channels follow the street layout: this allows location of the street axes. Adam points out that the city of Timgad (Algeria), once the ancient city of Thamugadi and agricultural centre of Numidia, was built in the year 100 on precise orthogonal axes and that a sewage network was placed under the axis of each street. The sewage system, still present today, has tunnels, which are 0.4 m wide and 0.8-1 m high. Each has a small well, which connected to the main, cardus maximus collector (Adam 1988, pg. 287).

After Rome was captured and partially destroyed by the Gauls in the V century B.C., Livy (Titus Livius) tells us of the harangue held by Camillus, following which, the Romans decided to rebuild their city. It appears that in their haste, the roads were not correctly aligned, and where the old sewers once passed through public land, they now ran under private houses (Livy, V, 55). The sewers of Ancient Rome are generally dated on the basis of their shape and the material used in their construction: «The first sewers were square, vaulted sewers, made of tuff; later concrete structures such as the *Cloaca Maxima*, retained this shape. Although the dimensions of sewers changed with the advent of earthenware materials (tiles and bricks), the roof covering generally consisted of two bricks, hence the name "*cappuccina*" The "covering" and base of the sewer were generally one and a half or two feet long, while the width of the conduits was generally of 45 cm (one foot and a half), 60 cm (two feet) or 120 cm (two x two feet)» (Moccheggiani Carpano 1985, pgs. 177-178).

More recent sewers were made of brick, with semi-circular or segmental arches, concave or trench brick bases and either with or without lateral embankments. Later still, egg-shaped sewers were developed, the minor arch of which, pointed downwards.

VII.7 - *Typology 3: Places of worship*

Structures built for religious purposes and regarded as holy, or natural cavities used for the same purposes.
«On the one hand, we have Man's need to practice his ritual traditions and his creed in meeting places or places of worship; on the other, we are more inclined to believe that the hunter satisfied a need for proprietary or scaramantic magic by 'visualising' the animal he intended to capture on cave walls» (Rossi-Osmida 1974, pg. 16). Be it that certain caves were effectively chosen as places of worship or be it that they became places of worship only subsequently to being used, we now know that many places, considered sacred by many religions, were created within natural cavities.

Whether due to a change in the concept or the basic idea of a cave or through making the most of certain situations rather than being simply subject to their effects or perhaps due to the very nature religion, various man-made structures have been uncovered along rocky flanks and crags. Some of these have even been found in natural rock recesses or underground. In this latter case, there are structures of essentially underground development and others, whose environments and even entire cultural complexes, were cut directly into the rock matrix (fig. VII.17).

Fig. VII.17. *Fresco in the sacred Specus of Subiaco (Rome) (photo C. Ninni).*

VII.7.1 - *Crypt*

The underground sections of a public building, normally a sacred or cemeterial building; in religious architecture, this is the only environment or complex of environments that develops in the area beneath a church.
In architecture, a crypt was originally a vaulted passageway, which was not necessarily underground. In Christian basilicas, the term was later applied to the area under the presbytery, where martyrs' tombs can often be found. The first examples being restricted to the pre-Roman period, with the advent of Roman architecture, the crypt takes on a specific character and becomes a room in itself. Despite their increase in size, crypts are primarily used for burial and ossuary purposes in subsequent periods (please refer to paragraphs VII.8.1, VII.8.2 and VII.9.4). There are many isolated crypts on private property, although these are more frequently found in cemeteries.

Viollet-le-Duc says the following of crypts: «Crute, croute, grotte. L'étymologie de cet mot (cacher) indique assez sa signification. Les premières cryptes ou grottes sacrées ont été taillées dans le roc ou maçonnées sous le sol. pour cacher aux yeux des profanes les tombeaux des martyrs; plus tard, au dessu de ces hypogées vénérés par les premiers crétiens, on leva des chapelles et de vastes églises; puis on établit des cryptes sous les édifices destinés au culte pour y renfermer les corps saints recueillis par la poété des fidèles» (Viollet-le-Duc 1996, pg. 378).

VII.7.2 - *Favissa*

Normally well-shaped environment reserved for votive objects; located on holy ground but outwith the sanctuary.
Possibly of Etruscan origin, the Latin term *favissa* generally referred to well-shaped cavities, used in the storage of votive objects as and when these became excessive in number. Disused objects and images of Gods who were no longer worshipped were also kept here.

VII.7.3 - *Holy well*

Vertical perforation in the ground, considered sacred in nature.
The term holy well refers to a series of man-made structures, with vertical axis, connected to particular places of worship. They sometimes reached a water source, which was considered to be sacred or to have health-giving properties whereas on other occasions they represented the conjunction with Mother Earth or with "underground dimensions". «And there is decisive proof that it was a God who gave human foolishness the following divination: in fact no-one in his right mind ever reaches true and inspired divination, if not when his rational control is hampered by sleep, by an illness or by a type of divine insanity. The rational man must instead reflect and recall that which the divination or his inspired nature tell him in his dreams or in his moments of wakefulness and rationally interpret the significance of the visions in order that he may establish how and in regard to whom they predict a negative or positive, future, past or present omen» (Plato, 71 e).

Well of destiny: at the Terrace of the Exedrae ("Terrazza degli Emicicli"), part of the lower structure of the ancient Paenestre Temple of the Primagenial Fortune in Rome, there is a *tholos* with a holy well, from which it seems that lots were drawn. «The overall depth of the well is 7.5 m, of which the first 5.4 m is lined by a wall of uncertain construction. The inferior section, of approximately 1.10 m, is lined by bevelled blocks of square tuff: the six rows end, higher up, in a band of thinner layers, which was originally the well's opening: there is in fact no doubt that this is part of an older section, which was later extended upwards with a concrete wall when the sanctuary's monumental transformation completely altered previous levels» (Coarelli 1987, pg. 50). Cicero tells of the oak lots (*sortes*), bearing ancient inscriptions in *De Divinatione* (Cicero, II 41, 85-6).

Well temple: this type of well consists of large steps and dates to the Nuragic period. It was utilised in the celebration of particular rites linked to the worship of water. One of the best examples is the Temple well of Santa Cristina at Paulilatino (Oristano), the long stairway of which, leads to the water source (Lilliu 1980, pg. 108).

VII.7.4 - *Mithraeum*

Normally an underground environment where religious ceremonies were held in honour of the Iranian god, Mithra.
In Latin these were known as *spelaeum* (cave) as Mithra was born in a cave and the original ceremony had to be held in a natural cavity. In the urban environment, or in the absence of natural cavities, a subterranean or semi-subterranean environment with no windows or with small skylights replaces the cave.

Normally of an elongated rectangular shape with staggered entrance on the longitudinal axis, worshippers would have gathered on the benches, which lined each wall. A small altar, in the shape of an aedicule, ara or steps, was at the far end and normally depicted the god Mithra as a young man with a Phrygian cap in the act of killing a bull (tauroctony) with a dagger as can be seen at the Mithraeum of St. Clement in Rome (Della Portella 1999, pg. 20 and pg. 43). The animal is the allegorical representation of cosmic and individual salvation. Other elements such as sculptures, bas-reliefs, and niches with frescoes and mosaics can sometimes be found, as can be seen in the Mithraeum of the Seven Gates in Ostia, where the mosaic-art symbologies depict initiation rites. An example of a cave mithraeum can be seen in Duino (Trieste).

VII.7.5 - *Rock hermitage*

Isolated environment generally cut directly into a rocky flank where one or more people retreat to conduct a religious lifestyle.
Environment created within a natural cavity or by adapting an existing underground structure. In many religions, a cave is considered the place of preference by men retreating to become hermits. «This rocky cave in the deserted valley / And my devout devotion / Portend the realisation of my dreams» (Milarepa, pg. 166). Once an underground structure has been visited by a 'sacred' or 'enlightened' person, it normally becomes a place of worship. Special structures can be created within it or it may become part of a larger, architectonic complex, which may also extend outwith the cavity.

Numerous rock-cut structures at the convent of San Cosimato convent in Rome, have been traditionally considered as hermitages; these are predominantly rock tombs and natural cavities, which have been converted into both places of worship and dwellings. St. Benedict of Nursia frequented the site at the beginning of the VI century and became leader of the *cenobium*. When, after a certain period of time, the monks decided that they were no longer going to obey his monastic rules and regulations, he somehow managed to escape being poisoned by them. Today, the Hermitage of St. Benedict is a series of three rock-cut structures, consisting of two small dwellings next to steps that lead to the last environment, which probably originated as a tomb and is now known as the Chapel of St. Benedict (Basilico, Lampugnani 2002, pgs. 82-86).

VII.7.6 - *Rocky place of worship*

Place of worship, partially cut into the rock or situated within a karst cavity.
Cut into the rock face or exploiting horizontal or sub-horizontal natural and artificial cavities, over time many places of worship were created as a testament to the many religious concepts.

The so-called Ajanta Caves (Maharastra, India) are in fact a monasterial Buddhist complex cut into a natural stepped rock terrace. Along the very sides of the ravine, originally consecrated to a Nagaraja or Serpent King, a community of Buddhist monks initiated the excavation of the *chaityas* (sanctuaries) and the *viharas* (monasteries) in the II century B.C. (Rowland 1963, pgs. 5-6). There are 29 cavities in all (five temples and twenty-four monasteries). In "Cave X", thought to be the oldest, there are wall paintings thought to date back to between the I century B.C. and the I century A.D. Sergeant tells us that the *chaityas* generally consist of high, vaulted chambers with circular apses housing rock-carved altars, at their far end. The *viharas* are predominantly quadrangular, central courts flanked on three sides by monastic cells, the façade of which consists of a pillar-lined walkway. The smaller *viharas* are simple colonnades, which open directly onto the cells (Sergeant 1914, pgs. 248-251).

Rock-cut church: building dedicated to Christian worship, cut directly into the rock face, created from adapting a natural cave or build within the cave itself. In the Mottola area in Puglia there are approximately thirty rock-cut churches, which were created following a precise liturgical plan, with the apse facing east; many of the churches still have their original frescoes. The area surrounding Matera also has many churches cut into the sides of ravines. The church of St. Peter in Princibus is situated below the entrenched village of Murgia Timone and is entirely cut into the limestone rock; some graffiti, including a horse, can be seen in a room with barrel vault (Belmonte 2001, pgs. 64-66). The church of St. Salvatore in Serino (Salerno), within which a large face is chiselled into the flowstone, was on the other hand, built inside a cave (D'Alessio 1993, pgs 57-69). In the Cappadocia region of Turkey, the particular geomorphological structure was suitable for the development of rock-cut and underground structures; the remains of hundreds of rock-cut churches can be seen in the area between Gulsehir, Kayseri and Nigde.

It should not be forgotten that many churches, whether or not dedicated to a particular saint, were originally pre-Christian sites, which were subsequently 'demonized' by Christianity. Several cavities were dedicated to St. Michael the Archangel whose teachings were first followed by eastern cultures at the beginning of Christianity and only later diffused to the west by the Byzantines. He was proclaimed by the Longobards as patron saint of their militia. The worship of saints is associated to mountainous areas and areas of numerous caves, in memory of the legend where Michael defeats Lucifer and banishes him to a cave (Lucrezi Berti 1973, pgs. 185-194). «Even further north, there are many rock-cut sanctuaries, dedicated to the archangel, as attested in Latium by the cave of St. Michael of Monte Tancia. The first known references date back to the VIII century, when Duke Ildebrando of Spoleto gave the caves structures to the Abbot of Fafa Probato» (Righetti 1987, pg. 490).

VII.7.7 - *Underground hermitage*

Isolated environment, generally cut directly into a rocky flank where one or more people retreat to conduct a religious life.
Similar to the rock hermitage, although not as common, the underground hermitage serves the same functions and can be either of the following: a natural cavity, an underground excavation or the adaptation of a pre-existing structure. For example, in Finale Ligure (Savona), the floor of the small Romanic Church of late antiquity *castrum* of St. Anthony of Perti, conceals the entrance of a natural cavity, which, first bypasses two small pools then leads onto a small room, which is partially lined with formations; on the right hand side, a flowstone with tiny pools appears artificially worn, so much so that its centre is almost smooth and a small pool, into which the drips collect, has been roughly carved in the flowstone 'mass'. As recently as the beginning of the 1980s, it was thought that a hermit once slept there.

VII.7.8 - *Underground place of worship*

Underground religious structure.
Various religions met in primarily underground environments; others simply found the underground environment more suitable and safer from a contingent point of view. It is not always possible to know what the celebrations of these particular religions entailed (fig. VII.18). In Sardinia there are several caves which were used for religious purposes: «The *Sa Domu and s'Orku-Urzulèi* (the ogre's lair) cave in the Olgiastra region was also considered sacred. Here long ago and in different times, three bronze figures were uncovered, a metal fusion cast and a high quality, intensely psychological paleosardinian plastic statuette known as "Dead man's mother", which others identify as a sort of "Nuragic Madonna", or symbol of "Mercy", cradling her dead son (Lilliu 1980, pg. 110).

In Rome in 1917, an underground structure, now known as the "Basilica Neopitagorica" was uncovered, although its original purpose remains unknown. The monument dates back to the I century A.D. and consists of a long, underground *dromos*, which leads to a basilical system with three naves and an end apse. The basilical structure became important in sacred Christian architecture from the IV century onwards. The floor is decorated with black and white tessera mosaic, while the walls and the stucco vaults have numerous insets with stucco representations of mythical scenes (Pavia 1999, pg. 114). The "*Dolcenum*", the place of worship dedicated to Giove Dolicheno, is also in Rome and is situated on the Aventino Hill. Originally a roofless, open-air building, it was modified and buried together with other public and private buildings following the city's decline.

The Santa Maria in Stelle (Verona) parish church has an underground section dating back to between the I and III centuries. This section is «linked to a spring in the hill behind the church by means of a vaulted channel, which is approximately 75 m long, 1.63 m high and 0.75 m wide and is interrupted by three settling wells» (Forlati 1962, pg. 245). It consists of an entrance hall (with robed statue) and tunnel leading to a quadrangular atrium and a large arch, before opening onto two semi-elliptic cells, on opposite sides, which receive light from two openings in the ceiling (now blocked). Both cells retain the original frescoes, one probably dating back to the III century, while the others depict more recent, Christian scenes; the cell on the right-hand side has a mosaic floor. «From the water and the inscription bearing the names of the people who built the structure, which appears on the architrave of the passage that channels water to the entrance, it is clear that this was originally a pagan temple. The inscription bears the following: *P. Pomponius Cornelianus et Julia Magia cum / Iuliano et Magiano filiis a solo fecerunt*» (Forlati 1962, pg. 254).

Underground church: dedicated to Christian worship, the church may have been built or created underground. For instance, the Sotterra church in Paola, Calabria, is entirely underground. Dating to around the IX-X century, its pronaos, nave with two lateral niches, iconostasis, presbytery and apse depict Christ with the Apostles. It is currently accessed via the Carmine Church, which lies above it (Verducci 1991). The St. Domenic Church in Narni (Terni), built in around the XII century and now site of the Municipal Library, has a single-nave underground church with a

Fig. VII.18. *«Chapelle Saint-Antoine, dans les salines de Wieliezka» (Badin 1876, pg. 157).*

semi-circular apse taken from Roman structures. The environment has several frescoes by different people over various historical periods (Nini 1997, pgs. 345-347). The underground Wieliezka Church in Poland, excavated within the salt mines, is also worthy of notice: «Une des curiosités de ces immenses èscavations, qui réfléchissent de tous côtés comme le cristal la clarté des lampes et des torches, est la chapelle Saint-Antoine, située au premier étage: cette chapelle est creusée dans la mine même et ne se compose que de sel; l'autel, les statues, les colonnes, le chaire, les ornements, tout est en sel» (Badin 1876, pg. 160).

Buildings cut and carved into the rock, are also considered underground places of worship; some striking examples are the monolithic Ethiopian churches of Lalibela. Nicolletti sub-divides these religious buildings, which he calls monoliths, into the following categories: monoliths, cave monoliths, semi-rock-cut monoliths, hypogeal and rock-cut environments and cave buildings (Nicoletti 1980, pgs. 331-332). The underground chambers of the Sphynx in El-Giza, Egypt should also be categorised under religious underground environments. Although their original purpose is unclear, they were most probably sacred in nature, given the monument under which they are located.

VII.8 - *Typology 4: funerary structures*

Structures or architectonic complexes for the storage of the human remains (corpses, bones, ashes) of one or more persons.

Throughout all eras, almost every civilisation felt the need to bury their dead. Each civilisation had its own 'burial system' depending on their particular religion and thus created primarily underground structures for the storage of the mortal remains. In view of the sheer numbers, this is the most common type of artificial cavity and it could be said that almost no anthropised place in the world does not have burials or funerary buildings.

Burials range from simple pits lined with stone layers through to monumental buildings such as the Pyramid of Giza in Egypt and the Pyramid of the Inscriptions at Palenque in Mexico or the megalithic tombs found in various areas of the Earth. There are many different types of sometimes complex architecture. Each one reflects a belief, a desire, a tradition, the available funding, the need to leave a mark, acquired technology and compliance with the law.

VII.8.1 - *Catacomb*

Series of underground tunnels and environments, sometimes on superimposed floors, used primarily as cemeteries by the ancient Christians and later as funerary places of worship.

In archaeological terminology, the term 'catacomb' refers to an underground Christian cemetery consisting of a structured system of passages, tunnels and cubicles, which are sometimes situated on different levels and are used for burial purposes. The burial sites are normally carved into the walls of the corridors themselves. The term may derive from a IV century toponym relating to a depression in the ground presenting large cavities, which was created by a stone quarry located at the IV mile of the Appian Way. Only during the Middle Ages (circa IX - X century) would paleo-Christian cemeteries, originally known by the generic name of *cryptae* be called catacombs.

«The policy of creating underground structures for funerary use was certainly not invented by Rome's first Christian communities: it was well diffused, as we know, throughout many civilisations and cultures in the ancient world, especially where the type of soil was easy to excavate and gave reliable 'support' to the underground structures. Still in the geographic context of Rome and Latium, medium to large hypogeal sepulchres were created by the Etruscans, the Sabines and by the very Romans. In this geographic area, underground burial was greatly facilitated by the excellent local tuff, which was easy to work and statically reliable» (Fiocchi Nicolai, Bisconti, Mazzoleni 1998, pg. 15).

Collective underground burial areas were originally used by all, including Christians. Only subsequently and very probably following improved organisation on the part of the Christians, were separate burial grounds created for the followers of St. Paul, in the name of Christ. This was also due to the demographic increase and the consequent need for additional cemetery space. The existence of underground cemeteries in the Roman Christian community dates back to between the end of the II century and the beginning of the IV century (Fiocchi Nicolai, Bisconti, Mazzoleni 1998, pg. 16. Pergola 1997, pg. 21). The burials of the so-called martyrs, bishops, popes and later the burials of those who were proclaimed 'saints', contribute to the increasing importance attributed to the catacombs, which became objects of devotion and were subject to significant monumental interventions.

Primarily the first part of the IV century, saw an increase in underground burials with the introduction of vast burial areas of varying planimetry, connecting stairwells, ventilation shafts and decorative frescoes and inscriptions. On the surface, Basilicas and reception areas for the devotees were built and it cannot be denied that the pilgrimages contributed to the global development of cemetery complexes. The tradition of catacomb burials came to an end in the V century and it was in this as well as the VI century that devotees would visit catacombs for primarily devotional purposes. «In relation to the planimetry of underground structures, two wide categories have been identified and there are also intermediate typologies: "closed" hypogea for a predetermined number of burials and which at the time of creation did not envisage future enlargement and "open" hypogea, created with several tens of tombs and with the possibility of further extending the initial tunnels or of creating new ones along the main axes» (Pergola 1997, pg. 60). To some extent, catacomb excavation originated from underground quarries and disused hydraulic systems such as cisterns and aqueducts. We owe the systematic study of catacombs to Antonio Bosio (1575-1629), through which Roman Christian Archaeology began.

The characteristic elements of the catacombs follow fixed criteria with some differences and diverse architectonic solutions appearing over time. The creation of an underground environment envisages an access ramp and this normally takes the form of either a rock-cut or masonry stairwell. The important aspect is that the tunnel be more or less straight with lateral branches. The tunnel often has *loculi* in its walls, one above the other and one next to the other, sometimes covering entire walls. A *loculo* can hold just one deceased (*loculo monosomo*), or two (*bisomo*) or three (*trisomo*), etc., and was sealed with marble or brick slabs or simple flat roof tiles. There are also the *cubicula*, their name deriving from the Roman bed chamber, which are quadrangular in shape and may be richly frescoed. An *arcosolio*, a is often to be found in the *cubicola*. Skylights are ventilation shafts through which some light may filter; initially these would also have been used for the evacuation of excavated material.

The catacombs of Larderia (IV-V century) in Cava d'Ispica (Ragusa), have both *loculi* cut into the walls and tombs set into the floor of the very corridors (Ardito 2003, pgs. 220-223). The catacomb of Santa Cristina in Bolsena (Viterbo) appears to date back to the beginning of the IV century and has been an accessible place of worship since the Middle Ages. As well as Roman tombs there is a chapel dedicated to St. Michael and a large underground room with floor tombs aptly named the "Longobard Sepulchre" (Carletti, Fiocchi Nicolai 1989, pgs. 15-24).

VII.8.2 - *Cemetery*

Burial site for both the tumulation and inhumation of the dead.
Generally consisting of pits excavated in the ground and sealed with stone slabs or simple mounds of earth, the cemetery could also contain chapels and monuments under which, chambers containing corpses wrapped in sheets or in wooden boxes or sarcophaguses could be found. For Christian burials, the term cemetery normally applies. Originally, the word sarcophagus referred to the limestone rock, which quickly consumed the corpses and only later took on the meaning of sepulchral stone, wood, earthenware or metal and generally monumental urn.

Crypt: this term refers to the sometimes monumental tombs underneath funerary chapels. This type of structure can be found in both cemeteries and private properties and is normally located in parks (please refer to paragraphs VII.7.1, VII.8.1 and VII.9.4).

VII.8.3 - *Columbarium*

A columbarium was used for the storage of cinerary urns.
From the Latin *columbarium* (meaning dovecote and group of sepulchres), the term indicates a type of structure for the storage of cinerary urns, the walls of which contain multiple rows of overlapping niches, reminiscent in appearance to a dovecote. The niches are either circular or quadrangular in shape and cinerary urns were bricked-in; they were used between the I century B.C. and the II century A.D. In modern-day cemeteries, the columbarium is a system of masonry compartments, on different levels, for temporary and permanent private burials.

VII.8.4 - *Domus de janas*

Type of Sardinian burial tomb in the form of a chamber and normally cut into the rock.
The *domus de janas* (House of the Fairies), is a type of ancient Sardinian burial. Developed in the Middle Neolithic (V millennium B.C. circa) and used until at least the First Bronze age, these underground structures were cut into any type of rock. Those in limestone and sandstone rock have an extended planimetry, while those in trachytic tufa are, although large, smaller than those excavated in granite and basalt (Atzeni 1985, pgs. 33-41. Lo Schiavo 1996, pg. 190). The tombs may be isolated or, as is more common, grouped together and consist of one or more communicating cells where bodies were placed in the foetal position surrounded by funerary gifts. Just over one thousand remain.

VII.8.5 - *Foiba*

Type of sinkhole into which, both the living and the dead were thrown.
Foibe, presenting a sinkhole at their base, are typical of Istrian karst. Although not strictly a type of burial, the *foiba* is linked to the civil and military excesses, which took place in Istria and on a lesser scale in the Triestine Karst from 8 September 1943 until shortly after the end of the Second World War. Many *foibe* still contain the remains of bodies.

The *Foiba* of Basovizza, near Trieste, is not a natural cavity, but a mining shaft, which was used after the First World War for the disposal of war material and again at the end of the Second World War in the slaughter of both civilians and soldiers. Sealed by a large memorial plaque, it is now a national monument.

VII.8.6 - *Morgue*

Particular collective burial structure located within a fortification.
When inside a besieged fortress, consideration must be given to the removal of corpses. This is particularly true during the summer months, as corpses can give rise to pestilence or at any rate spread disease in the operating garrison. Special underground rooms, aptly known as morgue (equivalent to the Italian term *obitorio*) were built within some XVIII century Savoia fortresses. The corpses of fallen soldiers were placed into these generally well-shaped structures equipped with special chutes to facilitate the introduction of the corpses. These large mortuaries have vents to extract decomposition gases.

VII.8.7 - *Necropolis*

Pre-Christian burial ground.
The term necropolis was used to refer to the underground sepulchres of Alexandria of Egypt. In archaeological terms it is reminiscent of pre-Christian burial grounds rather than Christian burials. «Of ancient burial grounds, those of the Etruscans were among the largest. As in Greece and other parts of Italy, the dead were buried outside but close to the city, on nearby slopes or in neighbouring plains» (Prayon 2000, pg. 335).

Cemetery: this term generally refers to Christian burial grounds; also known as graveyard. Its purpose is to serve as a burial site for both the tumulation and inhumation of the dead.

VII.8.8 - *Ossuary*

Its purpose is to collect and store the human remains, following their exhumation.
Sometimes subterranean or semi-subterranean environment or building, in which exhumed human remains or remains recovered from the ground following war or natural causes, are collectively stored. The structure may be either part of a cemetery or independent and may even take the form of a monument.

Memorial: this is a chapel-shaped structure or monument, which generally contains the remains and memories of those who died during war. The military memorial at Timau (Udine) is the 1937 reconstruction of the Crucifix Sanctuary, erected in 1284 on the ancient Monte Croce Carnico pass. It was designed by the architect Giovanni Greppi and by the sculptor Giannino Castiglione. «The Memorial holds the remains of 1,644 soldiers who died in the 1915, 1916 and 1917 war campaign, of which, 232 remain unknown. The remains were removed from various war cemeteries in the Upper But area. Internal and external (under the colonnades) *loculi* were created in the arches of the Temple and the names of the fallen were engraved on a bronze plaque» (Ministry of Defence 1977, pg. 31).

VII.8.9 - *Tomb*

Any type of burial place for the storage for human remains.
Generic term indicating any type of burial place for human remains (ashes, bones or corpses), of a specific type: rock-cut or underground tombs, well, tunnel, pit, chamber, *loculi* (*Columbarium*), cube aedicule, hut, house, soil, stone circle, raised tombs etc. In various civilisations, burials can point more or less in any of the cardinal points, depending to popular beliefs relating to the location of the afterworld (fig. VII.19). Three '*cappuccina*' tombs constructed from flat roof tiles were uncovered under the central nave of the church of San Salvatore in Brescia. «They had unfortunately been violated and their walls were decorated with white weaving atop different coloured

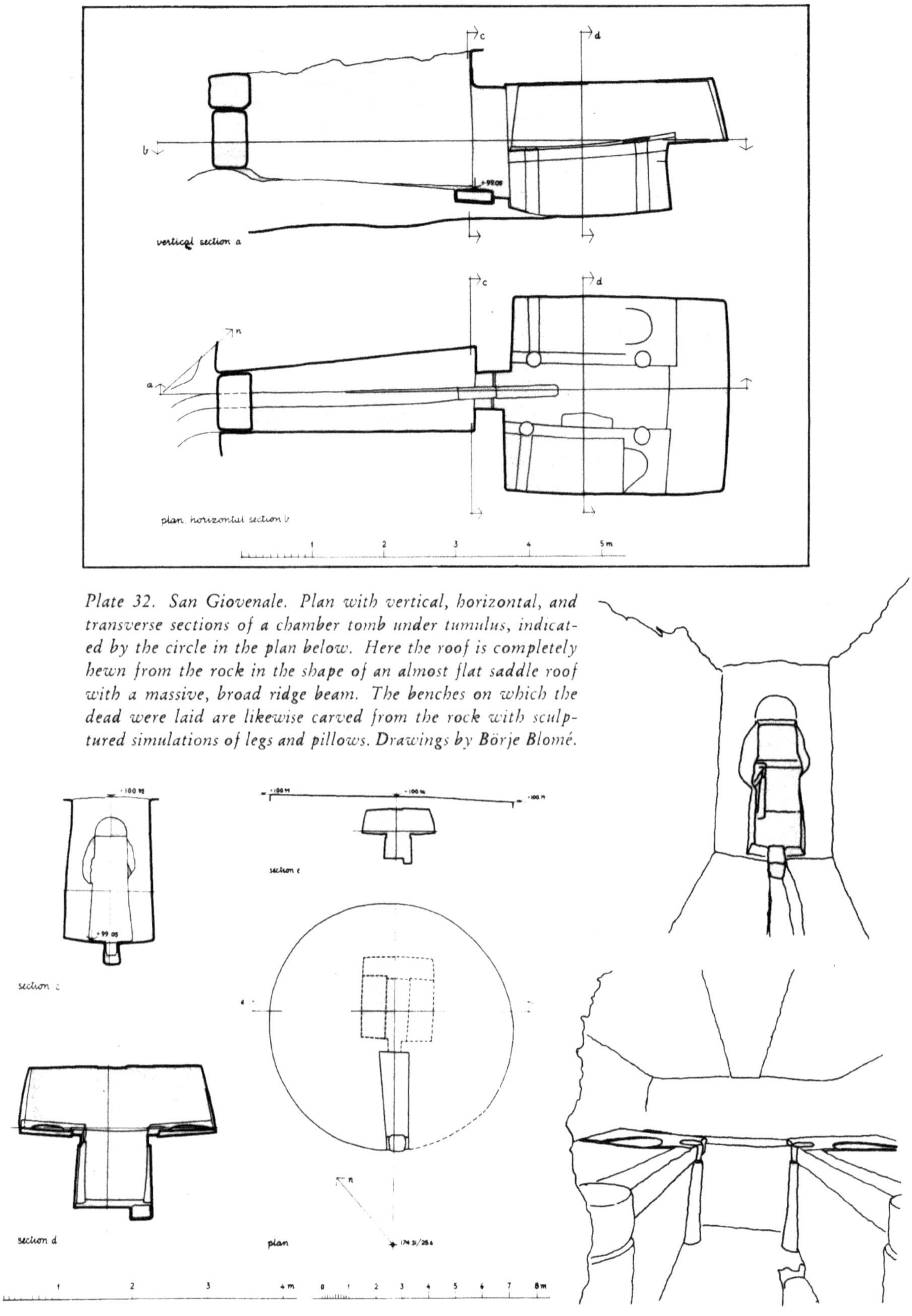

Fig. VII.19. *«Plate 32. San Giovenale. Plan with vertical, horizontal, and transverse section of a chamber tomb under tumulus, indicated by the circle in the plan below. Here the roof is completely hewn form the rock in the shape of an almost flat saddle roof with a massive, broad ridge beam. The benches on which the dead were laid are likewise carved from the rock with sculptured simulations of legs and pillows. Drawings by Börje Blomé» (Welin 1962, pg. 285).*

fasciae. Decorated tombs from the IV century to the XII century have been uncovered in Lombardy; the weave motif was common throughout these centuries therefore the tombs may even be Roman in origin» (Panazza 1962, pg. 63).

As previously mentioned, there are various types of tomb, among which:
chamber tomb with dromos: with an open entrance passage, leading to the underground chamber;
chamber tomb with machicolation entrance: with an entrance shaft and underground chamber;
corridor tomb with sepulchral recess: the entrance to this type of tomb is cut into the side of a sunken corridor and leads to the *loculo*;
Montanaro type tomb with double niche: with an access well and two sepulchral recesses;
Narce type, niche tomb: with a well access and sepulchral niche;
thólos tomb: type of corridor tomb with an underground passage known as a *dròmos*, its sepulchral chamber has a cupola made of protruding layers or segments and sometimes has a central pillar.

Mastaba: Egyptian funerary structure for officials and ministers; in its most basic form it is a tronco-conical pyramid-shaped building, with a rectangular base. It has a false door and its summit opens onto a well, which vertically intercepts it and leads to an underground chamber, sometimes 15-20 m deep.

Mausoleum: this is a large sepulchral structure of which, one example is the *underground chamber mausoleum*. The name derives from the Mausoleum of Halicarnassus, a sepulchral monument erected by Artemisia II in memory of her husband and for her own burial.

Temple tomb: situated in the territory of modern day Jordan, between the Dead Sea and the Red Sea, the city of Perta, known to Arabs as Wadi Musa (Moses Valley), is know primarily for its red and ochre monumental tombs, cut into the sandstone rock and which characterise the entire complex. The area has been inhabited since the Palaeolithic period and gained importance following Nabatean settlement in the area around the IV century B.C. Of the various types of tomb, the most important is the *temple tomb*, with its split-level façade; the different types of sepulchral chambers are then cut into the rock.

Tumulus tomb: also known as *burial mound*, this is a type of tomb used by some ancient civilisations. It can consist of a stone corridor and a chamber, built with large slabs and covered by soil or stone rubble; it can be circular, conical or dome shaped and be of monumental dimension, such as the Newgrange and Knowth tumulus in the country of Meath in Ireland (Eogan 1986, pgs. 11-29). Numerous examples of tumulus tombs have been provided by the Etruscans, both surface tombs and ones cut into the rock with an overlying layer of soil; among the most interesting, are those of the Banditaccia Necropolis in Cerveteri (Viterbo), the first phase of which, dates back to the VII century B.C. (Cristofani 1991, pgs. 45-72). Several tombs in Tarquinia (fig. VII.20), cut into the carbonate rock (Tarquninian Macco, or fossiliferous calcarenite) retain their original artwork: «Of great ancient art, only Etruscan and Italic artwork remains in abundance in its original form. This is due to the fact that, in the Italic world and particularly in Tarquinia, art decorated not only temples and public and private buildings but also decorated sepulchral chambers in order to recreate – in a purely symbolic sense – funeral ceremonies or scenes depicting the most significant aspects of life, around the dead» (Moretti 1974, pg. 9).

Several studies have been conducted in Italy in relation to the funerary gifts found in tombs and the decline of this tradition from around the IV century onwards (Gastaldo 1998, pgs. 15-33). Another interesting point is that the Roman tradition, and before that, the Etruscan tradition of separating the living and the dead into two distinct areas, was progressively disregarded and *intra muros* burials became increasingly popular. We have witnessed other phenomena, such as martyral basilicas becoming monastic sites, in turn giving rise to urban centres: «The resting places of the dead do not intrude on those of the living; indeed, these large religious and cemeterial complexes are the foundation of the inhabited areas, which in this very period began to resemble independent villages» (Giuntella 1998, pg. 62).

VII.9 - *Typology 5: civil structures*

Environments or structures for various purposes, strictly connected to the everyday life, finances and social life of human communities.

Having gained inner knowledge, Man began to modify the environment he lived in by building rock shelters, huts, houses, palaces, factories and roadways, sometimes using bold architecture. All of this, in his search for structural and cultural stability. Public construction often exploited the underground, creating both underground and semi-subterranean structures using criteria still used today, although today different types of material and machinery are

Fig. VII.20. *Tarquinia (Viterbo): Tomb of the* Aninas *- No. 5051, III - II century B.C. Showing* Charun *(photo G. Padovan).*

used in large-scale construction (fig. VII.21). Such structures were used, and are still used, in everyday life and served to develop an underground solution for community life and to resolve viability problems.

With the advent and development of the first structured communities, it is likely that the search for spaces in rocky crags and underground would have been simultaneous. Nicoletti underlines how «cave architecture imposes linguistic connotations on nature; geological formations and terrain viability are selected as already falling under architecture and thus become part of a project» (Nicoletti 1980, pg. 63). Dictated by practicality, by the impossibility of obtaining better solutions for the technological level of the time or through having reached an excellent system, integrated with the surrounding eco-system, Man created a wide range of "negative architecture", primarily through the removal of material. Such architecture does not necessarily require the use of wood, the extraction of stone or the manufacture of bricks; where this may be interpreted as "poor" architecture it is equally true that the same brick can create both architectural masterpieces and squalid suburbs. There are some examples where organisational peaks, in the broadest sense of the term, were reached. Peaks that is almost inconceivable from a modern viewpoint, which required different, if not superior pace, awareness and education. Civil structures built on the surface, but which are now underground are also included in this typology.

For example, in Libya, Chiauzzi came across five different types of underground dwelling units: «natural caves, other semi-natural shelters, wall hypogea, well hypogea and also ditch hypogea» (Chiauzzi 2003, pg. 171). Court hypogea, in particular, provided better shelter from the variable temperatures and were weather-proof. These were preferred to brick, surface dwellings, however not everyone could afford this type of dwelling. Excavation was expensive and many cubic metres of material had to be removed.

VII.9.1 - *Artificial grotto*

A structure resembling a natural cave, generally erected in public gardens for ornamental purposes and as a backdrop for statues, fountains and water features.
The use or artificial grottoes was established during the Renaissance period and is still popular today in the adornment of public gardens, avenues and thickets. The Great Grotto of Boboli Gardens in Florence, designed by Bernardo Buontalenti is worthy of notice. «With the Great Grotto, an articulated 'mysterious cave' at the end of the Vasarian Corridor in the Pitti area, Buontalenti reproposes the central structure theme, but with a particular naturalistic slant. Under his leadership, construction began in August 1583. The lower part of the façade was built following Vasari's design, between 1556 and 1560; between 1583 and 1585, Pietro di Tommaso Matti designed the first chamber; in 1585, Michaelangelo's 'Prisoners' were exhibited here and from 1586-87, Bernardino Pochetti carried out the frescoes; in 1587, Giovanni del Tadda continues decoration of the façade and Vincenzo de' Rossi positioned the statue group of Paride and Elena» (Fara 1988, pg. 196).

At the turn of the Nineteenth century, Ercole Silva talks of natural caves in the introduction to his treatise on the construction of the artificial grottoes in gardens: «Caverns are large, empty cavities with precipices that descend to the bowels of the earth. They are formed by erosion, in much the same way that abysses and chasms are formed by the eruption of volcanoes, by the action of water, of underground vapour and of earthquakes. They are not suitable for the gardens» (Silva 2002, pg. 157). Silva has been described as «the true theorist of the new art of landscape gardening in Italy» (Nenci 2002, pg. V). He believes that rocks, covered in moss and vegetation, could be used to soften the construction of such grottoes: «Above all, an artificial grotto must be situated in such a location as we are accustomed to seeing it in nature; it should rest on a hill, a cliff or be set among craggy rocks and streams, in deep, dark cavities. There is nothing less natural than a man-made grotto situated on a plain or in an open, isolated or badly positioned site, where it immediately catches the eye» (Silva 2002, pgs. 159-160).

VII.9.2 - *Butto*

Normally open-air waste disposal pit, sometimes found in natural or artificial caves.
A relatively deep, pit or well-shaped excavation, the *butto* was used for the storage of solid inorganic waste. It could be an unlined structure, cut into the rock or could exploit pre-existing cavities such as disused wells and cisterns or natural fissures in the rock, which may even have been widened and adapted. *Butti* normally contain important information and can turn out to be interesting structures. At Montelupo (Florence), a large quantity of XIV-XVI ceramics, were uncovered in an underground room used by ceramics factories as waste disposal pits. These ceramics now constitute the main part of the local "Archaeological and ceramics museum".

Fig. VII.21. *Road in cutting (via cava) near Pittigliano (Grosseto) (photo G. Padovan).*

VII.9.3 - *Cellar*

Partially or entirely underground room generally used for the storage of wine.
A cellar must have a low, constant temperature. This is achieved by creating partial or full rock cellars or by building an underground or semi-subterranean room with thick masonry walls and ceiling for good thermal insulation. Sufficient ventilation will be provided by small, suitably placed openings, sometimes of the “gola di lupo” type. Layers of waterproofing material and dry stone drains can be used to prevent water infiltration and humidity.

The room is used in the storage of wine, beer and distilled alcoholic beverages. By extension, the term also refers to a basement, or the lower floor of a building, which is either fully or partially underground and used as a warehouse or store-room for foodstuffs. Some cellars may have a well and an opening in the ceiling in correspondence to the well’s opening providing direct access to the above floor (fig. VII.22). There may also be a *butto*, a soakaway or a sewage system for the discharge of water, including water used in the maintenance of the room itself or of the equipment and machinery contained within the room.

VII.9.4 - *Crypt*

Originally a covered subterranean or semi-subterranean passage.
This was originally a simple covered passage, which was not necessarily underground. Generically speaking, the term *crypt* now refers to the underground sections of a public building used primarily for holy or cemeterial purposes, but also used for civil purposes, which relate to the purpose of the building itself. The same term also referred to the primitive catacombs and later to the subterranean or semi-subterranean section, directly beneath the presbytery of Christian basilicas, which often holds a martyr's tomb (please refer to paragraphs VI.7.1, VI.8.1 and VI.8.2).

VII.9.5 - *Cryptoportico*

Covered portico, with masonry arch, normally illuminated by slits in the side of the arch and used in Roman architecture. Cryptoporticos are longitudinal, subterranean or semi-subterranean architectonic structures, which are illuminated by small openings of the “gola di lupo” type. These were covered passages leading to the various parts of a building or connecting two separate building. These were present in both public and private buildings and were in use from the Late-Republican period to the end of the Imperial Period. «An important functional ‘curiosity’ of cryptoporticos was that they had hidden entrances, where two branches met, with entrance steps, which were both low in number and small in dimension. These were used to ensure a constantly low temperature within the passages» (Luciani 1984, pgs. 148-149). Cryptoporticos worthy of mention are those at Villa Adriana in Tivoli and those in Aosta and Vicenza. The cryptoportico in Arles in France dates back to between the end of the I century B.C. and the beginning of the I century A.D., while the *cryptoportico forense* in Smirne, dates back to the II century A.D.

In Rome, the Palatine cryptoportico linked the Domus Aurea to the palaces of Augustus, Caligula and Tiberius. Its three sections combine to form a divisory wall: «The principal and best conserved section, 109.1 m long with a constant width of 4.1 m and an arch of variable height, ranging between 3.05 m at the base and 4.4 m at the summit of the first section and 2.75 m and 4.1 m respectively in the second section (...). A brick structure with a thick coating and barrel vault, the only apertures are on the wall facing the *Fontana dei platani* (Plane Fountain) and the *ninfeo de li spechi*. There are eighteen ‘*gola di lupo*’ openings, four of which are closed (one is in between two protruding pillars) and fourteen of which are open; their openings are of 1.4 m and they are positioned approximately 3 m one from the other» (Luciani 1984, pg. 150).

VII.9.6 - *Dovecot*

Structure intended to house pigeons or doves, sometimes cut into rocky flanks and, or more rarely, created underground.
Also known as *dovecot* or *columbarium*, this structure built to house pigeons or doves generally consists of many small niches cut into the walls, each one containing a nest. Some types of columbarium may be cut into rocky flanks or created in underground environments or even in pre-existing underground structures, such as the Torre Pinta Hypogeum. The Cave of the Bandits in Tarquinia is a rock columbarium, which has been re-utilised as a pigeon farm and for the collection of guano to be used as fertilizer; a large window was built in the chamber’s external wall for illumination and ventilation purposes and to allow the birds access.

Fig. VII.22. *Sandstone cellar in Ceva (Cuneo) (photo G. Padovan).*

VII.9.7 - *Granary pit*

An underground area, generally used for the storage of grain.
A granary pit is normally a hemispherical, coated rock excavation; when created in the ground, it must be kept dry with a masonry coating. Its access is air-tight and due to the lack of ventilation, the carbon dioxide released from the cereal creates a protective environment. For example, near Pitigliano (Grosseto) and in Tarquinia (Viterbo) there are vast, egg-shaped ditches, which due their positioning and lack of supply channels are assumed to be granaries.

Food preservation areas were created underground using the same isolation criteria; in Sirolo, the so-called *grain pits* or *underground silos* are still intact. In the Romagna region (Italy), such silos are known as *granili* and just like the examples in Sirolo, their truncated cone is either cut into sandstone or marl or made with bricks; chutes with trapdoor closure link the *granili* to the surface (Recanatini 1997, pg. 180). In some regions of Italy, apples were stored in underground ditches and covered with hay until the latter part of the last century; the ditches were sometimes line with clay.

VII.9.8 - *Jails*

Place where people deprived of personal freedom are confined.
Building, environment or series of environments for the confinement of people awaiting sentencing or serving a sentence. Jails may be underground or may be created in the lower floors of structures used for other purposes. In Narni (Terni), at a time yet to be established, the large room under the apse of the Church of St. Domenic was separated and transformed into a detention centre; the cell graffiti includes two dates: 1759, carved under a monogram and 1809 (Nini 1997, pgs. 347-348). In everyday language the term *prison* if more common.

VII.9.9 - *Mushroom cultivation*

A structure for the cultivation of edible mushrooms, normally contained within a cave or tunnel.
Undergrounds structure, cave or semi-subterranean room for the cultivation of mushrooms for commercial use; these are often natural or artificial cavities such as disused mines or quarries, which have been adapted for this purpose. Edible mushrooms are cultivated on beds of manure and inoculated with mushroom spores from pure cultures. Temperature, humidity and ventilation require careful attention.

VII.9.10 - *Nymphaeum*

A building with particular architectonic features, containing a fountain.
A nymphaeum was originally a sanctuary consecrated to the nymphs. In Hellenistic and Roman times, a nymphaeum was a rectangular, circular, elliptic or often apse-shaped structure with niches, columns and a fountain. In Renaissance and Baroque times, nymphaea were also created in semi-subterranean or underground rooms in the gardens of villas. They would contain a fountain and often reproduced the inside of a cave.

«This representation of the caves consecrated to the Nymphs evokes that atmosphere of charm, which the Greeks so generously lavished on all things. These sacred places inspired faith without fear. Not yet incorporated in gardens, which in that country never deviated from the mainstream layout, they instead constituted separate structures. Their collocation alongside lakes, rivers, mountains and forests made nymphaea picturesque and poetic buildings» (Silva 2002, pg. 158).

VII.9.11 - *Pedestrian tunnel*

A constant and/or variable section structure, through which the continuity of a journey is ensured..
Pedestrian routes can consist of sections that are cut into the rock and sections that are cut and covered; where the tunnel is small, it is sometimes known as an *advancement passage.* Pedestrian tunnels can be found both in and outwith urban areas. These may be for private or public use (fig. VII.23). One particular type of pedestrian tunnel, for hunting use, was uncovered at an abandoned villa near Rota Imagna (Bergamo): outside the building is the entrance to a tunnel with several branches, leading to small bird shooting posts. The structure was excavated in alluvial deposits and is unlined.

Fig. VII.23. Aquae Olle *Hypogeal Stairway (CA 02066 LA RM) in San Cosimato (Rome). The tunnel currently leads to both the Hermitage of St. Benedict, carved into the rock face from a hypogeal tomb and to the subterranean section of aqua Claudia. The roadway was undoubtedly made in Roman times and served the aqueducts, which in this section extend beyond the gorge of the River Aniene and are carved into the overhanging travertine wall (photo G. Padovan).*

VII.9.12 - *Powder magazine*

Room, or rooms, for the storage of powder, artifices and explosives in general.
Rooms for the storage of explosive material are required for civil as well as military purposes. The location, construction and the distribution of munitions and explosives are regulated by stringent laws, which divide explosives into groups according to the way they react in the event of combustion or explosion. A powder magazine can be both an underground or semi-subterranean structure. For particular explosives, it may also consist of small barracks separated one from the other by strips of land.

Mining plants include a storage area for material used in the charging of mines. Where this area is situated underground, it is accessed by a single, straight corridor with one or two right-angled curves. Such curves serve to dampen the effects of accidental explosion within the tunnel. Explosives can also be kept in small, open-air magazines, so long as these are enclosed in a grilled, metal structure to act as a "Faraday cage"; the same method was used by military gunpowder magazines in the XIX and XX centuries.

VII.9.13 - *Railway tunnel*

A constant-section excavation ensuring the continuity of a railway track.
The excavation of a railway tunnel allows mountain ridges and reliefs to be crossed without the need for complicated bypasses. A tunnel can be straight or curved and have slightly inclined tracks and a wide angled curve. The excavation of long horizontal sections is normally avoided to eliminate water stagnation; the tracks always follow a slight inclination.

There are several types of tunnel:
ramp tunnel, used where entrances are situated at different altitudes;
helical tunnel, the same as the previous but used in mountainous areas where there are steep slopes;
overflow tunnel, when the entrances are at roughly the same altitude, the tunnel is excavated from each end, following a slight upwards slope in order that water can be easily discharged from each section.

As with the other types of tunnel, appropriate equipment should be used to plot the tunnel's axis on the ground (external marking). Once the exact position and altitude of the entrances has been established, advancement can take place from either end, following the internal plotting in order to correct any slope or direction errors. Before the advent of modern machinery, *an advancement passage* was excavated and equipped with temporary structures before the actual tunnel was created.

The constant-section tunnel was coated in masonry and more recently with reinforced concrete and even pre-fabricated roof supports. Railways were originally known as railtracks that is roads having tracks: a decisive element for both its concept and future development. The concept was developed in mines. Agricola's "*De re metallica*" already refers to a mining wagon, the front wheels of which were smaller than the rear and which had a metal pivot fixed vertically to the front of it, of roughly the same height as the rear wheels. The transport wagon was hand-pushed along the timber tracks and the pivot was used for steering (Agricola 1621, VI, pg. 113). More than a century after the treatise was first published, Della Fratta published "*Pratica minerale*", which refers to an identical wagon and describes its use as follows: two parallel planks, one clearly anchored to the ground by means of a structure, are placed in the mineral clearance tunnel for the wagon to slide over (Della Fratta 1678, pg.19). «It is surrounded by metal sheeting (the wagon; n.e.s.) and by strong, nailed bars having iron pegs at their base, around the heads onto which wooden wheels are attached, thus ensuring that these are secure; the smaller wheels are placed at the front, a circular peg between them; when the abovementioned wagon is driven or pushed through the channel or gap, the circular peg remains between the aforementioned planks on the mine's floor. This keeps the wagon above the ground, allows water to flow in the channel and ensures that the wagon is straight» (Della Fratta 1678, pg. 19).

In England, the Woodhead tunnel, on the Manchester-Sheffield line, was built in 1839-45; 6.3 km long, it has five ventilation shafts, which are up to 180 m. deep. The Fréjus railway tunnel (also known as the Mont Cenis Tunnel), was built between 1857 and 1870 and officially opened in 1871. It is a single section, double track tunnel and it is 13636 m long. «The construction of the Mont Cenis tunnel was complex, requiring much research on the part of the best technicians of the time. For example, in order to avoid having to extract water, the two tunnels under construction were made to incline slightly upwards at both the Piedmont and French entrances. But it was primarily the drilling machines, invented by engineers Germano Sommelier, Sebastiano Grandis and Severino Grattone, which allowed the successful completion of this great undertaking: such machinery used hydraulic energy to compress air and create the drilling movement, which was then used to make bore holes inside the tunnel; the mountain could now be perforated, without the usual vertical ventilation shafts» (Maggi 2003, pg. 58).

The double track Semplon railway tunnel is 19824 m long. The two tunnels were completed in 1906 and 1921 respectively. A very complex and difficult task, 58 men lost their lives during its construction. «It was thought that temperatures would be of around 30°-35°. The actual temperature turned out to be 52°. From km 8.9 to km 9.2 of the second tunnel, the ground was raised by lateral pressure, thus preventing the outflow of water (...). At 4,428 km from the south entrance, there were such extreme thrusts that the 0.40 by 0.40 m oak planks began bending and breaking like twigs (...). In some parts of the tunnel, metal girders with 40 cm I beams, which could easily support the strut of a seven-floor house were used. But even these would bend (...). In September 1902, explosives created another water source with a jet of 1118 litres per minute! The watercourse were so close to each other that 17 were found in a single 176 m section» (de Rizzardi 1956, pgs. 13-15).

Railways are classified according to their characteristics: on their importance, gauge, number of tracks, adhesion, traction, etc. These include:
- primary railway: with heavy traffic;
- secondary railway: with limited traffic;
- standard gauge railway: with 1435 mm tracks in straight sections and 1465 mm tracks in curved sections;
- narrow gauge railway: with inferior track width;
- adhesion railway: where wagons move through friction between the driving wheels and the tracks;
- rack and adhesion railway: where natural adhesion is insufficient, driving pinions mesh with a rack, placed between the two tracks (rack railway).

'Special railways' include various types of railways, with or without tunnels:
- cable railway: for steep inclines, this generally follows a short, rectilinear course;
- underground railway: for the rapid transport of a high number of passengers, usually within a city.

VII.9.14 - *Road in cutting*

Road in cutting consisting of a roadway below the natural ground level.
The cutting of soil or rock for the creation of roadways over orographic obstacles to the surrounding areas and of link-roads to connect such areas. This method was utilised in the past and is still used today. This type of road has lateral escarpments, the slopes of which depend on the characteristics of the surrounding land and its stratigraphy. In pre-modern times road construction roads in cutting were used for pedestrian and animal transit while the larger ones were used for sleighs and carts. The roadway and, more rarely, the rock face can have channels for the collection and discharge of meteoric or seepage water. The road in cutting can be considerable in size, extending several hundred metres with a depth sometimes in excess of 15-20 m.

Cutting: a type of *road in cutting*, consisting of a single vertical cutting in the rock face and of a horizontal cutting for the roadway; 'L excavation' allowed rocky spurs to be overcome without the need for trenches or tunnels (Coralini 1997, pg. 294).

Platform road in cutting: another type of road in cutting, which does not always involve the vertical cutting of the rock face, the partial roadbed of which is surrounded by protruding wooden, scaffolding.

Via cava: also known as cutting, this type of road can be for both pedestrians or vehicles or, due to its limited width may be for pedestrians only; it is made by cutting the rock at a depths of up to and beyond 15 m. Western-central Italy has a unique concentration of such structures. Created from the VII-VI centuries B.C., the majority are attributed to the Etruscans and the Falisci. Generally created in tuff, they are particularly common in the Pitigliano, Morronaccio and Poggio Buco areas; many serve sacred areas and necropolis.

VII.9.15 - *Road tunnel*

Normally a constant-section tunnel, this may be masonry lined and ensures the continuation of a road.
Logical deduction, leads us to believe that mineral extraction, once again, led to the idea of creating passages underground using acquired excavation techniques. Just like aqueducts, the tunnel axis was first plotted on the ground and then excavated. Shafts, inclines or openings were sometimes created in order to reach the necessary depth, for the removal of extracted material and for ventilation purposes. The *Crypta Neapolitana* at Piedigrotta

(Naples) has two inclined wells (inclines); the Cocceio Cave, also in the Neapolitan area, is served by wells, two of which are inclined and by an underground passage (Busana, Basso 1997, pgs.136-139). Other examples of road tunnels are provided by the Seiano Cava in Posillipo (Naples), St. Mary's Tunnel in Ponza (Latina) and the Furlo Tunnel (*Petra pertusa or Forulus*) in Fermignano (Pesaro), opened by Vespasiano in 77 A.D. and at the site of via Flaminia, still used today for vehicle transit. The size of Roman road tunnels depends on the type of road, on whether the road was for heavy or limited traffic, on the stability of the land and also on their current condition (Busana, Basso 1997, pgs. 158).

Italy has the Monte Viso Tunnel, also known at the "Salt Tunnel" or "*Pertus d'la Traversetta*", which is passes under the Punta Traversette of the Monviso Mountain Group at an altitude of 2882 m a.s.l. It connects the Po valley, once part of the Marquisate of Saluzzo, with the Guil valley in France, on the other side of the watershed. Excavations began in 1479 and were completed the following year. The tunnel, just over eighty metres long, approximately three metres wide and two metres high is entirely cut into the rock. It was mainly used in the transit of salt from France to Marquisate, thus eliminating the need to cross the dangerous Traversette pass. Over time it was closed several times on account of the various conflicts and landslides, which damaged parts of the route. Its umpteenth reopening was celebrated in 1907. It was again closed in 1973, so that the detritus blocking the entrances could be removed (Valbusa 1907. Amoretti, Gallo-Orsi 1973, pgs. 227-260).

Contingent factors, such as the need to retain specialised workshops in Italy, thus avoiding their transfer to Germany, resulted in the Gargano-Riva (Brescia) road tunnels being transformed into a plant for the production of war materials between 1943 and 1945. The old road, known as the "Western Gardesana", followed the ancient "Bènaco" along the banks of Lake Garda and seventy tunnel sections, totalling 7182 m were used to house Breda, FIAT and Caproni workshops (Cocconcelli 2002, pgs. 74-96).

One particular road tunnel is the Domusnovas Cave (Cagliari). This is a natural cavity, adapted into a carriageable road, the two entrances of which, present traces of ancient walls thus showing that it was once fortified. The Jenolan Cave in Australia and the Mas-d'Azil Caves in France are two road tunnels, which were created from natural cavities (Floris 1995, pgs. 63-71). Numerous traces of the Magdalenian culture (14000 - 19000 B.C.) were found in the latter, which was later used as a Christian place of worship; in 1625, approximately 2000 Huguenots defended themselves inside the cavity against the king's troops. It is a natural tunnel measuring 420 m in length, with an average height of 60-65 m and a width of 50 m; a road has been built within it.

VII.9.16 - *Rock apiary*

Place where the beehives of honeybees are kept.
The term rock apiary refers to a place where large numbers of beehives are kept within the same cavity. The cavity generally consists of a rock shelter or of a rock-cut flank, which can be sealed to create a rock environment suitable for beekeeping. The seal has both and entrance and several openings. The openings link directly to ledges, holding earthenware tubes containing the honeycombs. «The cylinder's open side (the hive's rear opening) would face towards the internal chamber (service chamber). It was closed with a wooden lid and sealed with propolis (from the bees) and wax (from man). The honey-comb cover (lid) would be removed when honey was collected. This type of honey comb could also be extended in length. Technical operations (inspections, fumigation, and honey collection) were carried out in the service chamber by the beekeeper at his own leisure» (Bixio 2002, pgs. 22-23).

VII.9.17 - *Rock dwelling*

Environment carved into a rocky flank or by cave adaptation (with horizontal or sub-horizontal development) or in a shelter under the rock to serve an individual and his family as a temporary or permanent refuge.
The term 'rocky' indicates a horizontal or sub-horizontal dwelling unit, created along rocky crags and which sometimes took advantage of rock roofs, recesses and even caves. Created in soft rock, such as certain tuff, carbonate rock or sandy limestone, this type of dwelling can consist of one or more rooms on two levels and have various elements, such as seats or wall cupboards, chiselled or modelled directly into the rock matrix; there may also be internal or external cisterns for the capture of rainwater or dripping water or wells for the capture of groundwater. Various examples of rock dwellings show the level of care taken in the excavation: openings providing light were positioned in such a way as to allow sunlight to reach the innermost recesses of the building. Numerous parts of Tuscany and Latium saw the adaptation of pre-existing rock tombs, normally of Etruscan origin, and on occasion their sheer concentration gave rise to articulated villages, known as *rock settlements*.

In the mid course of the River Fiora in the Latium area the landscape is characterised by tuffaceous plateaus, etched by watercourses. There is sometimes a rocky promontory at its confluence: «an artificial ditch separates the spur from the plains behind it, thus completely isolating the crag on all sides. This dwelling model is repeatedly found throughout the territory and in the current towns, at the summit of which the Fortress and the Church are normally found; houses are to be found on the highest plains, while artificial caves, used as cellars, as shelters for small animals and pigs, as storage areas for equipment and currently as garages for vehicles and tractors are carved into the rock (…). Almost all the above features can be found at the Sorgenti della Nova settlement, which here precedes, to the end of the second millennium B.C., a type of dwelling, which until now was thought to have been developed in Etruscan or mediaeval times. The difference being that masonry houses did not yet exist here and that artificial tufa caves as well as wooden and straw huts were used as dwellings» (Negroni Cotacchio 1981, pg. 159).

Sealed-up caves: an underground environment with one surface masonry face, often constituting the sole means of access. This type of cave results from the sealing of a natural cavity (or an adapted cavity), of a prominent rock shelter or simply through the excavation of an artificial cavity for dwelling purposes (Laureano 1993, pgs. 111-114 and pg. 142).

VII.9.18 - *Rock settlement*

Urban centre, the dwelling units of which are either partially or completely cut into a rocky flank.
A central settlement created along rocky crags either at the foot or half-way up the crest of hills or mountains, which presents signs of construction structure, organisation of spaces, of services, of work and production areas, of roads and control of territory control is classified as a rock settlement. «A rock settlement is one of the most popular dwelling models of late antiquity. Such settlements are still popular today in those Mediterranean regions presenting suitable geological conditions» (Brogiolo 1996, pg. 243).

There are numerous rock settlements in Sicily. Ancient documents on the subject are available from Arabian sources on Sicilian domination: «Ibn al-Atir explains that in 841, a group of knights plundered the "Fortezza di Grotte" (Cave Fortress), which took its name from the forty or so surrounding caves and, which the Amari chose to identify with the Municipality of Grotte Agrigento. A Byzantine letter also mentions communities sheltering in the mountain gullies and another X century source refers to the numerous hermits or religious prophets (*theoptikoi*), in Sicily at that time» (Uggeri 1974, pg. 158). The ravines, characteristic erosion valleys of the Murge in Apulia and Lucania are striking examples, deeply furrow the limestone plateaus, rich in karst phenomena such as dolines, sinkholes and caves. The ecosystem develops from the adaptation of caves or simple rock recesses and goes on to assume a terraced structure, such as in Matera for example, with superimposed levels where the roof of one hypogeum becomes the road or the hanging garden opposite the other hypogea (Laureano 1993, pgs. 109-121).

Following their closure in the 1970s and their inhabitants being transferred to modern houses, the "Sassi", picturesque rock districts of Matera, were included in the UNESCO World Heritage List. «During the course of the objective discovery of individual dwellings in the "Sassi" districts, we decided upon their classification on the basis of those interventions which were deemed necessary and this should not be underestimated. In particular:
a. the definitive closure of true *caves* was deemed necessary;
b. the restructure of the so-called *bassi*, in better condition than previous dwellings, was proposed;
c. and finally, it was proposed that the few large houses along the Gravina be adapted in line with the size of the household» (Marselli 2003, pg. 93).

In both Guermessa and Chenini in Tunisia, rock settlement are integrated into the terraced morphology of the mountains and consist of several rock-cut rooms with front masonry facades, which often take the form of a rectangular enclosure. There are many examples of rock settlements in the Cappadocia region of Turkey, including the Göreme settlement, cut into the stone cones formed by erosion of natural agents.

Cliff dwellings: these are pre-Columbian settlements which developed in large horizontal or subhorizontal rifts between Mexico and the U.S.A. «The Navajo National Monument is in the Kayenta region, in North Western Arizona at approximately 145 miles from Mesa Verde. It is famous for its three spectacular settlements located under enormous rock cavities: Inscription House, Betatakin and Keet Seel» (Gleria 1995, pg. 111). In the Navajo language, "*Betatakin*" means "*ledge house*". Similar settlements are also to be found in other geographic areas, such as in the Bandiagara cliffs in Mali.

VII.9.19 - *Sirocco Chamber*

Artificial cavity, used during the hottest hours of the day and diffused throughout Sicily.
«If we were to summarize the features of a Sirocco Chamber, two words would suffice: cave and source» (Todaro 1988, pg. 55). This type of underground environment, is generally quadrangular and with a single chamber. Aquifers are sometimes intercepted during excavation or alternatively, hydraulic channels are created. The chamber is equipped with an access ramp and a roof ventilation shaft, which also serves for illumination purposes and seats cut into the rock matrix. In Palermo, sirocco chambers were very popular from the XVI century onwards, where it seems that these were used on particularly hot days and for the preservation of food.

VII.9.20 - *Souterrain*

Room or rooms, specifically for civil use and built under the ground surface.
There are many examples of basements for civil use, either purpose built or created within pre-existing cavities. Every factory probably has its own rock or underground variant and many ancient public and private structures have suitable rooms for various purposes. For this very reason, this wide range of underground structures has been catalogued under the same sub-typology. Future explorations will undoubtedly lead to a more accurate and detailed subdivision. The washery is part of the mining plant and consists of several areas for crushing and mineral concentration through flotation or hydroseparator treatment. Although some parts may be underground, washeries are generally to be found on the surface. For instance, two washeries in the former Gavorrano coal field in Tuscany were build on large artificial terraces with internal spillway channels and rock chutes; in particular, the spillway channel of the larger washery is a quadrangular vertical shaft, approximately 15 m in depth, with an inclined quadrangular channel placed a few metres below its entrance.

Hypocaust: ancient Roman central heating system for both public (public baths) and private use. Hot air from a furnace was channelled under the floor and in the walls of the room to be heated. The temperature was regulated by increasing or decreasing the fire within the furnace. The floor was raised with masonry or brick pillars (*suspensurae*), thus creating an air chamber.

Underground car-park: very common, particularly in the last few decades, for parking of motor vehicles. Shelters for small coaster boats are sometimes created by cutting the rocky banks or exploiting large niches or marine, lake or river caves.

VII.9.21 - *Storeroom*

Room or rooms for the storage of various products and materials.
In this specific case, warehouse can refer to a subterranean, semi-subterranean or rock-carved storeroom and may even refer to the adaptation of a pre-existing cavity. Consisting of one or more rooms, the storeroom can be used for the storage of various products and materials. The underground rooms of the Roman temple in Brescia, reutilised for the construction of the Lookout Tower in the Viscount part of the Castle, serve as storerooms and are equipped with cylindrical stone containers for the storage of oil and dry foodstuffs (Breda 1986, pg. 48).

Horreum: In ancient Rome, this term referred to both private and public storerooms; the building could be either subterranean or semi-subterranean.

Depots: in industrial sites this is the group or rooms used for the storage of raw materials and manufactured products, while in business installations, the warehouse is used for the storage of merchandise.

VII.9.22 - *Underground dwelling*

Environment created underground to provide an individual and his family with temporary or permanent shelter.
The term underground dwelling refers to a temporary or permanent shelter, excavated under the surrounding natural surface level. The many types of underground environments used in daily life primarily depend on the consolidation and hardness of the rock matrix. Shanxi in China has dwellings with court and tunnel wells, which are still in use today.

Damùs: type of underground Libyan dwelling, used both for dwelling and industrial purposes. It has an entrance passage, known as *sghìfa*, which follows a steep underground descent and then curves to lead into the well court.

«Certain *dawamìs* (plural of *damùs*: n.e.s.) are vaster and deeper. Their courts are 10 m long, they have a large central well with large edges and numerous rooms, which are sometimes on two levels. The entrance arches, are reinforced in masonry and thus last longer. Certain parts of the rough, red court walls are whitewashed» (Chiauzzi 2003, pg. 185).

Matmata - troglodyte dwellings: type of dwelling deriving its name from the Jebel region of Tunisia; similar to the *damùs*. It consists of tunnelled steps leading to the base of a type of large well, which can be of various dimensions. Along its walls are the entrances to one or more levels of underground dwellings, equipped with hydraulic systems, mills and granaries.

Underground pit: Libya also provides examples of single-room, generally temporary dwellings used by farmers and shepherds. «A circular pit, one metre deep or even slightly more, by three to five meters wide is excavated in the ground. An external soil or stone border, similar to a wall and just under half a metre high, is built around it. A small opening serves as entrance. The border supports strong, closely-knit branches, which serve as a base for the roof. The branches are covered by a 30-40 cm layer of soil: this is the roof. The well-pressed soil goes from the roof to the ground and resembles a slightly flattened cupola. It encloses and thus strengthens the wall, filling the gap around it» (Chiauzzi 2003, pgs. 191-192).

Vicinanza: a type of underground dwelling, built until relatively recent times, was developed in Massafra (Taranto). It has yet to be established exactly when these were first developed. A downwards ramp, generally with steps, was excavated into the ground. The tunnel widens horizontally at a depth of 4-6 m forming a sort of quadrangular court known as *zoccata (*also quarry or *tufara*). The underground dwellings composed of one or more rooms are accessed via entrances in the vertical walls (*zoccata* façade). A rainwater cistern, laundry tank (pila) and drainage channel can normally be found in the courtyard. Such a complex is known as a *vicinanza*.

VII.9.23 - *Underground oil mill*

A type of underground installation for olive preparation.
Olive mills are sometimes built underground, by cutting rooms into the rock and by using pre-existing structures or natural cavities. There are no construction costs involved when creating underground structures and it is easy to keep the environments warm, dry and at a constant temperature; oil tends in fact to solidify at temperatures of around 6°, thus olive pressing must take place in a non-humid, mild environment. There are numerous underground olive presses.

Trappeto (“*trappitu*”): particularly popular in the south of Italy, this is tool is used to crush solid materials; especially used in olive preparation; the room in which the olives are pressed is known by the same name. The underground structure could have more than one room and was often equipped with tanks, store-rooms and cisterns. Both ‘Calabrian’ and ‘Genoan’ presses would be used and the pomace was collected in a cistern, which was also cut into the rock, known as *inferno* (“*nfiernu*”).

VII.9.24 - *Underground palmento*

A wide, shallow basin used for grape-pressing and must fermentation; by extension, the term also refers to the underground environment in which the basin was excavated.
The *palmento* is a masonry basin, which in underground environments is normally cut into the rock and rendered water-tight. A few examples of underground *palmenti* have been uncovered in the Montalè area (Sassari). These structures were either built *ex novo* or reutilised existing *domus de janas* or small rock-cut churches. The characteristics of such structures is homogenous «with a plant consisting of grape-pressing basins and their relative collection basins cut into the floor surface and with room for one or more screw presses, with their own collection basins. Each underground structure has between one and three pressing basins; these are cut into the rock and are approximately 2 m in diameter with an average height of 1.2 m; they are internally lined with a type of *opus signinum* (lime mortar with an aggregate of coarse pieces of broken terracotta)» (Rovina 1997, pg. 254). They generally have access steps and a cistern for the collection of rainwater.

VII.9.25 - *Underground settlement*

Underground settlements are normally more concentrated than rock settlements, in that they are fully underground and have, generally protected openings, providing access to the surface.

Such settlements can hold hundreds or thousands of people in a structure not dissimilar to a beehive, in an environment, the daily structure of which is more rigorous than in any other type of settlement. Life would have been centred around a marked structure and careful discipline; otherwise permanency and community life would have been impossible. The settlements would have had all the structures necessary for daily life, such as water supply systems consisting of wells and cisterns, waste disposal systems, public and private areas, work areas and places of worship. Ventilation, generally consisting of shaft ventilation, would have been of vital primary importance.

Generally speaking, the organizational structure of underground settlements is not solely dictated by defensive or financial motives or by the difficulty in obtaining or importing building materials. Climatic adaptation for thermal insulation, almost impossible by other means, also plays its part. Motivations and applications are undoubtedly subject to regional variation, however they share common principles. The complexities arising in the research and understanding of many such structures is given by the very nature of the territory and by their state of abandonment. They may have been abandoned spontaneously, following famine or due to water supply issues or even following epidemics, internal conflicts, and political or even wartime factors. The lack of maintenance therefore led to their obliteration by natural and progressive burial and structural collapse. Underground systems may have been occupied during different and distinct periods of time, in which case pre-existing structures may have been adapted, their spaces being amplified or reduced. The resulting stratification is difficult to interpret in terms of providing an accurate picture of the man-made complex.

In Turkey there are vast underground such as Ani in Armenia, Sivasa, Kaymakli and Derinkuyu in Cappadocia: «The sheer number of ancient hypogea in Turkey is surprising. In their research conducted between 1967 and 1973, the German scholar, Martin Urban and Ömer Demir, tourist curator of the underground city of Derinkuyu, located 39 underground settlements. However, according to the local inhabitants, for each Cappadocian village there is a corresponding structure carved in volcanic deposits. Yörükoglu (1988), an archaeologist from Kayseri, provides us with a list of 121 underground settlements, sub-divided by province: 4 in Yozgat, 5 in Kirsheir, 23 in Kayseri, 26 in Nevsheir and 64 in Nigde (the new province of Aksaray was recently de-merged from the latter). Nevertheless, he believes the actual number of such sites to be nearer 400. (...) Urban highlights the inadequacy of the term "underground city", as only larger structures may be strictly interpreted as urban settlements, the remaining structures not fulfilling the criteria of permanent settlements» (Bixio 1995, pg. 25). Rubble was removed from the underground city of Derinkuyu and eight levels were explored. However, scholars believe that there are further levels and that the structures may extend to a depth of 80 m. With various services, the environments are linked by corridors, stairways and inclines; its entrances are internally protected by a mill-stone door system, consisting of stone wheels resting in special manoeuvre areas, which were rolled across until the entrance was completely blocked. This system was used throughout Cappadocia although there are also instances of its use in other regions. Opinions in regard to the dating of the Anatolia underground settlements are conflicting: there are those who believe that the 'phenomenon' may have existed as far back as pre-history and those who believe that it has its origin in the Hittite period and lasted for centuries.

In his description of the homeward march of the Ten Thousand Greek Oplites after the battle of Cunaxa (401 B.C.), Xenophon states that there is an underground village in the area where there are hot springs and that the houses, excavated in the ground, have well-shaped entrances. They are sufficiently spacious and have excavated passages providing shelter for the animals, while there are stairways for the men. Before taking their leave and through a Persian-speaking interpreter, Xenophon and Chirisophus ask the head of the community, what the name of the land is and are told that it is Armenia (Xenophon, IV, 5, 25-26 e 34).

More of a curiosity, the "underground city" of Damanhur in Val Chiusella (Piedmont) may well be of architectonic and archaeological interest in a few centuries time. Established in 1975, the community excavated a small underground city with tunnels and temples and decorated it with frescoes, marble and sculptures.

VII.10 - *Typology 6: military structures*

Structures or environments for various purposes, strictly connected to wartime defence and offence.
Although castles hold undeniable appeal, it was the collective notion of their 'secret underground passages' that first led to their exploration (figs. VII.24 and VII.25). Its very 'genre' is ascribable to the type of activity dealt with in speleology. For example, in relation to the position occupied by the ancient *Praeneste,* Strabo tells us that it not only had a natural defence but that it had underground walkways in all directions leading to the plains, which were both utilised as 'secret passages' and for the purposes of water supply (Strabo, V, 11). The point of the exploration is therefore to document other types of underground structures: wells, cisterns, warehouses, prisons, underground

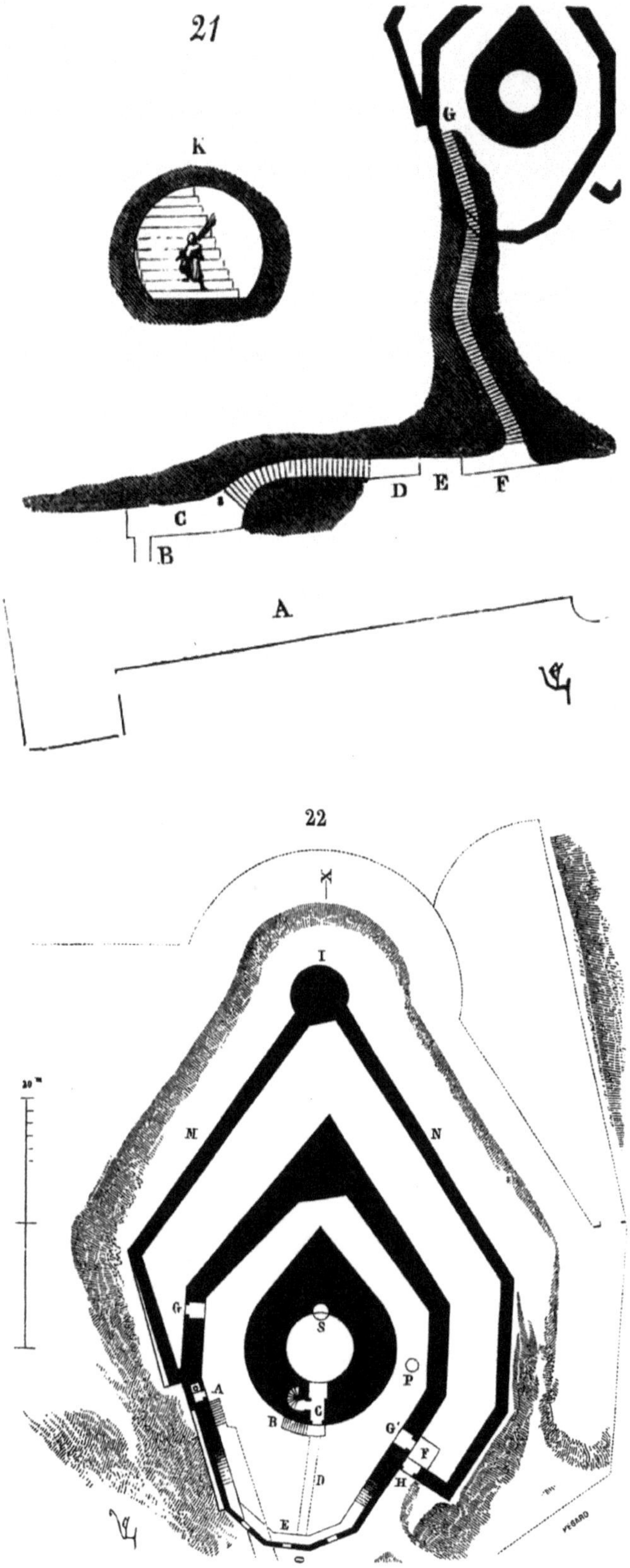

Fig. VII.24. *Secret passage in La Roche-Guyon Castle, France. Map number 21 details: A. clearing defended by walls in front of the castle; B. small drawbridge; C. platform cut into the hillside; D. second platform; E. rock tunnel communicating with the next platform by means of a timber bridge; F. third platform; G. ascending underground area leading into the second Donjon wall; K transversal section of the underground room «Les souterrains taillés dans le roc existent encore, et s'ils ne sont point des antres antiques, s'ils ne descendent pas aux enfers, ils datent d'une époque assez éloignée» (Viollet-le-Duc, pg. 413). Map number 22 details: A. entrance to the second Donjon wall; B. stairway; C. postern; D. small bridge; E. chemin-de-ronde; F. small ditch; G and G' posterns; H. postern; I. spur protection tower; M. and N. enceinte wall; P. well; S. cold meat and sausage preservation silos. (Viollet-le-Duc 1996, pgs. 413-414).*

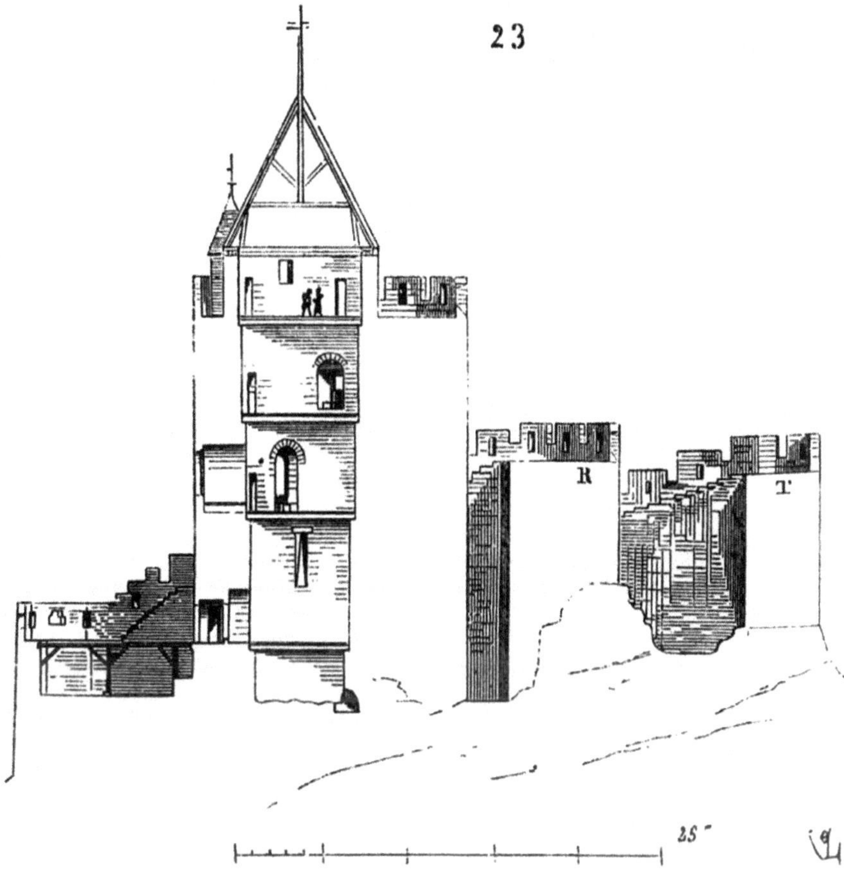

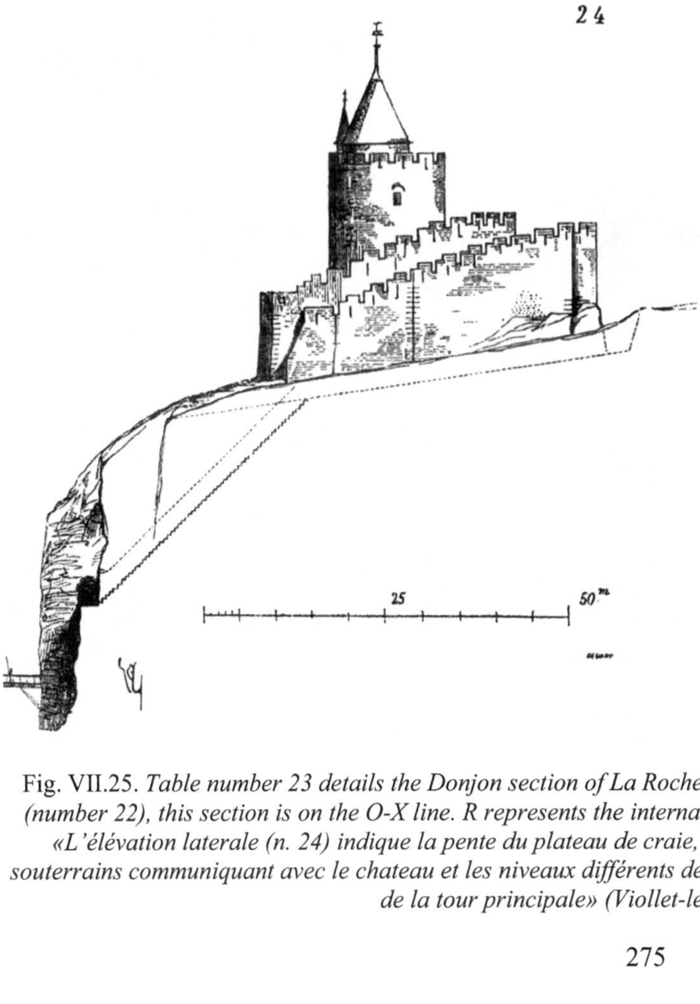

Fig. VII.25. *Table number 23 details the Donjon section of La Roche-Guyon Castle in France. Having seen the previous table (number 22), this section is on the O-X line. R represents the internal parapet and T represents the external chemin-de-ronde. «L'élévation laterale (n. 24) indique la pente du plateau de craie, son escarpement fait à main d'homme, la position des souterrains communiquant avec le chateau et les niveaux différents des parapets des deux chemises, ainsi que le commandement de la tour principale» (Viollet-le-Duc 1996, pg. 414).*

passages and connecting tunnels. Without of course calling off exploration and cognitive activity in those environments that only appear to be underground.

We shall now attempt to explain the abovementioned concepts. In order for a fortified enceinte to survive a siege, a few 'systems' be in place. One such system is the need for a water supply. Without water there can be no life. It therefore follows that without water there can be no defence. As well as quenching thirst, water extinguishes fires and ensures a level of hygiene sufficient to prevent the spread of disease. Such systems consist of wells for the extraction of water from aquifers, of aqueducts for the continuous supply of drinking water (normally stored in underground or semi-subterranean tanks) and of cisterns for the collection and storage of meteoric water. «Their cannons have breached out walls. They have taken the great iron gate off its hinges and dragged it to the encampment. But this was all in vain. Thirst overwhelms us. The two wells we excavated do not provide us with sufficient water. Now that the battle has ended, water is reserved for many the injured» (Kadarè 1993, pg. 197).

It should be taken into account that a defensive perimeter must provide all necessary services and adequate material reserves. Space must therefore be managed wisely and underground areas should be created, for purposes other than mere bomb shelters. There are thus areas for the storage of foodstuffs and areas used as military quarters or in the storage of munitions (artillery magazines, gunpowder magazines). Some systems provide for the use structures and not necessarily underground structures within the bastioned enceinte, whereas other systems provide exclusively for the use underground structures.

Defensive installations are perfected over time through the use of various building materials and the introduction of counterforts, towers, ditches and avant-corps. Slow yet constant, the change in defensive solutions in a sense results from the application of new technologies, only second to financial commitment and the time available for their construction. Innovations also arise from evolving warfare techniques; the results attained often rendering the specific type of fortification ineffective. It should also be taken into account that when building military structures in different places and following different methods, upgrades must be made and any failure should be seen as a learning curve. Had this been the case, Man would have abandoned the so-called 'art of war' long ago, in favour of a pacifist culture. Machiavelli observes that the princes built fortifications to provide safe shelters: «In order to hold their states more securely, it was customary among princes to build fortresses to serve as bridle and bit to those who might plot against them, and to provide refuge from immediate attack»; however in conclusion he affirms that: «I shall praise he who builds fortresses and he who does not, and I shall fault whomever places his trust in the fortresses caring little about being hated by the people» (Machiavelli, XX, pgs. 106-108).

Prior to the use of firearms, underground works inside the walls were not strictly necessary for defence purposes. Following their introduction, underground works immediately became a part of the fortification's defence. In bastioned buildings, countermines and demolition installations generally constitute the most important defensive element. It should also be taken into account that over time and following partial destruction and burial or subsequent urban repairs, raised sections could end up below the level of adjacent ground. The XV century "Ghirlanda" (Garland) curtain wall, protecting Milan Castle along its 'country' façade is one example worthy of notice: partially demolished at the end of the XIX century, the underground tunnels, corridors and casemates within the scarp wall and on the first floor of the curtain are still intact.

Defence works have been created almost everywhere, using different types of materials and sometimes using existing works. They can be used in the defence of houses, in territory control or in the defence of obligatory passages through valleys or along rivers. Various types of fortification are used in the defence of ports and the hills are guarded by towers and forts. In the XVII century "La Serenissima" (the most serene) Republic of Venice has the "octagons" built in the Venetian lagoon. These were isolated, octagonal fortifications, equipped with artillery and positioned in the lagoon to guard waterways such as the Malamocco and Spignon Canals (Concina 1990, pg. 258). «The walls are effectively a technical, military, financial, social, political, judicial, symbolic and ideological phenomenon. They define the outside and the inside and the dialectic relations between the city and its surroundings: outskirts, county and distant areas linked by roads and by imagination. The walls are an essential part of the urban image» (Le Goff 1989, pg. 1).

Let us now take a brief look at the development of fortifications. As a rough (and in some aspect arbitrary) guide, the development of dwellings and the first dry-stone city walls could be attributed to the so-called "Neolithic period". Sardinian Nuraghe, Roman castrum, Irish dún, Scottish broch, Slovenian castellier, Persian kārwānsarāy, Venetian covelo: scores of books could be written on the military or at any rate defence works, which preceded the use of cannons and arquebuses. But let us go back to the most famous type of fortification, that which is evocative of romantic poetry, noble gestures and brutal revenge: the castle. The basis of this defensive work is the creation of a high and apparently insurmountable obstacle such as the curtain wall, in the face of impetuous enemy charge. Cutting weapons, thrusting weapons and throwing weapons are used in battle. Tension and torsion weapons such as ballistas and catapults alongside mangonels, bores and battering rams saw great development. The first cannons, also used to demolish defensive wall structures, make an appearance. Their undoubted advantage is that their range is far superior to that of the usual throwing weapons. The Guelph or Ghibelline battlements of towers and curtains were inadequate against new siege warfare techniques. Works are made with lower and thicker walls, to better withstand cannon fire. Ditches are systematically equipped with counterscarp walls and additional works, thus paving the way for the development of the "bastioned front fortification" of Italian origin.

In the latter part of the XV century, Antonio Averlino, known as the "Filarete" introduces a star-shaped fortress formed by the intersection of two squares at a 45° angle in his "Sforzinda" treatise. Throughout the XVI century, European military engineering was developed by famous figures such as Francesco di Giorgio Martini, Giuliano da Sangallo, Leonardo da Vinci, Niccolò Machiavelli, Michelangelo Buonarroti, Antonio da Sangallo the Younger, Giulio Savorgnano and Nicolò Tartaglia The talented Albrecht Dürer also contributed. As did Francesco de' Marchi. «A large number of systems are described in Marchi's treatise published in 1599: He provides a general outline of the external works he called *pontoni*, such as the demilunes, the lunette, the tenaille and the counterguard, which were popular in the VII and VIII centuries» (Fara 1989, pg. 159).

Star-shaped fortification plans are based on the application of mathematical theories and take into account both cannon range and the need to eliminate 'dead angles', or the points which cannot be reached by shells. However, one of the primary defence systems of a bastioned stronghold is its underground countermine system. From the end of the XVI and the XVIII centuries, fortifications are systematically equipped with underground tunnels, usually underneath the primary defensive perimeter. In the event of siege, their purpose is to identify and intercept enemy excavations and interrupt their advancement by underground combat or by using explosives to destroy the attack passage.

The words of Galileo Galilei provide clarification on how a fortification should be taken and how curtains and bastioned defences can be rendered useless: «It would appear that there are five primary means by which fortresses may be taken and conquered, these being:
- the battery, when an artilleryman opens a wall from afar and access to the fortress is gained by means of the opening thus created;
- the hoe, by digging close to the wall with iron bars, picks and other tools and thus destroying it;
- the third being escalade, where ladders are used to scale the walls;
- the fourth being the use of mines, their explosion in an underground cavity destroying the walls in an instant;
- the fifth and final being the siege, whereby having removed all possible assistance, the defenders are driven by hunger to surrender» (Pellicanò 2000, pg. 100).

The wartime experience of the XVIII century, brought with it the need for a permanent system of countermine tunnels, which would become an effective, if costly, wartime tool. Tunnels were created using the cut and cover method or were simply excavated in the ground (fig. VII.26). They generally had a masonry facing and a vaulted roof to protect them from water infiltration and humidity, which were pre-requisites for the use of black powder.

Other essential elements were casemated works, chiefly used in the protection of artilleries and communication tunnels for the quick movement of soldiers, even under enemy fire. Covered roads were also built along the external perimeter, marked by the positioning of the glacis and ditch counterscarps, where tunnels with loopholes became more and more frequent to keep the ditch itself under control. There is no shortage of underground communication works.

«Regardless of the pre-existence of ancient circuits, modern-day military architecture stems from a geometric system in which any change affects the systems, as happens in a magnetic field; this system directly relates to the prospective system» (Fara 1988, pg. 94). «The first *modern* fortified works appeared at a time when artilleries were already fairly advanced, although rather late in terms of Brunelleschi, despite the common cultural base» (Fara 1989, pg. 81).

One type of fortification with a specific purpose was the so-called trade castle or fort, a manifestation of European colonialism. «Fort construction chronology and their dislocation mark the phases of the various European settlements. The change of typologies, of defence works first directed towards an unknown land seen as hostile, were increasingly directed towards the sea and were indicative of the change in African relations, relationships which had

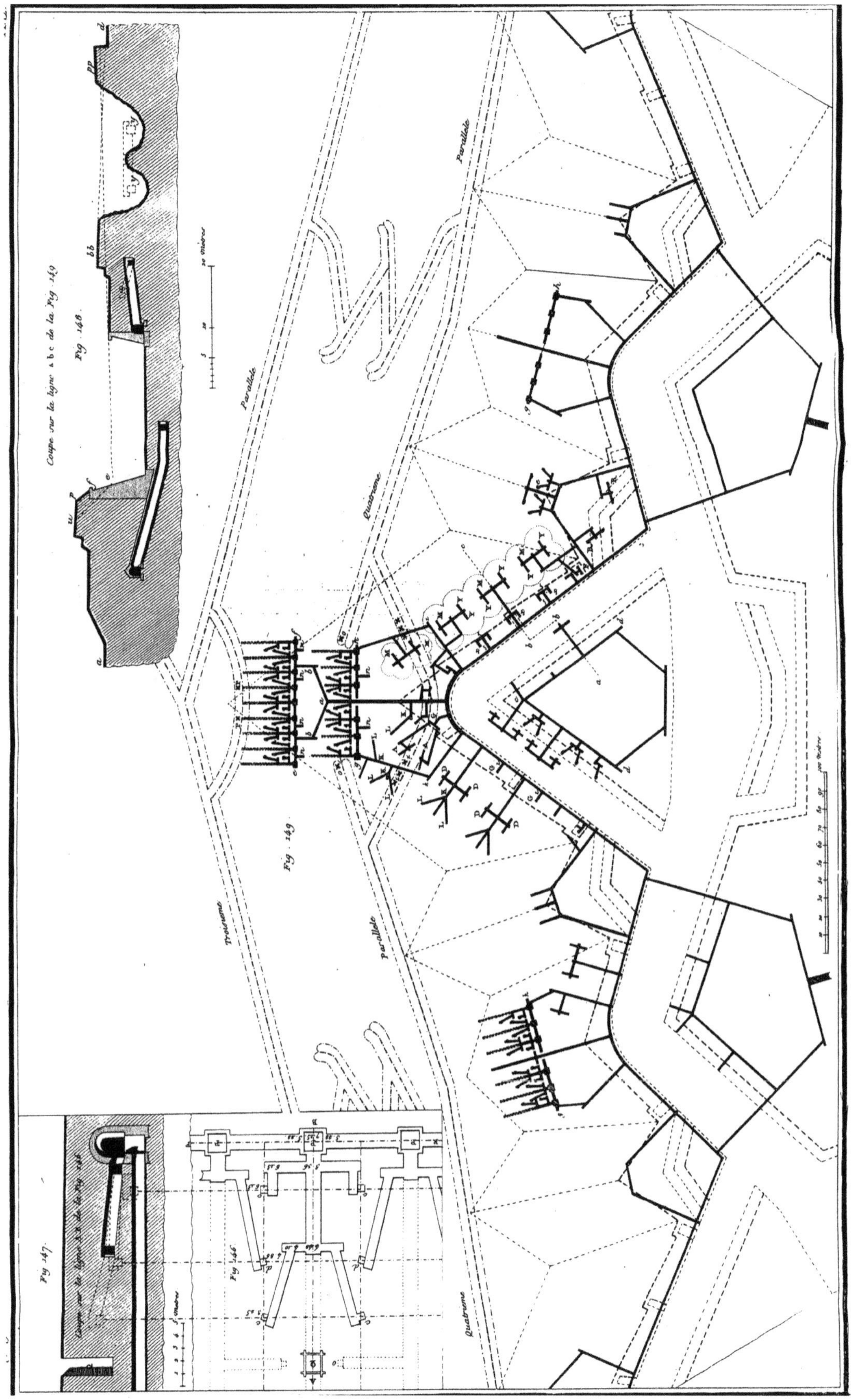

Fig. VII.26. *Bastioned fortification countermine system (Gillot 1805, pl. 12).*

certainly not become more kindly or fair but which continued to be based on corruption, in its every form; and above all, this was indicative of the battle between the European powers» (Bassani 1990, pg. 2).

With improved artillery and the systematic use of mortars firing large explosive projectiles (XVII-XVIII centuries), external works were increasingly (ravelins, counterguards, hornworks, crownworks, caponiers, lunettes etc.) were increasingly adopted, thus extending the defensive perimeter in order to keep enemy batteries as far as possible from the main fortification as well as to combat infantry attacks, with their increasingly accurate and quick-loading guns. An excellent and almost integral example of a star fort, in use until the beginning of the XIX century with its relative extensions and improvements is the Citadel of Alessandria, designed by Giuseppe Francesco Ignazio Bertola in 1727.

In France we have the Blaise-Françoise de Pagan (1604-1665), deemed by many as a brilliant and innovative military engineer, whose research was the source of much inspiration. In his time, Sébastian Le Prestre, Seigneur of Vauban (1633-1707), Marshal of France and Royal Engineering official was the master of military architecture and siege strategies; his treatises would become famous. Bernard Forest de Bélidor (1697-1761), Bengt Wilhelm Carlsberg (1696-1778) and Marc-René de Montalembert (1714-1800) should also be mentioned. The Montalembert school of thought prevailed, even in respect of its successors, representing the European military architecture model for the XIX century (Fara 1989, pgs. 243-248). Menno van Coehoorn (1641-1704) marked a turning point in Dutch fortification techniques, capitalizing to the best of his abilities on the possibilities offered by a terrain with a water table in close proximity to the surface (Ponzio 1997, pg. 32).

In the "Strongholds" paragraph of his work, Carl Philipp Gottlieb von Clausewitz observes that until such a time as armies became a permanent fixture, the primary task of castles and fortified cities was to protect the inhabitants and the feudatory or local squire. Such a purpose was subject to gradual modification until fortresses, if suitably positioned, assumed the function of directly guarding the territory. They could thus even become «a means of conducting war in a more co-ordinated manner» (von Clausewitz, VI, X). Their strategic value would influence military campaigns for the conquering one or more strongholds rather than the destruction of enemy forces. In some measure, this makes us lose sight of the original task and leads to the concept of fortresses without cities and inhabitants, just like the primitive defence systems (von Clausewitz, VI, X).

The end of the XVIII century saw the end of 'modern' fortifications with bastioned front. The introduction of rifled barrels and breech loading, the use of ogival projectiles with more effective launch loads, ensure that artillery in the second part of the XIX century, benefits from an increased range, thus becoming more accurate and destructive. This leads to the rapid modification of not only the concept of fortification, but of the application of new defensive systems, increasingly equipped with semi-subterranean and subterranean works for the protection of artilleries, garrison soldiers and logistic services. Structured defence is implemented with the creation of fort 'enceintes', around the primary stronghold or "body of the place". Armoured casemates, revolving steel gun turrets and cupolas (including retracting turrets/cupolas) become increasingly popular and as do "cave works" or those works carved in the rock. Trenched fields, rings of detached forts with armoured forts and permanent field defences were rapidly developed throughout the European continent. In the latter part of the XIX century, Major William Palliser designed a projectile, the conical iron tip of which became sufficiently hard, after specific treatment, to pierce armour. Through the very same system, the metallurgist, Hermann Gruson von Magdeburg-Buckau, realised that extremely resistant iron sheets, suitable for armoury, could be produced. Henri-Alexis Brialmont (1821-1903), a Belgian engineer, subsequently designed several fortification systems, which would later become known as: "Brialmont forts".

At the beginning of the First World War, fortifications consisted of redoubts, forts, blockades and strongholds and some areas had trenched fields while others had bastioned fronts. However, "permanent fortifications" were not as successful as initially anticipated and could be destroyed by modern artillery. On the other hand, the more efficient and flexible field fortification gained importance. Initially, field fortifications consisted of a line of resistance centres connected by trenches and protected by barbed wire entanglement. From 1917-1918 lookout and listening towers were positioned along the front and these were linked to the unsheltered resistance line, which lay behind. Behind this were various lines of deep trenches, with bomb-proof shelters and command posts and even casemated command posts. Along the mountain front, rock-cut shelters and posts were widely used as were natural cavities.

The experience of the First World War, the ballistic development of heavy artillery, the inclusion of aviation as a siege weapon and the use of tracked vehicles, determined an almost total lack of works with elevated features which were easily identifiable above the ground. Defence works were created underground, their emerging parts buried and externally protected by anti-tank obstacles (ditches and "Dragon's teeth"), mine fields and light infantry defence. Trenched fields, armoured forts and more importantly "defence lines" were thus created with forward position works (observation posts, barbed wire entanglement and anti-tank obstacles) and rearward position resistance works (artilleries, machine guns and anti-tank weapons), which were sometimes supported by different types of underground casemated systems.

Between World War I and World War II, the Italian Alpine Wall was mobilized with chiefly light to medium-light artillery (47 mm and 75 mm machine guns), thus surrounding the Alps: from Ventimiglia to Fiume, omitting the area along the Helvetic border, already protected by the Cadorna Line. Many of the fortifications are now to be found on French or Slovenian territory. Other European and extra-European nations also adopted permanent works, the most famous being the Maginot Line in France.

During the Second World War, the new permanent fortifications again proved to be ineffective and the heavy use of aerial forces further limited their resistance. The most famous defence line is the Atlantic Wall, the main part of which was built along the coastal area nearest to England. Other defensive lines include the Siegfried Line (Westwall), the Stalin Line, the Czechoslovakian Line, the Gothic Line, etc. A large number of underground installations were developed to protect war production plants. Works carved in mountain reliefs were widely and effectively used as they provided excellent protection to both men and vehicles.

Subsequent events would confirm the longstanding preference for field fortifications and basic underground shelters. Shelters, underground fortifications and missile installations were built in anti-nuclear, anti-chemical and anti-bacterial underground silos. In post-war Italy, certain fortifications on the Austrian border and part of the Alpine Wall were reconsidered, and more were built to block a hypothetical Soviet invasion. In Friuli-Venezia Giulia, a defensive line was mobilized along the banks of the River Tagliamento as far as the sea and a second line was mobilized on the border of ex-Yugoslavia (now Slovenia) and along the course of the Isonzo River. Underground and semi-subterranean fortifications were built as were concrete positions with steel demi-crown and armour for cannons, light artillery works and lookout posts. To speed up the implementation of defences, «the shells of Sherman tanks set in cement with armoured cupola» were also used (Cappellano 2003, pg. 5). Tank turrets, their primary armament removed, were used for automatic weapons. Today, almost all the works are in disuse.

In Tunisia, numerous fortifications bear witness to the French occupation of Bizerte. Military installations can be broken down into three periods: the1881 to 1918 period, the period between the two world wars and the 1943-1963 period (Moulins 2005). Forts and trenched fields were used during the Franco-Vietnamese war in Indochina. Dien Bien Phu, pivot of the French defence, was conquered after a lengthy battle. Extensive trenches characterised the final phases of the Korean War. The Americans introduced "fire support bases" during the Vietnam War: these consisted of a series of light and heavy artillery positions enclosed in a perimeter of barbed wire entanglement. The Viet Minh and Viet Cong excavated extensive underground systems for use as shelters, storage rooms and hospitals. These also served as temporary hiding places to escape from shellfire (Rottman 2006).

VII.10.1 - *Artillery magazine*

In permanent of field fortifications, this is a small room for the storage of munitions.
A small, bomb-proof room situated in proximity to the defensive line, sometimes between emplacements and sometimes used as a transverse. Its purpose was to supply munitions to the line sections.

Expense magazine: a small room in the terreplein of a traverse situated between two emplacements or artillery units, already used in bastioned XVIII century fortifications. It serves as a munitions and artillery magazine.

VII.10.2 - *Bastion*

Fortified work consisting of a terreplain enclosed in a thick polygonal support wall perimeter, also known as a rampart.
In a defensive system, a bastion is built to defend and provide corner support to the curtains or to provide additional strength to straight sections, such as the walls around Lucca. Usually pentagonal in shape, a bastion has four external sides: two faces forming a salient angle and two flanks, connected to the curtains. The demi-bastion has only one flank, one face and a second flank on the capital of the salient angle. The profile of the external wall consists of two sections, a scarped lower section and a vertical upper section, separated by a large horizontal band. A bastion's defensive system was completed by solid defence shelters built on top of the walls, known as merlons, between which cannons could be fired. It normally had re-entering flanks, with rounded angles (round or square orillons), in which artilleries were positioned to guard the areas opposite the curtains). These batteries could be placed on

multiple, overlapping levels and could be positioned within casemates or within barbettes and were given different names according to their position.

The first bastions appeared in Italy at the end of the XV century but were only systematically used during the course of the following century. They remained in use, with relative upgrades and modifications to improve their defensive capacity, throughout the XVIII century. In the XIX century, the advent of more powerful fire arms resulted in bastions being gradually replaced by other systems.

Various types of environment were created within or under the bastions. There may be communication and service tunnels leading to artillery posts, such as those of the Grosseto bastions or of the Astagno Hill citadel in Ancona, the construction of which was overseen by Antonio da Sangallo the Younger until 1534. There are also many examples of bastions granting access to countermine systems.

Sortie: passage or small tunnel giving access to the ditch and used to laterally surprise an enemy. The *sortie* is normally accessed from within the bastion, as exemplified by the Venetian belt of Bergamo.

VII.10.3 - *Battery*

In fortifications, this is the part of the work, which holds defence artillery.
The term battery refers to a terreplein or reinforced concrete permanent post, sheltering permanent or field artillery. In this specific case, we shall consider those batteries which are not part of a fort but which are part of the defensive enceinte, of a wall, of a general defence system. Such batteries take on the name of the section to which they belong, such as mountain battery, anti-aircraft battery etc.

If the knowledge that coastal defence could be ensured by the Royal Navy prevailed in England for many years, the development of navies in the post-war period, particularly on the part of France, led to a programme of coastal fortifications in the latter part of the XIX century. As for Italian XX century coastal batteries, these normally consisted of four cannons, of the same calibre, anchored to semi-subterranean concrete emplacements by means of an iron carriage with mechanical rotation system. Each emplacement was equipped with its own artillery magazines. During the Second World War, the Piombino (Livorno) coastal battery artillery magazine was situated in the "G. Sommi Picenardi" Battery ammunitions depots at Punta Falcone: cannon loads were stored in an underground gunpowder magazine while projectiles were stored in a camouflaged wood and metal sheet external artillery magazine (Dondoli 1997, pg. 47).

In 1941, the three 40.6 cm SCK/34 cannons arming the "Groβdeutschland" battery in the sector of Danzig, were dismantled and transferred to Noires-Mottes, near Calais, as reinforcement to the Atlantikwall. Special casemates were built to house the three large pieces of artillery. Each casemate was organized in the following way: «Chaque casemate d'artillerie, mesurant 50 mètres de longueur, 30 mètres de largeur et 17 metres de hauteur pour une masse de 17.000 m^3, est organisée sur le même modèle de plan type S 262, dont le programme de construction est instauré à la fin de l'année 41. Elles comportent chacune trois niveaux inférieur, un niveau intermédiaire et un niveau supérieur. Dans ce dernier se trouvent situées les différentes entrées possibles, l'une mixte matériel et hommes et l'autre pour le personnel» (Chazette 2001, pg. 21).

The area was enclosed in a trenched field with anti-tank ditches, additional defence works and casemated services. The complex was inaugurated on 19 September 1942 and given the name of "Lindemann Battery" in honour of the late Kapitän zur See Ernst Lindemann, Commander of the Bismarck battleship. The Atlantikwall consisted of many other batteries, among which the "Oldenburg", situated east of Calais and the "Siegfried", later to become known as "Todt" and which is now a museum (Charzette 2004 a, pgs. 20-22). Heavy batteries on railway wagons such as the E.722 battery at Cherbourg, complete with two 28 cm Kurze Bruno Kanone cannons, were also utilised for defensive purposes (Chazette 2004 b, pgs. 22-23).

VII.10.4 - *Bonnet*

Military ditch defence work, equipped with gunholes suitable for various types of weapon.
In XIX and XX century fortifications, the term refers to a casemated work with gun or machine gun and rapid fire cannon positions, which transversally protrudes over the ditch. It is used in the protection (flanking) of the fort or fortified work and over the years perfected the use of oblique fire to sweep over the ditch. The type of bonnet depends on its position.

Rear bonnet: serves to protect the area behind a fortified work (rear) and faces into the fortified square. Externally, this work remains opposite the supposed enemy attack directive (attack front).

Counterscarp bonnet: serves to defend the ditch and can be constructed in counterscarp wall salients. This type of work communicates with the counterscarp gallery by means of covered walkways.

VII.10.5 - *Castle*

Mediaeval fortified work, generally surrounded by walls with towers and having various rooms and which was used either to house and defend rich or noble owners or as a territory control point.
A castle built in an elevated position may have internal underground works such as wells, cisterns, storerooms and communication passages. Occasionally, especially when in a high position, a castle's perimeter defence can be provided by the rocky flanks into which it is carved. One example is the *rock castle* of Sperlinga (Enna) in Sicily. Built on a sandstone spur, on the remains of previous rock settlements possibly dating to the XII-VIII centuries B.C., the castle first appears in literature dating to the XI century. Several of its rooms are carved into the rock and a rock village lies south of the castle.

Ricetto: a typically mediaeval enceinte wall with towers for the protection of houses, stables and warehouses and used as a shelter by those living in the countryside in times of peril. The ricetto in Candelo (Biella) is well preserved.

Stronghold (rocca): built in an elevated position, this is generally smaller yet more solid than a typical castle. This particular type of stronghold was established in the Renaissance period.

«Thus Parzival continued on his journey and briskly trotted down the road as far as the ditch. Here the bridge was raised and the secure and stalwart castle towered upright as if created on a lathe. Only the wind or one capable of flight could have possibly gained access to the castle and no damage could certainly be caused by ground assault. There were towers and high rooms in great numbers and marvellous fortifications. Should all the world's armies surround the castle for as many as thirty years, those within its walls would still not have given a single loaf of bread to be freed of them» (von Eschenbach, V, 226).

VII.10.6 - *Caponier*

Additional fortification typical of bastioned work, used in the flanking of ditches; not necessarily positioned in the ditch scarp it could be of any shape and size.
A caponier normally takes on the form of a protected open-air post (also known as caponniere). Sometimes, especially with the development of offensive and defensive systems, it developed into a casemate. Normally used by riflemen, the caponier served in defence of the bottom of the ditch by means of enfilade fire or, where there was no ditch, in the defence of the curtain wall. It may communicate with other works by means of underground tunnels.

At the beginning of the XIX century, the Prussians implement a series of works where caponiers took on a specific role in the defence of forts and strongholds. Construction of Fort Poznan (Poland) commenced in 1828 and followed the "Prussian system". Here, the military quarters and the storerooms consist of casemates protected by a thick layer of soil. The ditch counterscarp is equipped with a tunnel from which the underground countermine works branch in addition to caponiers, the passages of which are always underground (Hogg 1982, pgs. 139-142).

VII.10.7 - *Casemate*

Initially, these defence works were built at the base of the external escarp for ditch defence purposes. They were later built inside the bastion itself and used to shelter cannons; by extension the term is used to describe any type of fortification providing shelter to fire arms. In the modern day, the term normally refers to an armoured cannon shelter: casemates with revolving steel cupolas are the basis of modern forts.
One of the requirements of a fortification is that it must be able to provide as much protection as possible to the garrison during combat and in particular, it must able to protect artilleries from enemy counterbattery fire. During the XVII and XVIII centuries, the use of mortars and the new "ricochet" bombing techniques (rebound), first used by the Vauban in the siege of the city of Ath in Belgium (XVIII century) make the need to provide defence works with bomb-proof rooms, impenetrable to enemy artillery fire, all the more evident. Environments sufficiently large to hold both cannons and their operators and with gunholes for external fire were created behind curtain walls and sometimes above the ramparts (as in the Citadel of Alessandria).

The term casemate was later used to describe any bomb-proof work, even those with only light artillery loopholes or those used as military quarters or storerooms. Today the extensive range of casemates includes the many types of fortification all over the world. There are long range cannon, anti-tank cannon, howitzer and machine gun casemates. These are normally deployed to control mountain passes, bridges, road tunnels, barracks and ammunitions depots. They can be part of blockades, trenched fields, valley forts, coastal defences or of defensive lines in general.

Block: the term was adopted in 1929 (and corresponds to the Italian term *malloppo*) to indicate an isolated casemate or a group of casemates within a single fortified building in French Maginot works (Gariglio, Minola 1994, pg. 276).

Blockhaus: This German term literally means a house made out of tree-trunks. By extension, the term also refers to a defensive work, originally constructed from tree trunks and surrounded by a small ditch or by accessory defences, for protection of a garrison. The first examples date back to the American War of Independence (1778). Since the XIX century, the term has been used to refer to a *guard-house*, or a generic casemated work.

Bunker: this is another German term which refers to a reinforced concrete casemate. This is normally an underground armoured dugout, with one or two entrances providing access to the surface or adjacent works. The term is also synonymous with reinforced concrete pillbox, a semi-subterranean work or a work having rooms beneath the surface section. This may be circular in shape with a rounded or flat ceiling and horizontal loopholes for firearms (fig. VII.27).

Malloppo: this is a modern, post World War I work is normally a visual combat position, with a loophole and either cave or casemated battery. The Alpine Wall, an Italian defensive line built in the period of time between the two world wars, provides many examples. Underground passages and/or ramps of stairs sometimes linked the *malloppi* to other parts of the work.

Tobruk e Ringstände: small casemated works. «Depuis les combats de Tobrouk qui se sont déroulés en Afrique du Nord en novembre 1942 et au cours desquels les forces de l'Axe encerclées durent s'enterrer dans la sable pour ne laisser apparaître que leurs tourelles défilées, le nom de Tobrouk va s'attribuer à un petit ouvrage bétonné enfoui dans le sol. Ce dernier, plus communément appelé Ringstände, est issu du programme des fortifications de campagne renforcées, donc bétonnés, codé dans la terminologie allemande Vf pour Verstärkt feldmässig» (Chazette 2004 c, pg. 1). A common and versatile model is the standard Vf 58 *Ringstände* model, from the well of which a soldier armed with a light machine gun could shoot or an observation binocular or a visual signal could emerge: «Doté d'un orifice de 80 cm de diamètre, il présente les caractéristiques suivantes: une fouille de 58 m^3 et une grosse masse de béton de 11,5 m^3 armée de 750 kg de tores en acier. Sa hauteur est de 2,75 m, pour une largeur de 2,36 m et une longueur de 3,79 mètres» (Chazette 2004, pgs. 1-7). Other types of *Ringstände* could hold a mortar (Vf 69) or a cannon such as the Vf 600v or the Vf 221. Others were set up to contain anti-aircraft pieces or flame-throwers while tank-turrets such as the Vf 236, Vf 237 etc. originating from the Vf 67 were positioned on top of others still.

7000-type work: also known as "Pariani Posts" in honour of the Italian general who instigated their creation in 1938, this is a casemate consisting of a single concrete block. Designed to withstand small and medium calibre weapons, it was armed with one or two machine guns and sometimes with 47/32 anti-tank guns, positioned behind metal armour set in concrete (Gariglio, Minola 1994, pg. 24).

«The Bunker is instinctively the terrifying curse of the war, which still chills our memory and which we would like to be free of. But is our attempt to forget, and the further destruction that this causes, the way forward? The last decade has seen the slow acknowledgement of issues, globally attested by World War II fortifications. In particular this is due the work of urbanist, Paul Virilio, theorist of the "oblique function" who conducted an enquiry on the impact of military space on the territory» (Uggeri 1976, pgs. 261-263). In the 1970s, Virilio organised the "Bunker Archéologie" at the Centre National d'Art et de Culture Georges Pompidou in Paris.

VII.10.8 - *Cave works*

Term referring to underground military works, used to store artillery and normally excavated into mountain-sides or rocky slopes; hence the term "cave battery".

Underground excavations for defensive purposes have taken place since antiquity. Underground works are sometimes carved into rocky flanks and even complete surface works. In the latter part of the XX century in particular, there is widespread use of works cut into reliefs, used to shelter light and heavy weaponry positions, observatories and

Fig. VII.27. *World War Two observation and/or defence bunkers in Monterosso al Mare (La Spezia) (photo A. Thum).*

logistic services. One example is the so-called "Cima Grappa cave fortress" (Veneto), completed by the Italians during the First World War. It consisted of a main tunnel of approximately 1,500 m with various branches leading to 23 pieces of artillery, machine guns and observatories. With electric generators, magazines, munitions depots and tanks of drinking water, it was equipped with ventilation system with special anti-gas filters, internal compartmentalization systems and anti-gas curtains to block the openings (Giardino 1929, pgs. 124-129). In certain cases, the photoelectric cells could be mounted on wheels or rails and positioned in tunnel sections for easy removal and re-entry in the event of enemy battery fire.

Cave battery: also known as *galleria cannoniera* (cannon tunnel), this type of battery consists of one or more artillery positions, which are generally cut directly into a rocky wall and can be either isolated or linked together by a tunnel, which may in turn be equipped with accessory works. An underground system described as a "cannon tunnel", built between 1916 and 1917 at an altitude of 223 m north of San Michele del Carso. It is an obtuse-angled tunnel, with six casemates for as many artillery pieces opening up along the same side of the tunnel; a passage, leading to an observation post, branches off from one of the casemates. On the other side are two entrance tunnels, which once led to the surface and a third tunnel, which links the main tunnel with an entrance tunnel (Stocker 1999, pgs. 252-256). The inside of the tunnel are in reinforced in masonry and concrete.

VII.10.9 - *Countermine*

Tunnel or underground passage for the interception and destruction of enemy mines; these were sometimes used against siege positions or enemy troops.
This has been the counter-measure of preference against enemy mines since antiquity. It is a tunnel or a simple underground passage, excavated towards a similar enemy work, in order to intercept, occupy and destroy it. This was generally achieved by setting its wooden supports on fire. Vitruvio writes that during the siege of Appollonia the architect Tifone from Alessandria ordered that various surface communicating tunnels be excavated within the city walls. The tunnels were to be of a length suitable for the shooting of arrows, in the (successful) attempt to intercept the tunnel used by the enemy to overcome the defences and conquer the city (Vitruvio, X, 10).

From 1564 to 1570, Galeazzo Alessi underlines the importance of countermine tunnels in his "*Libro di Fortificatione in modo di Compendio*": «*Il Castriotto uuole che le contramine p[er] la importanza loro, si faccino ne luoghi asciutti, a tutti i corpi de Baloardi, Cau[alie]ri e Piatteforme: et il Maggi approua tanto q[esta] opinione che dice essere necessarie ancora ne luoghi di acqua, p[er]chè dice ancor quelli potersi minare se ben hoggi li moderni, come il Cau[alie]re Paciotto no[n] le usa più solo p[er] fuggire la spesa et p[er] il tempo, che mi corre nel farla*» («The Catriotto states that because of their importance countermine tunnels should be created in dry places, beneath the Ramparts, Cavaliers and Platforms: and the Maggi is in agreement with this opinion and adds that countermine tunnels are also necessary in those places where there is water and that although modern men such as Knight Paciotto no longer built these on account of the expense and the time involved in their construction, these areas can still be mined») (Coppa 1999, pg. 76). The siege of the Famagosta Fortress (Cyprus), ending with the return of the Venetian garrison (1571) was characterised by intense use of mines and countermines. Following this event, it was clear to European military forces how their defence works should be equipped in advance with countermine tunnels rather than during the course of a siege.

From the end of the XVI and the beginning of the XVIII centuries, fortifications are systematically equipped with underground tunnels, usually directly below the primary defensive perimeter. In the event of siege, their purpose was to identify and intercept enemy excavations and interrupt their advancement by underground combat or by using explosives to destroy the attack passage. During the XVIII century the wartime experience ensured that a permanent system of countermine tunnels, beneath and under a fortification itself, provided an advantageous and lasting defence and became an efficient, if costly, instrument of war.

The engineer, Giuseppe Formenti states the following in relation to the risk of mine attacks on the Milazzo Fortress in Sicily (manuscript No. 5524 of the National Library of Vienna): «*La ciudad baxa merece toda reflexión porque una vez perdida, se hallaría la ciudad cerrada y el cabo en mucho aprieto: porque por la demasiada altura de las murallas y cercanía de las casas hasta el pie de ellas, quedaría el enemigo cubierto y no obstante ser el sitio de peña fuerte, podría obrar con hornillos en las murallas, donde no estan contraminadas a demans que no tendría la ciudad cerrada ni el castillo por donde ser socorridos*» (Formenti 1705).

The tunnels were built using the cut and cover method or were tunnelled underground; they generally had a masonry facing and a vaulted roof to protect them from water infiltration and humidity, both necessary for the use of black powder.

«In seeking to hinder British, Australian and Canadian tunnellers, all of whom were playing their part beneath the ridge by the end of 1916, German efforts were not restricted to countermining. Low flying aircraft were sent out to examine the forward areas for any signs of tunnelling works, communicating all sightings of blue (grey-green to Germans) clay to the artillery» (Barton, Doyle, Vandewalle 2004, pg. 177).

Magisterial gallery or counterscarp gallery: approximately 1.80 m high and one metre wide, this tunnel is the basis of most countermine systems. It develops around the shape of the fortress directly under the main enceinte of the body of the place or immediately beyond the counterscarp wall of the main ditch (widely used strategy). This subterranean system, leading to the *capital tunnels*, i.e. outworks perpendicular to the fortifications' perimeter, can thus be accessed from the bottom of the ditch. Tunnels from the body of the place sometimes pass under the ditch and lead to the external underground system. Right-angled *mine tunnels* or *mine branches*, on average 1.2 - 1.7 m high and 0.7 - 1 m wide, to contain the blast wave caused by explosion of the mine, branch off from the capital tunnels. These tunnels end in demolition chambers, small chambers into which demolition charges (mines) are placed. The positioning of these subterranean defences envisages possible enemy approach to outwork defences and the body of the place.

Capital and counterscarp tunnels and demolition chambers are similar in size throughout Europe, although smaller ones have been known such as those of the Verrua Savoia Fortress in Piedmont (Padovan D., Padovan G., Bordignon, Ottino 1997, pgs. 187-208). Such uniformity is due to the war experience gained and to the advantages of military treatises on underground war (Amoretti 1965, pgs. 39-52. Duffy 1996, pgs. 82-83).

The systems are positioned at a depth of approximately 3-4 m, although they can sometimes descend to even 10-15 m and extend for several kilometres. They generally have ventilation shafts or pipes to ensure air circulation. Where such systems do not provide sufficient air exchange, a forge bellow is used to introduce air through tin or wooden pipes. The shafts can also be used to supply surface garrisons and as a means of verbal communication for the co-ordination of mine action in respect of events on the besieging field. The underground passages also have drainage

wells, which collect any water infiltrations (Amoretti 1965, pgs. 57-102). In French mining jargon, an underground mine is known as a "camouflet".
Demolition chamber: masonry or dugout chamber containing an explosive demolition charge. The demolition chamber has multiple purposes:
- to eliminate an enemy mine by causing it to collapse through the explosion of an underground charge;
- to destroy enemy siege works by blocking entrance to the passage with mounds of soil and then causing an explosion, which thus vents upwards and creates a crater on the surface;
- to destroy the works of a fortress under siege and now under enemy occupation (demolition passage or tunnel).

During the French siege in 1706, the countermine system of the Turin Citadel is articulated in *lower tunnels* situated at a depth of approximately 14 m and *upper tunnels* at a depth of approximately 6 m.
- from inside the Citadel and more specifically from the centre of the three external-facing bastions and the two curtains (these inclusive), branch the *lower capital tunnels*, one for each of the aforementioned works, which push to beyond the furthest outwork defences and have several mine branches;
- behind the first glacis along the counterscarp wall is a *magisterial tunnel* from which numerous mine branches branch off;- at the point where the two systems meet, there are five communication steps: one of these was blown by the miner Pietro Micca (known as "Passapertutt" [Pass-everywhere]) to block the irruption of French soldiers who had gained access through the Demilune's upper capital entrance (Amoretti 1996).

VII.10.10 - *Counterscarp gallery*

This work is set into and runs along the length of the ditch counterscarp wall.
Normally equipped with loopholes, the counterscarp gallery provides cover for the troops inside the passage, allowing them to attack any enemies with the ditch with 'reverse fire'. The term reverse means that the shot is not fired from the Body of the Place in an outwards direction, but is fired inwards, from the counterscarp wall. The counterscarp gallery may have avant-corps, casemates and caponiers.

The counterscarp gallery of Milan Castle, named "secret inside road" by Leonardo da Vinci, is one example (da Vinci, manuscript B – folio 36 verso). In brick or rare stone tunnels with barrel vault, the tunnel has a hundred or so double-embrasure large corner apertures, communication areas with ditch ravelins and many tunnels leading to the external *Ghirlanda* (Garland) (Padovan 1996, pgs. 64-75). The Bastion of Saint Ignatius in the Demonto Fort (Cuneo), was defended by a dry ditch with counterscarp gallery, an integral 11.36 m section of which, presenting three deep loopholes overlooking the ditch and the entrance sortie (buried) survives. A loophole and countermine tunnel can be seen in an uncovered section of the same stone, masonry lined tunnel. Beneath the tunnel is a hydraulic conduit for the discharge of water from the ditch (Padovan 2003, pgs. 320-333).

VII.10.11 - *Cupola*

Type of hemispherical armoured casemate, inclusive of those which spin on their own axis.
Also known as gun turret, the cupola is made of metal and is armed with a howitzer or cannon; alternatively, it can house machine guns or photoelectric cells. It can also be used as an armoured observation post to direct artillery fire. It is more commonly hemispherical in shape although it comes in many shapes according to its purpose. Its element can be of the "disappearing" variety, that is, it has systems which enable it to retract inside the casemate thus avoiding both visual exposure and enemy fire. It is generally the main active element of modern forts. Whether isolated or linked to defence works, it is an integral part of a fixed defence as are the *wall*, the *centre of fire*, the *blockade*, etc.

VII.10.12 - *Defensive Tambour*

Bomb-proof defensive work with a purpose similar to that of the bonnet.
This is a cylindrical casemate (hence its name), in the magisterial wall of a fortified work, protecting the ditches and glacis by means of artillery fire. The tambour may be in a detached position from where it provides defence of the rear of a work.

VII.10.13 - *Demolition passage*

This refers to an underground work or work built within the defence perimeter when it has become indefensible.

In bastioned fortifications, the demolition passage is created below accessory works such as tenaille lines, counterguards and ravelins. It may be part of a system consisting of a main tunnel allowing rapid access to demolition passages with chambers. As previously mentioned, its purpose is that of rendering useless that which has become indefensible.

Fougasse: very similar to a countermine, a fougasse is less deep. A fougasse is placed beneath the glacis in traverses and demilunes, at a depth of no more than 4 m below the natural ground level; its purpose is to detonate in the front of advancing enemy infantry. In XIX and XX century works, a fougasse is part of various systems used in the demolition of pedestrian, road and railway systems.

VII.10.14 - *Demolition tunnel*

Demolition work within defence works or under roads.
Underground demolition works, for the obstruction of roadways were created in addition to fortifications systems, particularly in the XIX an XX centuries. Demolition tunnels and demolition chambers completed the defences of blockade fortifications and road cuttings; similar works were also placed inside railway and road tunnels.

The demolition tunnel could be part of the bastion's very structure and was used both in its defence and partial demolition thus allowing a second, rear-lying bastioned front to be created. A very specific and rare typology, this is generally a large tunnel (of up to 6 m in height and 6 m wide), which follows the internal outline of the bastion's two external faces. It is equipped with small wells in its arch and externally branching demolition passages. Where the bastion is partially demolished by battery fire or mine explosion, the bore holes are detonated to 'knock' the two external faces of the bastion into the ditch, thus creating a salient angle (outward, enemy-facing angle) and a new ditch (the tunnel itself, uncovered) with full scarp wall (internal breastwork). It also serves as a bomb-proof shelter and as a passageway from one side of the bastion to the other.

VII.10.15 - *Fort*

Defensive work of limited dimensions, containing only military works within its walls.
A fort is a work which may be isolated and is utilised in the surveillance and defence of a locality or an obligatory passage. It may also be part of a trenched field or a fortified region. Once its defensive function becomes redundant, the work is often used for other purposes, as a depot, gunpowder magazine or detention centre. In Italy, XX century forts are known as *armoured batteries, armoured forts* or simply *works*.

Fuentes Fort, a bastioned hillside fort, is the largest of its type in Lombardy. It was built in defence of the territories lying north of the Duchy of Milan at the beginning of the XVII century, upon the orders of Pedro Enriquez Azevedo, Count of Fuentes. It blocked the raids of the Grigioni of Valtellina as well as possible French and Venetian attacks (Giussani 1903. Padovan 1997, pgs. 293-298). Built from local stone, its defensive perimeter, including the tenaille, which houses one of the gates, remain intact. Fort George in Inverness, Scotland is one of the best conserved bastioned forts in Great Britain. The Vauban-style fort was built at the far end of a small peninsula by William Skinner between the years of 1747 and 1769.

Between the end of the XIX century and the beginning of the XX century, the Italian-Austrian border saw the construction of many fortifications. Although for different purposes, both countries had armoured batteries at the core of their defence system. Italian forts built during this period were always casemated. The forts were built on a rectangular ledge of bomb-proof concrete, which is 10-15 m wide and 60-80 m long. This type of complex normally has two storeys and is built on a previously cut rock ledge. Its surface, level and concealed surface in respect of the nearby soil, only the metal cupolas of well installations emerge. The armour normally consists of 4 to 6 medium calibre cannons placed on a single line parallel to the battery's largest axis. The weapons are fixed to revolving carriages and steel cupolas with 360° rotation are firmly attached to the carriage itself (Minola, Ronco 1998, pg. 69. Girotto 2002, pgs. 127-128). The forts have machine gun positions and are surrounded by ditches and barbed wire entanglement.

Armoured batteries were widely used during the First World War. Modern day armoured batteries are primarily masonry, stone or armoured sheet works and the more recent models are generally concrete and reinforced concrete buried works with artilleries placed under retracting or more commonly, armoured cupolas. Fort Lusardi (Lecco) is the only Italian armoured battery to survive with its full heavy weaponry. Also known as "North Montecchio Fort", it is part of the Cadorna Line and its preparation works began in 1911. There were some improvements made in 1916.

The fort has four 149/35 cannons with revolving cupolas ("S" installation, Schneider) (Flocchini 1994, pg. 48). It has underground weapons depots and water storage cisterns (figs. VII.28 and VII.29). All other forts, such as the Austrian Three Stones Fort guarding the Val Parola (Alto Adige), were destroyed during the war, demolished for their metal or have had their artillery removed.

At the turn of the XX century, Switzerland was also protected by various fortifications. There is the Saint-Maurice Blockade (Valais) consisting of Fort Savatan, Fort Dailly and the "Batterie du Mex" with eight small calibre cave cannons, while the Gondo Blockade (Semplon) consists of Fort Gondo, the Figenen Battery and small blockhäuser and shelters. There is also the Trenched Field of San Gottardo, the central Reduit of Andermatt and the Trenched Field of Bellinzona. All these systems are equipped with forts, batteries and emplacements (Ascoli, Russo 1999, pgs. 211-213).

An old Russian proverb says: «He who has Przemyśl has Galicia». Galicia means "industry" and Przemyśl became the main point of Austrian defence in Galicia, and is now considered one of the most imposing fortifications in Europe. A 48 km ring of nineteen forts and twenty-three redoubts provided external protection to the fortified city. The design of the forts followed the "Brialmont system", with structured subterranean installations. During the First World War it is besieged by the Russians and is conquered on the 22 March 1915, due to the lack of food (Hogg 1982, pgs. 195-196).

Fort Eben-Emael was built in Belgium between 1932 and 1935 and lies north of Liège, close to the Dutch border. The fort controlled the Albert Channel, the River Maas, westbound roads from Maastricht and the canal's bridges. Considered impregnable, partly on account of the Albert Canal, which acts as an enormous ditch, it has casemated batteries and steel cupolas, two of which are retracting anti-aircraft defences. A defensive ring of fortlets and casemates surround the fort and there are false cupolas to fool the enemy. Anti-gas systems have also been installed. On 10 May 1940, Eben-Emael was conquered in just a few hours by German Assault Group Granite paratroopers, who gained access to the fort by means of DFS 230 assault gliders (Setti 2004, pgs. 96-129).

The fortified complex of Boden, on the River Lule in Sweden, is an example of a trenched field consisting of a crown of armoured forts and batteries, linked to casemated defence works and infantry trenches. Work commenced at the turn of the XX century and some works remained in use until recent times. In 1988, the Boden complex was included in the Cultural Heritage conservation list. Rödberget Fort was opened to the public and a reception area created nearby. In relation to Fort Mjösjöberget: «the bulk of its surviving outfit will remain on-site and the fort will be sealed off until such a time as it may explored in the not too distant future, by archaeologists expert in the field of military architecture» (von Kartaschew 2002, pg. 16).

Blockade or *road block*: a blockade or fortified blockade generally controlling access to a valley road carriage or obligatory passage. The Tombion Blockade was built in 1885 and is a road block controlling the carriageway from Valsugana to Bassano del Grappa via Cismon. «The blockade consists of two single-storey casemated batteries, in perpendicular position to the road; of a two-storey building serving as military quarters placed between the two and of a hidden two-storey, parallel to the road section. The latter was equipped with a lightning conductor (Faraday cage) to protect the powders it stores as well as the office-dwelling-armoury area (...). Water supply was guaranteed by the perennial spring inside the work itself (southern side of the south court) although there was also a large cistern, with double filtering system and a capacity of 80 m^3, was also used to collect rainwater from the horizontal surfaces of roofs within the complex (Girotto 2002, pgs. 88-89).

Fortress: fortification work, with basic, continuous enceinte and additional internal or external works; the term is sometimes synonymous with stronghold. One example is that of the city-fortress of Palma, modern-day Palmanova (Udine), the construction of which began on 5 October 1593 under the "La Serenissima Republic of Venice". Completed in 1623, its nonagon structure corresponds to a bastioned star work with nine ramparts in the first enceinte. A second enceinte was again built by "La Serenissima" from 1658 and 1690 and a third was built during the Napoleonic period, from 1806 to 1809, (Di Sopra 2003, pgs. 11-20). «It is thought that Palma was conceived as the "ideal fortress" rather than as the "ideal city" as its design almost exclusively caters for military needs» (Bertossi 1993, pg. 29).

Fortification: this term refers to the construction of defence works, including the pentagonal, walled *bastioned-front fortification* with ditch, scarp and counterscarp. These may be subdivided as follows: *permanent fortification, semi-permanent fortification and field fortification.*

Fig. VII.28. *Fort Lusardi (Montecchio Nord), in Colico (Lecco). The only Italian First World War fort to survive intact. In the foreground is one of the four revolving turrets ("S" installation, Schneider), armed with a 149/35 cannon (photo G. Padovan).*

Fig. VII.29. *South Battery of Fort Osoppo (Udine) built at the beginning of the XX century. It consisted of four 149 A "Grillo" armoured cannons (Fiala 1988, pg. 249. Club Alpinistico Triestino 1994, pgs. 34-37). A speleologist inside one of the four well installations with no cupolae or artillery pieces (photo G. Padovan).*

Fortlet: a small defence work normally armed with light weapons and small calibre artillery. One example is the Adda Fortlet, a XVI century work which protected the final section of the River Adda (Lombardy) before its emission into Lake Como. The fortlet is an accessory work of the Fort of Fuentes fortified system.

Stronghold: this refers to a fortified work, generally from the late Renaissance period, which more or less surrounds a village or city, its type of defensive systems depending on its period of construction. This work was used to shelter garrisons defending an area and as a logistics base.

VII.10.16 - *Fortified cave*

Term referring to natural cavities, of which, at least the entrances are protected by a defence work.
Over time, natural cavities, particularly those with wide entrances and horizontal structure, were often equipped with fortification works. These may have a basic closure wall or may be articulated defence works in the tunnel below the parapet. There are many fortified caves in the karst area of Hong Lin (China). The large entrance of the Shui Xiang Dong cave, located directly beneath the village of Hong Lin, is approximately 120 m high and is crossed by a small, external watercourse. The cave is partially blocked by an imposing dry wall created from local, grey-coloured limestone chips, often set in irregular rows. The upper part of the best conserved section of this defence work has square and rectangular loopholes (Buzio, Mengoli, Zorzin 2005, pgs. 131-132).

The Castel Lueghi Cave in Slovenia, better known as Predjama, is set out on four levels and extends for approximately 5 km. The upper level, known as Erazmova jama (Erasmus Cave), holds the remains of a XIII century fortress. The entrance to a new castle (Predjamnski grad), conserved and perfectly visible to this very day, was built in the latter part of the XVI century (Aallen, Strinati 1976, pg. 208).

Corona (crown) and *covalo* (or *covelo*): these terms refer to castles built in a cave or in large rock shelters (Gorfer 1985, pgs. 178-179), such as San Gottardo Castle in Mezzocorona (Trento). In the Istrian karst area, “Tabors” were built on top of mountain reliefs and even in caves. These were makeshift but efficient defensive emplacements. They were used between the XV and XVI centuries in the protection of both goods and persons from Turkic raids (Radacich 1993, pgs. 11-20).

VII.10.17 - *Gun powder magazine*

Room, or rooms, for the storage of gunpowder and munitions.
Given the intrinsic danger of the stored material and the destructive effects that accidental explosion would have on surrounding buildings, a gunpowder magazine must have specific characteristics. Some fortifications or military complexes have underground gunpowder magazines, while in others the magazines are on the flank or in the flank of another fortification. They generally have wooden floors, air gaps to provide thermal and protection against humidity and groundwater flow and sometimes have a small guard-house. Some XIX and XX century examples also have a “Faraday cage”.

The gunpowder magazine at Fort Lusardi in Colico (see paragraph IV.10.15) was built inside the rocky relief behind the armoured battery. There are six quadrangular artillery magazines, each of different dimensions, on each side of its long corridor. The walkway has a dehumidification system. «The dehumidifier, consisting of galvanised metal sheets, runs along the ceiling of the gunpowder magazine, following its lines, curves and arches. Small bricks hold the metal sheet a few centimetres from the ceiling and air circulates within the gap, cooling the metal, which in turn condenses the humidity. The droplets of water slide down the dehumidifier’s smooth surface and are channelled into gutters, which divert the water to the ground. A cement channel then carries the water along the corridor where it is discharged into special drains, which remove the water from the fortified work» (Cassinelli 2002, pg. 29).

VII.10.18 - *Mine*

Underground passage providing access to a fortified work. This normally refers to both a defensive and offensive work, placed under either permanent or field works, with a bore hole at its summit for its future demolition.

The aim of a siege is the conquest of the fortress or walled city by surrender of its occupants or by direct assault. In the case of the latter, the defensive perimeter can be scaled or alternatively a gap can be created (breach) with siege weapons or by undermining the support stones at the foot of the wall, causing it to collapse.

In his chapter entitled «*De cuniculis, per quos aut murus defoditur, aut civitas penetratur*», Vegezio tells us that a type of hidden, underground siege, is known as *cuniculus* and derives from rabbits, which dig burrows in the ground. One system is for soldiers to gain access to the besieged city by means of an underground passage (mine) and to then open the gates. However, an underground passage can also be taken under the foundations where a chamber with underpinned arch is created and supported by wooden frames, to be later set alight. Once the support is removed, the wall collapses (Vegezio, IV, 24).

These specific siege techniques have been known and used since antiquity. Titus Livius refers to these very techniques in his rendition of the Roman conquest of Veii, following the siege of the city using field works: «*Operum fuit omnium longe maximum ac laboriosissimum cuniculus in arcem hostium agi coeptus*» («The most important and most demanding of these works was the creation of a tunnel, which we carved towards the enemy stronghold») (Livius, V, 19, 10). Upon mobilization of the tunnel, the legionnaires attack various sections of the walls to attract the attention of the defenders, thus allowing 'special troops' to emerge from the passage. «*Haec precatus, superante multitudine ab omnibus locis urbem agreditur, quo minor ab cuniculo ingruentis periculi sensus esset. Veientes ignari se iam a suis uatibus, iam ab externis oraculis proditos, iam in partem praedae suae uocatos deos, alios uotis ex urbe sua euocatos hostium templa nouasque sedes spectare, seque ultimum illum siem agere, nihil minus timentes quam subrutis cuniculo moenibus arcem iam plenam hostium esse, in muros pro se quisque armati discurrunt, mirantes quidnam id esset quod cum tot per dies nemo se ab stationibus Romanus mouisset, tum uelut repentino icti furore improuidi currerent ad muros* (...). *Cuniculus delectis militibus eo tempore plenus, in aede Iunonis quae in Veientana arce erat armatos repente edidit, et pars auersos in muris inuadunt hostes, pars claustra portarum reuellunt, pars cum ex tectis saxa tegulaeque a mulieribus ac seruitiis iacerunt, inferunt ignes*» («Having thus prayed, mighty forces attacked the city from all angles, in order that the impending danger from the tunnel would be not be noted. The Veiians, unaware of the betrayal on the part of their diviners and foreign oracles did not suspect that the gods had already been summoned to partake in the loot and that others, evoked by prayers from their city, were casting their eyes upon enemy temples and new sites. Nor did they suspect that that which they were now living was their final day. Fearing all but the possibility of the stronghold being overcome by enemies whom had gained access through a tunnel under its walls, they ran in arms, each one unto himself, scattering themselves here and there on the bastions, unable to understand for what possible reason, after so many days in which not a single Roman had moved from the outposts, they now ran blindly, as if overcome by sudden rage, towards the walls (…). The tunnel, which in that moment was full of select soldiers, suddenly unleashed a swarm of armed soldiers into the Temple of Juno in the Veii stronghold: some of them charge the walls surprising their enemy from behind, others remove the gate chains while others still lit the fire as women and servants threw stones and tiles from the walls») (Livius, V, 21, 4-7, 10-11).

In the IV century, the city of Dura Europos (Siria), held by the Romans, is conquered by a Persian army: archaeological investigations carried out on the remains of the fortress have provided an image of a true underground battlefield consisting of demolition passages and countermine tunnels (Bonetto 1997, pgs. 337-398. Please refer to: Mesnil du Buisson 1939 and 1944).

The excavation of passages could also serve other purposes. During the siege of Uxellodunum, the fortified city of the Carduci, Caius Julius Caesar had a passage built to drain the spring emerging just outside the city walls «*ad postremum cuniculis uenae fontis intercisae sunt atque auersae*» («the veins supplying the spring were blocked or diverted by means of underground tunnels causing the spring to dry up») (Caesar, VIII, 43). The episode is also quoted by Frontino ("*Stratagemata*") in a paragraph dedicated to the methodologies of spring deviation and pollution: "*De fluminum deriuatione et uitiatione aquarum*" (Frontino, *Str.*, III, 2).

From the XII-XIV centuries, city walls are characterised by curtains with scarped base, better able to absorb a "mine attack". In a letter proposing his services to Ludovico il Moro, Leonardo da Vinci affirms his ability to "bring down" (collapse) any stronghold or other fortress without resorting to bombardment, with the exception of strongholds «*non fusse fondata in su el saxo*» («built on solid rock») (da Vinci, C.A., 391 r.a.). In later centuries, artilleries in special "breach batteries" have the task of effecting breaches. The system was frequently expensive in terms of both men and weapons and required long periods of time. Where there were no significant results, mines were utilised following one of the below methods:

- *Deep mine:* in this case the wall to be mined is approached by underground means; an underground passage is excavated, with wooden, even pre-fabricated reinforcements. This may have a continuous series of right angles in order that an explosion's blast wave cannot erupt along the passage itself. One or more bore holes are prepared under the curtain to be destroyed. Once the explosive has been placed, the passage is filled with soil, so that the explosion blasts upwards, causing far more damage than a mine attack. In the latter part of the XVIII century, Forest de Belidor, a French engineer, formulates new concepts and applies the so-called "Globe of Compression", a type of 'supermine' (Gillot 1805) with devastating effect.

- *Mine attack:* the section of curtain to be mined is reached by open-air approach. Once the external wall facing has been undermined, a small hole known as a bore hole or demolition chamber is cut into its depth and is filled with explosive. So long as these are sufficiently powerful and well placed, the detonation of two or three bore holes causes a lot of damage. Open-air approach renders this method particularly rapid but exposes excavation personnel to serious risks, which may influence its successful outcome.

Mining Companies and special artillery divisions consisting of civilians employed in mines or caves operated the mines. They normally worked in teams of four or more people:
- the first person cuts the ground with has "pickaxe";
- the second collects the loose soil;
- the third moves the containers of soil to the entrance;
- the third conceals the leaf mould which would alert the defenders and lead to them digging a countermine.

The carpenters prepare the frames and the axes for reinforcement of the passage. A close-knit team of miners can dig a 4-5 m passage in twenty-four hours, even under a ditch full of water. So long, that is, that the excavation is not being carried out in compact rock. In which case, progression is significantly slower.

Demolition shaft: this is a vertical excavation, generally used in fortifications or military works, allowing access to the area in which a *bore hole or demolition chamber* is to be created.

It was only with the development of artilleries, especially large calibre artillery (mid XIX century) that mine techniques was temporarily abandoned. There was a brief ceasefire in the Russian-Japanese War (1904) when General Kiten Maresuke Nogi besieged the Russian Port Arthur in Manchuria (China). Following disastrous frontal attacks and while waiting for adequate artillery, General Nogi resorts to rational siege strategies: approach trenches and mines.

During the First World War the use of mines and countermines was an attempt to split front staticity based on the defensive barbed wire entanglement-trench-machinegun trinomial and the use of artillery. Their widespread use and tragic consequences primarily affected the Italo-Austrian mountain front. At Lagazuoi Piccoli in the Dolomites, the Italians managed to take hold of the Martini Ledge and an underground war between the two contenders thus began. On the flank of the Tofana di Rozes, in the immediate vicinity, the Italians cut into the hard dolomite, creating a tunnel leading to beneath the Austrian Castelletto positions (Viazzi 1976, pgs. 160-178) and decide to excavate only one demolition chamber: «Initially, Tissi had planned two chambers. Malvezzi felt pushed for time on account of the presumed countermine and made do with one. It was 5.00x5.50 m, with a central height of 2.30 m (63 m^3), just the right size to accommodate the 35,000 kg of explosives which finally arrived» (Striffler 1994, pg. 270).

«Although mining had been going on for nine months before I took over the appointment of Inspector of Mines, I did not realise that geology had anyting to do with military mining. We were not mining very deeply in those days, and naturally if came to a wet place, if the Germans were mining we continued, and if it was wet we pumped; but there was no geological science so far as that was concerned. It was when we came to deep mining that its importance was obvious. BRIGADIER-GENERAL R N HARVEY, 1919» (Barton, Doyle, Vandewalle 2004, pg. 71).

VII.10.19 - *Postern*

This is a small defiled, open gate in a hidden location, situated far from the main gate and used only under specific circumstance.

In bastioned works the term normally refers to an open gate in the part of the bastion covered by the orillion; also known as a false gate, secondary or sortie gate, its use is of easy interpretation. By extension, the entire tunnel, providing access to the gate, whether this be underground or within the walls themselves, is known by the same name.

Eurialo Castle in Syracuse has a well-structured example of underground "dynamic defence" works. Built by Dionysus between 402 and 397 B.C., it sits at the top of the great walls that close the Epipole terrace and controls the road, which once connected Syracuse with the inland parts of the island. With tenaille works and ditches, it was designed for sorties and counter-attacks. A series of tunnels and postems allowed the enemy to be surprised both laterally and from behind in advanced phases of attack. Built along a western-eastern axis, its west apex (the most vulnerable point) has three ditches, the middle one of which is relatively wide (Mauceri 1939. Cassi Ramelli 1964, pgs. 41-46). A tunnel connected to the lookout tower's avant-corps runs parallel to the rear-lying ditch, into which it opens with numerous sorties. A tunnel completely flanks the advanced work near the ditch barrier wall and joins the gate protection fort (positioned in the tenaille) and the two postems in front of it, concealed by transversal walls. Another tunnel section with postems runs under the north walls.

VII.10.20 - *Ravelin*

Work separated from the enceinte, built inside a castle or fortified work.
The ravelin was primarily used in mediaeval times to protect the castle gate and therefore its entrance. An underground room and a stairway (now buried) leading to the counterscarp gallery have been uncovered in the Porta Comasina Ravelin of Porta Giovia Castle in Milan. (Padovan 1996, pgs. 132-133). In bastioned fortifications this work, consisting of two faces or two faces and two flanks, was placed in front of the curtain. It could house casemated rooms, communication tunnels and entrances to countermine tunnels or demolition works.

VII.10.21 - *Redoubt*

Small fortified work, whether isolated or part of a defensive system.
A relatively unimportant military work, the redoubt may be isolated or part of a larger defensive system. Gariglio and Minola list three types:
- small bastion also known as a lunette, generally found at the foot of the glacis.
- small quadrangular or irregular shaped fort for the reinforcement of trenches or trenched fields;
- small fort providing bridge, lock protection etc.; masonry and/or casemated fort (Gariglio, Minola 1994, I, pg. 276).

Casemated, masonry redoubts are known by the term guard-house or fortlet. This term is sometimes used to refer to a "fortified defensive line". In the past, these were sometimes a fortresses' last line of defence. The Marie-Thérèse Redoubt is a guard-house situated to the left of the Arc, south east of Avrieux in France, built to block the Mont Cenis road. Constructed by the Reign of Piedmont and Sardinia, between 1819 and 1825, it is part of the Esseillon barrier. This perpendicular, "Montalembert" stronghold is in the shape of a horse-shoe, is structured on three levels and has a counterscarp gallery: «To the left of the sleeper bridge, a stone stairway on the high counterscarp wall leads to the ditch. A defensive tunnel, with vertical loopholes to batter the ditch with enfilade fire, was built inside the wall. One of the loopholes in the north side has three embrasures, which all face different directions, thus allowing different areas to be covered with fire. Two loop-holed masonry traverses connect the tunnel to the west wing of the Redoubt. Sixty metres of underground tunnel, branching off from the left passage, connects the Marie-Thérèse Redoubt to a recently restored guard-house, situated within the national road » (Gariglio, Minola 1994, I, pg. 219).

VII.10.22 - *Reduit*

Small fortified work, whether isolated or part of a defensive system.
Like the *redoubt,* this is a relatively unimportant military work, which can be either isolated or part of a larger defensive system. Gariglio and Minola list three types:
- small ravelin or demilune, inside a larger ravelin or demilune;
- temporary or permanent work, where the combatants retreat after a first defence;
- in fortified cities this is a small citadel situated opposite to the true citadel; it is normally a rear bastioned fortification (Gariglio, Minola 1994, I, pg. 278).

In some cases the term 'reduit' refers to an area fortified by forts, field batteries or semi-permanent works. The area protected may contain armament or other types of production centres.

VII.10.23 - *Road Tunnel*

Underground carriageway, specifically used for passage to military works.
Road tunnels were normally built in mountain areas to serve both permanent and field military works. During the First World War, it became necessary to supply the front lines, scattered on mountain reliefs.

The most well-known tunnel is the "Road of the 52 Tunnels", built in 1916 by Italian soldiers in order to send supplies to the positions on the Pasubio Mountain (Veneto slope); the road leads from Bocchetta Campiglia to Porte del Pasubio and is approximately 6330 m in length, 2300 m of which is tunnelled (Pieropan 1978, pg. 23). There are artillery embutements at certain points and the eighteenth tunnel has five small wells leading to bore holes for the eventual obstruction of the roadway (Gattera 1995, pg. 30). Other systems, such as military depots, present many examples of cut and cover tunnels.

VII.10.24 - *Shelter*

Normally underground shelter, built and equipped to protect both people and materials; also known as "refuge".
This can be constructed from reclaimed materials or existing works such as basements and cellars can be used. Alternatively, a shelter can be planned and built for specific purposes like generic bomb shelters or atomic shelters. Some XVII-XIX century works have rooms or external or semi-subterranean works which are "bomb-proof".

Air-raid shelter: also known as bomb shelter, this is generally a casemated work built to protect military and civil personnel (figs. VII.30 and VII.31). Between World War I and II and during the course of the latter, shelters were built under public buildings and factories, for primarily civilian use. They had blastproof and shrapnel-proof walls, reinforced doors and ventilation systems, etc. (fig. VII.32). In 1933 in Milan, a first, experimental, air-raid shelter was built inside a public building in Piazza Ascoli. This building is now the site of the "Virgilio" Institute (Thum 2005, pgs. 71-80). Many shelters were created in cellars and basements: with basic underpinned ceilings. These were basically designed to withstand not only the bombs dropped by planes but also the collapse of the overlying building. Investigations for the research and census of both civilian and military air-raid shelters began in Trieste several years ago (Bigi, Gobessi 2003). Details of bombings on the city and the re-utilisation of old hydraulic storage works were also documented (Bigi, Gobessi, Radacich 2004).

Bomb-proof shelter: generic, generally casemated work for the protection of both men and materials from artillery fire and subsequently from air-raids. These were built everywhere and many forms exist: open-air, basement and underground.

Hochbunker: this is a particular reinforced concrete building in the shape of a large pencil, built on the surface, with one or more underground rooms. There were various types however each model had the same external feature.

VII.10.25 - *Souterrain*

Room or rooms, built below the ground's surface and specifically used for military purposes.
Military works can have various subterranean rooms for various purposes. During the course of time their purposes can change due to new logistic or defensive needs. Souterrains can be used for the storage of wood, coal, weapons and barrack materials. They can also be used as military or cantonment quarters in times of need. Under the buildings which flank the place of arms in the XVII century Fuentes Fort (Lecco), there are underground rooms which were once used as stables, depots and cellars.

VII.10.26 - *Traditore*

Battery situated in a defiled position or casemated accessory building to a fortress.
The term *traditore* refers to the barbette battery (open air) or casemated battery, hidden by the orillion and positioned in the bastion's sunken flank. Its purpose was to flank the curtain and the adjacent bastion front and to control the area in front of the ditch. In modern fortifications this is a casemate separate to the main body of the fortification, which is generally armed with machine guns and rapid fire cannons, for defilade enemy fire. It was sometimes positioned at the entrance or could be positioned to control an especially delicate section around the fort. Fort Busa Verle, built at the turn of the XX century by the Austrians on the Vezzena Plateau (Veneto) had a "*traditor*" with two 80 mm mod. 9 cannons in embrasures (Gorfer 1987, pg. 620).

VII.10.27 - *Trench*

Field fortification work consisting of a rectangular or, more frequently, trapezoidal longitudinal excavation.
In the siege of bastioned fortifications, the approach trench was the dugout walkway from the contravallation line to the stronghold under siege. Sébastian Le Prestre, Seigneur of Vauban (1633-1707), observes that a stronghold attack depends on the pickaxe-shovel binomial for the excavation. The word "excavation" ("*sape*") refers to the head of a

Fig. VII. 30. *Fanni shelter and military depot in via Don Bosco in Cagliari; used during the Second World War. The main tunnel is carved in limestone rock and has a cement base and levees (photo G. Padovan).*

Fig. VII.31. *Fanni shelter and military depot in via Don Bosco in Cagliari; used during the Second World War. This tunnel section is lined in reinforced concrete (photo G Padovan).*

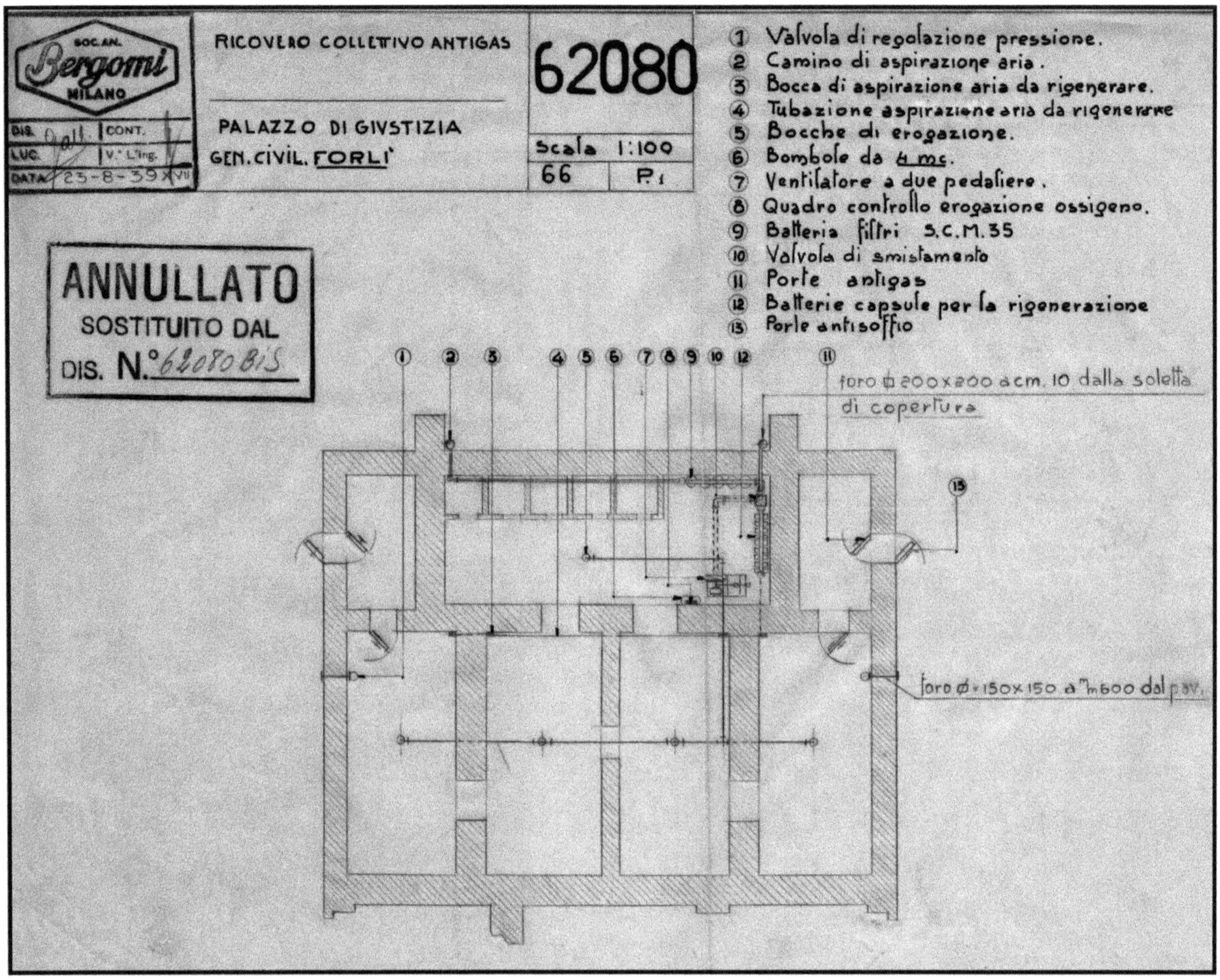

Fig. VII.32. *Milan S.A.B. (Società Anonima Bergomi) air-raid shelter project, 1939 (Archive A. Thum).*

trench which is extended and made deeper, little by little, by day and by night thus allowing men, equipment and artilleries to get closer to the place to be besieged and limiting losses. In the chapter entitled "*De la sape*" ("*Excavation*"), Vauban thus explains the abovementioned concept «Nous entendons par la sape la tête d'une tranchée poussée pied a pied, qui va jour et nuit également» (Vauban 1742, pg. 67).

In both the American Civil War and the Boer War, trenches were of great important in allowing soldiers to dodge rifle-fire. During the First World War (1914-1918), initial manoeuvre warfare tactics develop into a harsher "trench warfare".

A typical Great War trench has the following primary elements:
- parapet, a wall facing the enemy;
- a platform for the shooters, elevated in respect of the area occupied by infantry.
- a walkway in the deeper part of the trench.

The walls could be protected by sand-bags, reinforced with timber planks and covered with boards, wattle or XPM expanded metal panels, like in British trenches. There is no shortage of wicker or metal gabion shelters. A small channel for water drainage was dug beneath the gabion and a wooden walkway was sometimes built to raise the ground surface. These walkways, which were parallel to the front-line, were not straight but were broken and were sometimes similar to gear-teeth in appearance. The parallel trench lines were connected one to the other by other trenches. They were protected by entanglements or reels of barbed wire. The sectors came in various shapes and sizes and differed one year to the next according to the new defensive and offensive tactics adopted.

Shelters for both troops and officials lined the flans of trenches. These could be vast, reinforced concrete works but were more often simple holes in the ground, which did not provide any guarantee of safety against artillery fire. Trenches on the mountain front were often carved directly into the rock and had shelters, where these were possible (please refer to: Barton, Doyle, Vandewalle 2004).

«Is this the front line? Yes Sir. They are all pits similar to this one», Carlo Salsa, Infantry Lieutenant in the First World War tells of his arrival at San Michele del Carso in 1915. This is how he describes the trench he is to occupy with his men: «In the low trackway the soldiers are forced to squat in the mud so as to avoid becoming targets; the uneven edges of the shelter barely covers the men's heads. There is no room for movement; this ditch we are in is crammed with soldiers, retracted legs, guns, heaps of munitions boxes and is overflowing with rubbish: everything is caked in mud as tenacious as birdlime. The edge of the trench is bulging with bodies all in disarray: one by one I am able to make out human forms» (Salsa 1982, pg. 65).

The use of trenches did not cease in 1918 with the end of hostilities. Although with different criteria, they are still used today, in the wars, which due to our inability to learn and apply the lessons learned, continue to inflict our planet.

VII.10.28 - *Tunnel*

In fortifications this is generally a passage carved into the city walls or underground and then covered.
A tunnel is a subterranean communication passage, built either underground or within curtain walls, allowing passage from one sector of the defence perimeter to another, out of enemy sight and out of the line of fire. A tunnel can has many purposes and is built with such purposes in mind. Reference can be made to Heinrich Schliemann and how he was struck by the cyclopitic walls of the city of Tyrins, in Argolis to the south west of Argos, mentioned in antiquity by Homer, Pausanias and Strabo. He provides a brief description of the tunnels with pointed vaults made from a system of large overhanging rocks: «The south walls have covered (underground) tunnels, of unique construction. The east wall contains two parallel corridors, one of which has six niches in its external wall. The south wall holds a 4 m long tunnel, at the centre of which is an enormous door jamb with a huge hole for the bolt. It is therefore assumed that, when necessary, the passage could be closed. These tunnels obviously provided a means of passage between the two towers or guard-houses» (Schliemann 1962, pgs. 68-69).

In more recent works and in specific situations, tunnels may be carved in the rock. Such tunnels generally lead to cave works or positions outwith the main body of the work or simply provide external access which in defilade position, permit sorties or allow reinforcements and provisions to reach the fortification. During the First World War, communication tunnels were built inside glaciers or icehouses although this was chiefly on the Italian-Austrian front. In Marmolada (Veneto), the Austrians equipped the "Ice City" with shelters, munitions and wood depots, observation posts and communication passages leading to cave works (Bartoli, Fornaro, Rotasso 1993).

Rifleman's tunnel: communication tunnel with loopholes for light artillery, with or without a bonnet. The Fenestrelle Stronghold (Piedmont), a colossus blockade work build and extended over several periods (XVII-XIX centuries), a *covered stairway* built within a 1500 m masonry tunnel with light artillery loopholes, links the two main forts and the less important works of the complex (Contino 1993, pgs. 57-59).

VII.10.29 - *War cave*

Term referring to natural cavities used and adapted to serve as shelters, depots etc. during the First World War.
During the Great War, both the Austrians and Italians made use of caves, primarily in the Kras region. They used both horizontal and vertical caves in which they built depots and quarters for the troops, sometimes over several storeys, with ventilation shafts, drinking water and adequately protected entrances. Other caves were adapted and used to house light weapons and batteries. This type of work has become known as *war cave* (Gariboldi 1926, pgs. 129-152). Approximately 200 such caves have been registered in the karst areas of Trieste and Gorizia alone.

The Visogliano Cave is currently situated in the municipal territory of Duino-Aurisina (Trieste), was used during the First World War, following internal adaptation. The walls used as Austro-Hungarian barracks are still visible to the right of the entrance. Stone steps lead to a small ledge containing the remains of a military construction (figs. VII.33, VII.34, VII.35, VII.36 and VII.37). In recent times, the material blocking the terminal section was removed in the hope that further sections would be uncovered (Gherlizza, Radacich 2005, pgs. 65-66).

Fig. VII.33. *Entrance to Visogliano Cave (photo M. Radacich).*

Fig. VII.34. *Visogliano Cave. A few solid steps lead into the cave (photo M. Radacich).*

Fig. VII.35. *Austro-Hungarian soldiers in Visogliano Cave (photo G. Kaloper – Archive cap. Ezio Cervia g.c.).*

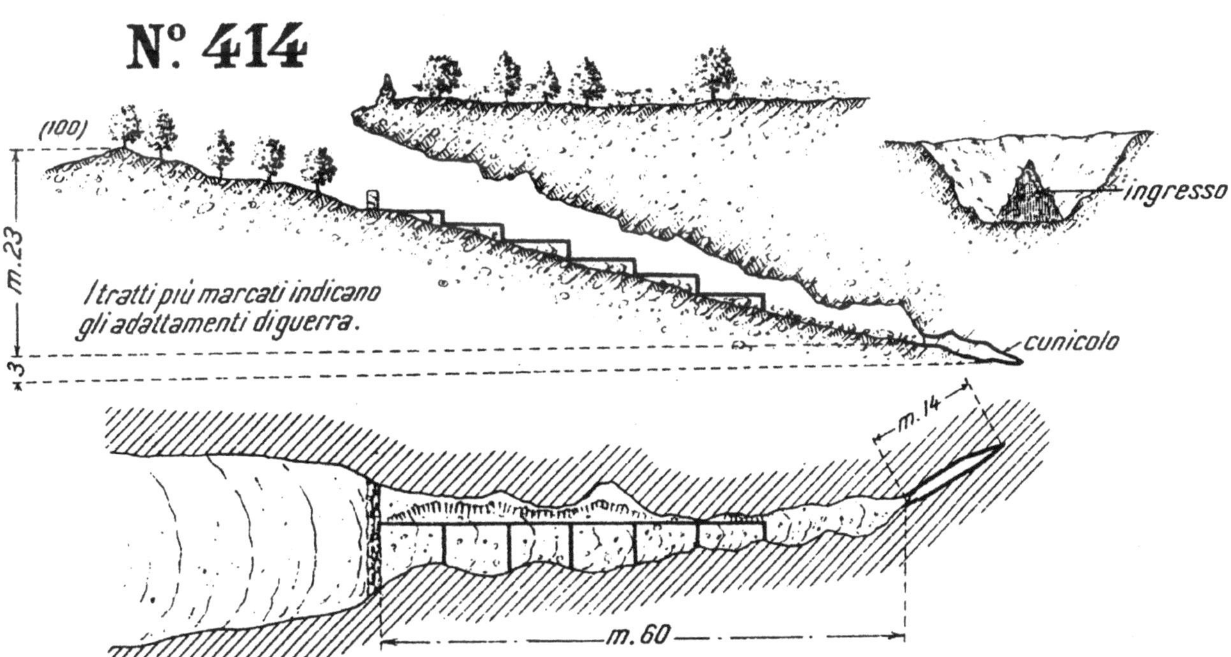

Fig. VII.36. *Period survey of Visogliano Cave (Club Alpinistico Triestino Archive).*

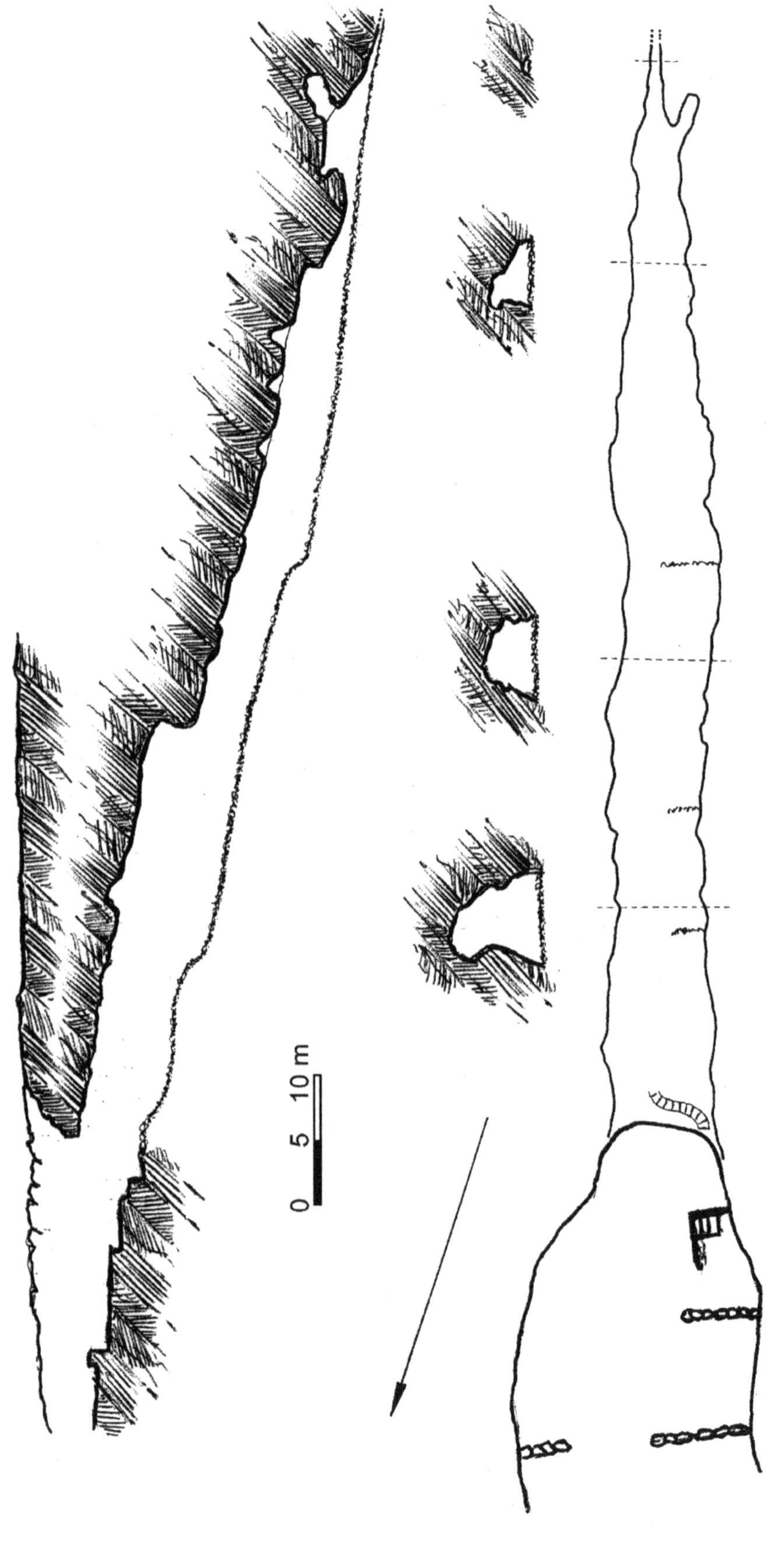

Fig. VII.37. *Present day survey of Visogliano Cave (Club Alpinistico Triestino Archive).*

Going on past 'experience', it was though that the use of natural cavities may also have envisaged atomic conflicts: «as t is often necessary to carry out pre-emptive adaptations, even during peacetime, especially as regards accessibility, additional entrances, ventilation and basic preparation, it is recommended that such works be disguised as works for tourist or other purposes» (Franzosini 1949).

VII.11 - *Typology 7: unidentified works*

Works of unknown purpose.

We sometimes come across underground works, the purposes of which are unknown. Due to inadequate or limited research, these works remain under the unidentified, 'unusual' or even 'mysterious' category. Others, despite in-depth research, may pose several unanswered questions, both in relation to their size and on account of the continuous structural interventions and adaptations, the total absence of internal and external elements or due to the lack of written documentation (fig. VII.39).

There are two "orientalizing" tumuli at the gates of Tarquinia (Viterbo): The drum of the larger tumulus holds the entrance to a lightly inclined underground passage, which descends beneath the funerary monument and leads to a quadrangular well, full of detritus and just under 2 m deep. Beyond this, the passage continues for a few more metres, curves and ends on the excavation front. Part of the drum was restored in restored in relatively recent times. The passage entrance was restored with exposed brick, thus it cannot now be established whether or not the work was built at the same time as the monument. That is, it cannot be established whether the drum was a functional part of the funerary work or whether it was created later for an as yet unknown purpose (fig. VII.40). Only complete removal of the mound of soil and a stratigraphic survey (at least covering the area opposite the entrance) can provide the information required for full comprehension. In any case, in its current state the underground passage remains classified an "unidentified work", the purpose of which is unknown.

Fig. VII.39. *Tuscania (Viterbo): subterranean named "Queen's tomb", but it is not a funerary structure (photo G. Padovan).*

Fig. VII.40. *"Antro delle Gallerie" (Gallery Cavern) in Valganna (Varese) Discovered at the end of the XIX century by Abbot Inganni of Milan. Its purpose is still unknown. The cavern extends for more than two kilometres. The photograph shows a communication shaft between two levels. The lower level is flooded (photo G. Padovan).*

CHAPTER VIII

BIBLIOGRAPHY

Aallen V., Strinati P 1976, *Guida alle grotte d'Europa*, Bologna.

Abignente G. 1888, *La proprietà del sottosuolo. Studio storico e giuridico*, Annali di Agricoltura, Ministero dell'Agricoltura Industria e Commercio, Roma.

Alfieri N. 2000, *Il nome "Topografia antica"*, in Bonora G., Dall'Aglio P. L., Patitucci S., Uggeri G. (edited by), *La topografia antica*, Bologna, pgs. 11-14.

Acquaro E., Francisi M. T., Mezzolani A. 2002, *Approvvigionamento idrico a* Tharros: *analisi e funzionalità conservative*, in Giorgetti D. and Riera I. (edited by), In binos actus lumina. *Atti del Convegno Internazionale di Studi su Metodologie per lo studio della scienza idraulica antica (Ravenna, 13-15 maggio 1999)*, La Spezia, pgs. 57-69.

Adam J.-P. 1988, *L'arte di costruire presso i romani*, Milano.

Agricola G. 1550, *De re metallica*, 1621 reprint, anastatic, Milano 1977.

Amoretti G. 1965, *Le gallerie di contromina della "Mezzaluna della Porta di Soccorso" della Cittadella di Torino*, in Presidenza dell'Accademia San Marciano - Torino (edited by), *Armi Antiche*, Torino, pgs. 57-102.

Amoretti G. 1996, *La verità storica su Pietro Micca*, Torino.

Amoretti G., Gallo-Orsi G. 1973, *Il Buco di Viso: primo traforo alpino*, in Associazione "Amici del Museo Pietro Micca e dell'Assedio di Torino del 1706" e del "Centre Culturel Franco-Italien" di Torino (edited by), *Di qua e di là dai monti*, Savigliano (Cuneo), pgs. 277-260.

Ardito F. 2003, *Italia sotterranea*, Torino.

Ascoli M., Russo F. 1999, *La difesa dell'arco alpino 1861-1940*, Stato Maggiore dell'Esercito, Ufficio Storico, Roma.

Ashby T. 1991, *Gli acquedotti dell'antica Roma*, Roma.

Associazione Subacquea "Orsa Minore" 1981, *Pozzi e cisterne medievali della città di Perugia. Ricerche subacquee e documentazione*, Quaderni Regione dell'Umbria, Regione Umbria.

Atzeni E. 1985, *Aspetti e sviluppi culturali del neolitico e della prima età dei metalli in Sardegna*, in VARIOUS AUTHORS (edited by), *Ichnussa*, Roma, pgs. 33-41.

Aquino C. 2005, *Studio del territorio dal punto di vista geologico, geomorfologico e idrogeologico per intercorrelazioni con le operazioni do speleologia in cavità artificiali*, in Padovan G. (edited by), *Archeologia del sottosuolo. Lettura e studio delle cavità artificiali*, Notebooks on Medieval Topography (British Archaeological Reports, International Series, S1416), Oxford, pgs. 137-150.

Badin A. D. 1876, *Grottes et cavernes*, Paris.

Bailly-Maitre M.-C., Bruno Dupraz J. 1994, *Brandes en Oisans. Le mine d'argent des Dauphins (XII-XIV s.) - Isère*, Documents d'Archéologie en Rhône-Alpes, No. 9, Lyon.

Balestracci D. 1987, *I bottini medievali di Siena*, Siena.

Barker P. 1977, *Tecniche dello scavo archeologico*, Milano.

Bartoli M., Fornaro M., Rotasso G. 1993, *La città di ghiaccio*, Trento.

Barton P., Doyle P., Vandewalle J. 2004, *Beneath Flanders Fields the Tunnellerd' War 1914-1918*, Staplehurst, Kent (UK).

Basilico R. 2005, *Il rilievo*, in Padovan G. (edited by), *Archeologia del sottosuolo. Lettura e studio delle cavità artificiali*, Notebooks on Medieval Topography (British Archaeological Reports, International Series, S1416), Oxford, pgs. 187-210.

Basilico R., Bianchi S., Ninni C., Padovan G. 2003, *Il pozzo del castello di Pavarolo*, in Anzanello E., Dal Cin F., Gasparetto P., Gava S. (edited by), *Montello 2002, Atti del 21° Incontro Internazionale di Speleologia (Nervesa della Battaglia, 1-3 novembre 2002)*, Villorba (Treviso), pgs. 277-292.

Basilico R., Casartelli M., Frignani F., Lampugnani M., Ninni C., Padovan G. 2006, *Tornando con Ashby a San Cosimato. Documentazione e studio delle opere idrauliche presenti nella gola dell'Aniene*, in Riera I (edited by), In binos actus lumina. *Rivista di Studi e Ricerche sull'idraulica storica e la storia della tecnica*, Atti del Convegno Internazionale di Studi Archeologia e Società. L'idraulica degli Antichi fra passato e futuro (Narni, 18-20 Ottobre 2001), La Spezia, pgs. 125-156.

Basilico R., Casini A., Padovan G. 2002, *La storia dell'acqua in un castello della Maremma Toscana: Campiglia Marittima (Livorno)*, in Club Alpinistico Triestino - Gruppo Grotte Sezione Ricerche e Studi su Cavità Artificiali (edited by), *Atti del V Convegno Nazionale sulle Cavità Artificiali. Osoppo 28 aprile -1 maggio 2001*, Trieste, pgs. 41-68.

Basilico R., Lampugnani M. 2002, *Il complesso conventuale eremitico di San Cosimato: un approccio critico*, in Club Alpinistico Triestino - Gruppo Grotte Sezione Ricerche e Studi su Cavità Artificiali (edited by), *Atti del V Convegno Nazionale sulle Cavità Artificiali. Osoppo 28 aprile -1 maggio 2001*, Trieste, pgs. 69-90.

Bassi P., Berto R., Perletti F. 1996, *Inquadramento geologico regionale del territorio di Milano*, in Padovan G. (edited by), *La fortezza celata. I sotterranei del Castello Sforzesco di Milano*, Vigevano (Pavia), pgs. 20-33.

Bassani E. 1990, *Utilità di una mostra*, in Pezzoli G. e Brena D. (edited by), *Forti e castelli di tratta. Storia e memoria di antichi insediamenti europei sulle coste dell'Africa nera*, Centro Studi Archeologia Africana, exhibition catalogue, pgs. 2-3.

Bavagnoli L. 2005, *Catasti passati e catasti futuri: l'introduzione dell'informatica*, in Padovan G. (edited by), *Archeologia del sottosuolo. Lettura e studio delle cavità artificiali*, Notebooks on Medieval Topography (British Archaeological Reports, International Series, S1416), Oxford, pgs. 355-358.

Bazzi F. 1907, *La Fortezza di Verrua nelle sue relazioni con la Casa dei Savoia, (1533-1706)*, Torino.

Belmonte B. 2001, *Al parco dei millenni*, Matera.

Benevolo L., Albrecht B. 2002, *Le origini dell'architettura*, Bari.

Benuzzi A., Relli G., Fortuzzi L. 2005, *La Linea Gotica Ritrovata. Bologna 1944-45*, Bologna.

Bertossi S. 1993, *Palmanova fortezza d'Europa 1593-1993*, Udine.

Bertulessi M., Padovan G. 2005, *La speleologia subacquea in cavità artificiali*, in Padovan G. (edited by), *Archeologia del sottosuolo. Lettura e studio delle cavità artificiali*, Notebooks on Medieval Topography (British Archaeological Reports, International Series, S1416), Oxford, pgs. 251-258.

Bertulessi M., Rota L. 2005, *Note sul lavoro di ricerca e rilievo*, in Brusetti L., Cogliati M., Gruppo Sommozzatori Almè (Bergamo) (edited by), *Labirinti sommersi. La cementeria del Fabbricone e le gallerie Pelucchi*, Comune di Olgiate Molgora, Pagine di Storia No. 3, Olgiate Molgora (Lecco), pg. 52.

Bianchi G. 1997, *Rocca San Silvestro e Campiglia Marittima: storia parallela di due insediamenti toscani attraverso la lettura delle strutture murarie*, in Gelichi S. (edited by), *I Congresso Nazionale di Archeologia Medievale*, Pisa, pgs. 437-445.

Bianco S. 1999, *Siris-Herakleia: l'uso dell'acqua nella città e nel territorio*, in Soprintendenza Archeologica della Basilicata (edited by), *Archeologia dell'acqua in Basilicata*, pgs. 73-86.

Bigi B., Gobessi M. 2003, *L'urlo della sirena. I ricoveri pubblici antiaerei a Trieste 1940-1945*, Trieste.

Bigi B., Gobessi M., Radacich M. 2004, *10 giugno 1944 ore 9.12 di un sabato mattina*, Trieste.

Biringuccio V. 1540, *De la Pirotechnia Libri X,* Venezia.

Bixio R. 1995, *La cultura rupestre nell'area mediterranea e in Cappadocia*, in Bertucci G., Bixio R., Traverso M. (edited by), *Le città sotterranee della Cappadocia*, Società Speleologica Italiana, pgs. 18-30.

Bixio R. 2002, *Apicoltura rupestre a Malta*, in Società Speleologica Italiana (edited by), *Opera Ipogea*, No. 3, Genova, pgs. 19-26.

Bixio R., Saj S. 1991, *La peste ed il bastione dell'Acquasola*, in Società Speleologica Italiana (edited by), *Speleologia*, No. 24, Milano, pgs. 5-7.

Boccacci M. 1984, *La Rocca di Campiglia Marittima: studi e ricerche*, Firenze.

Bodini G. 2002, *Antichi sistemi di irrigazione nell'arco alpino*, Ivrea.

Bonetto J. 1997, *La Guerra Sotterranea: i passaggi ipogei nella poliorcetica greca e romana: fonti e archeologia,* in Busana M. S. (edited by), Via per Montes Excisa. *Strade in galleria e passaggi sotterranei nell'Italia romana,* Roma, pgs. 337-398.

Bonvesin da la Riva, De magnalibus Mediolani, Corti M. (edited by), Milano 1997.

Bordignon L. 2000, *Tra i resti di un'architettura fortificata. Progetto per la fortezza di Verrua Savoia*, Politecnico di Torino. Facoltà di Architettura, Tesi di Laurea anno accademico 1999-2000.

Botturi G., Pareccini R. 1991, *Antichi acquedotti del territorio bresciano*, Milano.

Bouillon M. 1972, *Découverte du monde souterrain*, Paris.

Bovini G. 1952, *Rassegna degli studi sulle catacombe e sui cimiteri "*sub divo*"*, Città del Vaticano.

Breda A. 1986, *Sul colle Cidneo i primi documenti dell'antica* "Brixia". *Esiti e programmi della ricerca archeologica in Castello*, in VARIOUS AUTHORS (edited by), *Il Castello di Brescia*, Brescia, pgs. 45-50.

Bregani E. R. 1999, *Istoplasmosi*, in Società Speleologica Italiana (edited by), *Speleologia*, Rivista della Società Speleologica Italiana, No. 40, Città di Castello, pgs. 96-100.

Bregani E. R., Tien T. V., Ceraldi T., Delfitto C., Figini G. 2000, *Istoplasmosi: una malattia non solo tropicale. Recenti Progressi*, in VARIOUS AUTHORS (edited by), *Medicina 2000*, 91 (7 - 8), pgs. 396-401.

Bregani R. 2005, *Malattie e prevenzione*, in Padovan G. (edited by), *Archeologia del sottosuolo. Lettura e studio delle cavità artificiali*, Notebooks on Medieval Topography (British Archaeological Reports, International Series, S1416), Oxford, pgs. 277-283.

Brogiolo G. P. 1996, *Il progetto "Insediamenti rupestri dei Berici"*, in VARIOUS AUTHORS (edited by), *La chiesa rupestre di S. Cassiano (Lumignano di Longare - Vicenza)*, Archeologia Medievale, XXIII, Florence, pgs. 243.

Burdy J. 1996, *L'aqueduc romain du Gier*, Oullins (France).

Busana M. S., Basso P. 1997, *Le strade in galleria nell'Italia romana*, in Busana M. S. (edited by), Via per montes excisa. *Strade in galleria e passaggi sotterranei nell'Italia romana*, Roma, pgs. 81-250.

Buzio A., Casini A., Padovan G. 2000, *Attività estrattive nelle Grigne. Alcune note riguardo la Grotta del Pallone e la Grotta Ferrera,* in Padovan G. and Riera I. (edited by), *Atti del XV Congresso di Speleologia Lombarda, S. Omobono Imagna Terme 2-3 ottobre 1999, volume III Speleologia in Cavità Artificiali*, Milano, pgs. 141-162.

Buzio A., Mengoli D., Zorzin R. 2005, *Le grotte fortificate di Hong Lin (Guizhou - Cina)*, in Padovan G. (edited by), *Archeologia del sottosuolo. Lettura e studio delle cavità artificiali*, Notebooks on Medieval Topography (British Archaeological Reports, International Series, S1416), Oxford, pgs. 131-136.

Caesar C. J., De bello gallico, translation by Brindesi F., Milano 1985.

Cagnana A. 2000, *Archeologia dei materiali da costruzione*, Mantova.

Cambi F. 2003, *Archeologia dei paesaggi antichi: fonti e diagnostica*, Roma.

Cambi F., Terrenato N. 2002, *Introduzione all'archeologia dei paesaggi*, Roma.

Canestrini G. n. d. (1945?), *Arte militare meccanica medievale*, Milano.

Capacci C. 1918, *Acquedotti ed acque potabili*, Milano.

Cappa G., Cigna A., De Michele E., Parea G. C. 1962, *Ricerche sugli aspetti del fenomeno carsico profondo nel Gruppo delle Grigne (Lombardia). IV. La caverna Ferrera di Mandello 1502 Lo*, in Museo Civico di Storia Naturale (edited by), *Atti della Società Italiana di Scienze Naturali e del Museo Civico di Storia Naturale di Milano*, vol. CI, Milano, pgs. 20-42.

Cappellano F. 2003, *Fronte ad est*, in *Storia Militare*, No. 114, year XI, Zingonia-Verdellino (Bergamo), pgs. 4-15.

Caramella L. (edited by) 1999, *Ghiacciaie*, Sesto Calende (Varese).

Caramellino C. 1987, *Verrua Savoia, immagini di una Fortezza*, Verrua Savoia.

Carandini A. 1996, *Storie della terra, manuale di scavo archeologico*, Torino.

Cardinale M., Castellani V., Vignati A. 1978, *L'emissario sotterraneo del lago di Albano, duemila anni dopo*, in *Quaderni del Museo di Speleologia "V. Rivera" L'Aquila*, year IV, No. 7-8, L'Aquila, pgs. 17-30.

Carletti C., Fiocchi Nicolai V. 1989, *La catacomba di S. Cristina a Bolsena*, Pontificia Commissione di Archeologia Sacra, Città del Vaticano.

Carrozzo M. T., Nuzzo L., Quarta T. 2002, *Metodi fisici per i beni culturali*, in Castellano A., Martini M., Sibilia E. (edited by), *Elementi di archeometria*, Milano, pgs. 317-342.

Casini A., Cascone G. 2000, *Un contributo alla definizione della metodologia di studio e di rilevamento delle attività minerarie d'età preindustriale*, in Padovan G. and Riera I. (edited by), *Atti del XV Congresso di Speleologia Lombarda (Sant'Omobono Imagna Terme, 2-3 ottobre 1999), Vol. 3 - Speleologia in Cavità Artificiali*, Milano, pgs. 93-122.

Casini A., Padovan G. 2002, *Speleologia in Cavità Artificiali: metodologia e strategie d'indagine*, in Club Alpinistico Triestino – Gruppo Grotte Sezione Ricerche e Studi su Cavità Artificiali (edited by), *Atti del V Convegno Nazionale sulle Cavità Artificiali. Osoppo 28 aprile -1 maggio 2001*, Trieste, pgs. 155-184.

Cassinelli S. 2002, *Forte Montecchio. Baluardo tra Alto Lario e Valtellina*, Daverio (Varese).

Cassi Ramelli A. 1964, *Dalle caverne ai rifugi blindati. Trenta secoli di architettura militare*, Milano.

Castellani V. 1999, *Civiltà dell'acqua. Archeologia del territorio da Roma Arcaica alle antiche civiltà mediterranee*, Roma.

Castellani V., Dragoni W. 1981, *L'emissario medioevale del lago Trasimeno*, in Società Speleologica Italiana (edited by), *Speleologia*, Rivista della Società Speleologica Italiana, No. 6, Milano, pgs. 38-40.

Castellani V., Dragoni W. 1991, *Opere arcaiche per il controllo del territorio: gli emissari sotterranei artificiali dei laghi albani*, in Bergamini M. (edited by), *Gli Etruschi maestri d'idraulica*, Perugia, pgs. 43-60.

Castoldi M. 1996, *Pozzi romani a Milano: dall'uso al disuso*, in Antico Gallina M. (edited by), *Acque interne: uso e gestione di una risorsa*, Milano, pgs. 113-122.

Cazzato U., Fagiolo M., Giusti M. A. 2001, *Atlante delle grotte e dei ninfei in Italia. Toscana, Lazio, Italia meridionale e isole*, Perugia.

Cazzato U., Fagiolo M., Giusti M. A. 2002, *Atlante delle grotte e dei ninfei in Italia. Italia settentrionale, Umbria, Marche*, Perugia.

Ceccarelli Lemut M. L. 1985, *La Rocca di S. Silvestro nel medioevo ed i suoi Signori*, in Francovich R. (edited by), *Un villaggio di minatori e fonditori di metallo nella Toscana del Medioevo: S. Silvestro (Campiglia Marittima)*, Archeologia Medievale XII, Firenze, pgs. 322-341.

Chazette A. 2001, *La Batterie Lindemann*, Paris.

Chazette A. 2004 a, *Les batteries côtières en France. Volume I. Les batteries lourdes de Marine*, Paris.

Chazette A. 2004 b, *Les batteries côtières en France. Volume II. Les batteries lourdes sur voie ferrée. Les batteries du 194 mm au 240 mm*, Paris.

Chazette A. (edited by) 2004 c, *Tobruks Typologie. Atlantikwall – Südwall*, Paris.

Cherubini C., Geminaro S. 1991, *Stabilità degli ipogei in rocce calcarenitiche*, in Paone R., Piciocchi C. (edited by), *3° International Symposium on underground quarries (Naples, 10-14 July 1991)*, Napoli, pgs. 100-115.

Cherubini C., Sgobba D. 1997, *Le cave sotterranee di tufo pugliesi: descrizione degli ipogei e valutazione di stabilità*, in Club Alpino Triestino - Gruppo Grotte Sezione Ricerche e Studi su Cavità Artificiali (edited by), *Atti del IV Convegno Nazionale sulle Cavità Artificiali, Osoppo 30/31 maggio - 1 giugno 1997*, Trieste, pgs. 51-68.

Chiauzzi G. L. 2003, *La "modernità": rimozione degli ipogei del Gébel Gariàn*, in Rami Ceci L. (edited by), *Sassi e templi. Il luogo antropologico tra cultura e ambiente*, Roma, pgs. 168-193.

Chiesi M. 2001, *Il Traversante di Trebbia*, in Società Speleologica Italiana, *Opera Ipogea*, year III, No. 1, Genua, pgs. 15-28.

Chimenti M., Consolini F. 1958, *Relazione tecnica sull'esplorazione dell'emissario del lago di Albano*, in Circolo Speleologico Romano (edited by), *Notiziario Circolo Speleologico Romano*, year VIII, No. 1, Roma, pgs. 20.

Cicero, *De divinatione*.

Club Alpinistico Triestino – Gruppo Grotte 1994, *Gli ipogei naturali e artificiali del Campo di Osoppo*, Trieste.

Coarelli F. 1987, *I santuari del Lazio in età repubblicana*, Roma.

Cocconcelli G. D. 2002, *Tunnel factories. Le officine aeronautiche Caproni e Fiat nell'Alto Garda 1943-1945*, Carugate (Milano).

Colombo G. 1933, *Manuale dell'ingegnere*, 60th-65th edition, Milano.

Comoli Mandracci V., Marotta A. 1995, *Tortona e il suo castello. Dal dominio spagnolo al periodo postunitario*, Alessandria.

Concina E. 1990, *Le fortificazioni lagunari fra il tardo Medioevo e il secolo XIX*, in VARIOUS AUTHORS (edited by), *La laguna di Venezia*, Verona, pgs. 249-272.

Contino T. 1993, *La piazzaforte di Fenestrelle*, Torino.

Contu E. 1985, *L'architettura nuragica*, in VARIOUS AUTHORS (edited by), *Ichnussa*, Milano, pgs. 115-129.

Coppa A. 1999, *Galeazzo Alessi. Trattato di fortificazione*, Milano Politecnico - Facoltà di Architettura, Quaderni del Dipartimento di Conservazione e Storia dell'Architettura, Milano.

Coralini A. 1997, Conplanatis montibus et caesis rupibus… *Vie in galleria, in tagliata e in trincea nel mondo romano al di fuori dell'Italia peninsulare*, in Busana M. S. (edited by), Via per montes excisa. *Strade in galleria e passaggi sotterranei nell'Italia romana*, Roma, pgs. 279-335.

Cosci M. 1988, *Fotointerpretazione archeologica*, Firenze.

Cristofani M. 1991, *Cerveteri. Tre itinerari archeologici*, Roma.

D'Alessio V. 1993, *Il culto di San Michele Arcangelo, santuari tra Salerno e Avellino*, Solofra (Avellino).

Dall'Aglio P. L. 2000, *Geomorfologia e topografia antica*, in Bonora G., Dall'Aglio P. L., Patitucci S., Uggeri G. (edited by), *La topografia antica*, Bologna, pgs. 178-192.

da Vinci L., *Manuscrit B de l'Institut de France*, in *Leonardo da Vinci. Scritti*, Recupero J. (edited by), Ariccia (Roma) 2002.

da Vinci L., *Codice Leicester*, in Regio Istituto Lombardo di Scienze e Lettere, *Codice di Leonardo da Vinci della Biblioteca di Lord Leicester in Holkam hall*, Milano 1909.

da Vinci L. *Codice Atlantico*, in Recupero J. (edited by), *Leonardo da Vinci. Scritti*, Ariccia (Roma) 2002.

Della Fratta e Montalbano M. A. 1678, *Pratica Minerale. Trattato*, Cima M. (edited by), Firenze 1985.

Della Portella I. 1999, *Roma sotterranea*, Verona.

Del Lungo S. 2001, *Toponimi in archeologia: la Provincia di Latina, Italia*, Notebooks on Medieval Topography, (British Archaeological Reports, International Series, S911), Oxford.

Del Lungo S. 2005, *Introduzione alla toponomastica delle cavità artificiali e naturali*, in Padovan G. (edited by), *Archeologia del sottosuolo. Lettura e studio delle cavità artificiali*, Notebooks on Medieval Topography (British Archaeological Reports, International Series, S1416), Oxford, pgs. 215-218.

Del Pelo Pardi G. 1943, *Bonifiche antichissime. La Malaria e i Cunicoli del Lazio*, in VARIOUS AUTHORS (edited by), *Atti della Reale Accademia dei Georgofili*, Firenze, pgs. 283-323.

de Lumley H. 2004, *Le Sol d'occupation acheuléen de l'unité archéostratigraphique UA 25 de la grotte du Lazaret. Nice, Alpes-Maritimes*, Aix-en Provence.

de Rizzardi A. 1956, *Cinquantenario del traforo del Sempione*, Museo Nazionale della Scienza e della Tecnica "Leonardo da Vinci", Milano.

Di Martino 2005, *I resti animali*, in Passi Pitcher L. (edited by), Hic est bonum comedere. *Stoviglie e vettovaglie rinascimentali di una guarnigione di Pizzighettone*, Cremona, pgs. 95-98.

Di Sopra L. 2003, *Palmanova città fortezza*, Udine.

Dolci M. 1958, *Esplorazioni dell'emissario del lago di Albano in comune di Castelgandolfo*, in Circolo Speleologico Romano (edited by), *Notiziario Circolo Speleologico Romano*, year VIII, No. 1, Roma, pgs. 17-19.

Domergue C. 1991, *Les mines de la péninsule ibérique dans l'antiquité romaine*, Collection de l'École Française de Rome, 127, Roma.

Dondoli A. 1997, *Aspetti tecnici e strutturali delle batterie costiere a Piombino nella Seconda Guerra Mondiale*, in Archivio Storico della Città di Piombino (edited by), *Le difese costiere a Piombino nei due conflitti mondiali*, documentary exhibition catalogue (Piombino 18 Novembre - 8 Dicembre 1997), Piombino (Livorno), pgs. 47-48.

Drower M. S. 1993, *Fornitura di acqua, irrigazione e agricoltura*, in VARIOUS AUTHORS, *La preistoria e gli antichi imperi*, Storia della tecnologia 1, tomo II, Torino, pgs. 528-566.

Duffy C. 1996, *Fire Fire & Stone*, London, pgs. 82-83.

Eogan G. 1986, *Knowth and the passage-tombs of Ireland*, London, pgs. 11-29.

Esposito G. 2005, *Alcune note sulla fotografia in ipogeo*, in Padovan G. (edited by), *Archeologia del sottosuolo. Lettura e studio delle cavità artificiali*, Notebooks on Medieval Topography (British Archaeological Reports, International Series, S1416), Oxford, pgs. 223-232.

Fara A. 1988, *Bernardo Buontalenti. L'architettura, la guerra e l'elemento geometrico*, Genova.

Fara A. 1989, *Il Sistema e la Città. Architettura fortificata dell'Europa moderna dai trattati alle realizzazioni 1464-1794*, Genova.

Fiala K. A. 1988, *Le fortificazioni permanenti della "Zona Carnia" e gli "Sbarramnenti Carinziani"*, in Schaumann W. (edited by), *Storia e itinerari nelle località della guerra: Alpi Carniche Orientali*, Bassano del Grappa (Vicenza).

Ficacci L. 2000, *Piranesi. The Complete Etchings*, Roma.

Finzi E. 1990, *L'impiego delle tecniche geofisiche per l'individuazione di preesistenze antropiche*, in Francovich R., Manacorda D. (edited by), *Lo scavo archeologico: dalla diagnosi all'edizione*, III Ciclo di Lezioni sulla Ricerca applicata in Archeologia, Certosa di Pontignano (Siena) 6-18 November 1989, Firenze, pgs. 169-202.

Fiocchi Nicolai V., Bisconti F., Mazzoleni D. 1998, *Le catacombe cristiane di Roma. Origini, sviluppo, apparati decorativi, documentazione epigrafica*, Regensburg (Germany).

Flocchini A. 1994, *I forti della Grande Guerra*, in *Storia Militare*, No. 13, Parma, pgs. 43-52.

Floris A. 1988, *Cagliari sotterranea*, Cagliari.

Floris A. 1995, *Guida alle grotte turistiche della Sardegna*, Cagliari.

Floris A., Padovan G. 1997, *Bibliografia delle Cavità Artificiali Italiane (primo contributo)*, in Pensabene G. (edited by), *Atti XVII Congresso Nazionale di Speleologia, Castelnuovo Garfagnana Settembre 1994*, II volume, Firenze, pgs. 79-174.

Forbes R. J. 1993, *Ingegneria idraulica e impianti sanitari*; in VARIOUS AUTHORS, *Storia della Tecnologia 2, Le civiltà mediterranee e il medioevo*, tomo II, Torino, pgs. 674-706.

Forlati Tamaro B. 1962, *L'ipogeo di S. Maria in Stelle (Verona)*, in VARIOUS AUTHORS (edited by), *Stucchi e mosaici alto medievali*, Atti dell'ottavo Congresso di studi sull'arte dell'alto Medioevo, Verona-Vicenza-Brescia 5-11 Ottobre 1959, vol. I, Milano, pgs. 245-259.

Formenti G. 1705, *Descripción de la Isla de Sicilia y sus costas maritimas*, Doufour L. (edited by), Siracusa 1991.

Franzosini M. 1949, *Utilizzazione delle cavità naturali ai fini militari*, reprint, excerpt from *L'Universo*, Rivista dell'Istituto Geografico Militare, year XXIX - No. 6, 1949.

Frare G. P. 1996, *Cave. Tecniche di coltivazione e recupero ambientale*, Treviso.

Frega G. 1984, *Lezioni di acquedotti e fognature*, Napoli.

Frignani F., Sassi G., Frignani F., Casini A., Padovan G. 2002, *Il rilevamento delle cave di gesso a Vezzano sul Crostolo (Reggio Emilia)*, in Club Alpinistico Triestino, Gruppo Grotte Sezione Ricerche e Studi su Cavità Artificiali (edited by), *Atti del V Convegno Nazionale sulle Cavità Artificiali. Osoppo 28 aprile -1 maggio 2001*, Trieste, pgs. 267-289.

Frignani F. 2005, *Alcune note di topografia: acquisizione e gestione delle informazioni sul territorio*, in Padovan G. (edited by), *Archeologia del sottosuolo. Lettura e studio delle cavità artificiali*, Notebooks on Medieval Topography (British Archaeological Reports, International Series, S1416), Oxford, pgs. 151-186.

Frontino S. G., *De aquaeductu urbis Romae*, Galli F. (edited by), Lecce 1997.

Frontino S. G., *Stratagemata*, Galli F. (edited by), Lecce 1999.

Furiesi A. 1998, *L'acqua a Volterra. Storia dell'approvvigionamento idrico della città*, Siena.

Gambarelli G. (edited by) 1995, *Pizzighettone città murata*, Pizzighettone (Cremona).

Gariboldi I. 1926, *Le grotte di guerra*, in Bertarelli L. V. e Boegan E. (edited by), *Duemila Grotte. Quarant'anni di esplorazioni nella Venezia Giulia*, Milano, pgs. 129-152.

Gariglio D. 1997, *Le Sentinelle di Pietra*, Cuneo.

Gariglio D., Minola M. 1994, *Le fortezze delle Alpi Occidentali. Dal Piccolo San Bernardo al Monginevro*, vol. I, Cuneo.

Gastaldo G. 1998, *I corredi funerari nelle tombe "tardo romane"*, in Brogiolo G. P., Wataghin G. C. (edited by), *Italia settentrionale*, in *Sepolture tra IV e VIII secolo. 7° Seminario sul tardo antico e l'alto medioevo in Italia centro settentrionale (Gardone Riviera 24-26 Ottobre 1996)*, Mantova, pgs. 15-59.

Gattera C. 1999, *Il Pasubio e la Strada delle 52 Gallerie*, Valdagno (Vicenza).

Gerbella L. 1947, *Arte Mineraria*, I-II, Milano.

Gherlizza F., Radacich M. 2005, *Grotte della Grande Guerra*, Trieste.

Giardino C. 1998, *I metalli nel mondo antico. Introduzione all'archeometallurgia*, Roma-Bari.

Giardino G. 1929, *Rievocazioni e riflessioni di guerra. II. L'armata del Grappa. La battaglia difensiva del 1918*, Milano.

Gibertini U. 2005, *Gas in ipogeo: tipologie, valutazioni, rischi e prevenzione*, in Padovan G. (edited by), *Archeologia del sottosuolo. Lettura e studio delle cavità artificiali*, Notebooks on Medieval Topography (British Archaeological Reports, International Series, S1416), Oxford, pgs. 265-276.

Gide A., *I sotterranei del Vaticano*, Gigliozzi G. (translated by), Roma 1995.

Gigante G. E. 2002, *Origine e significato dell'archeometria e delle scienze applicate alla conservazione*, in Castellano A., Martini M., Sibilia E. (edited by), *Elementi di archeometria*, Milano, pgs. 9-19.

Gillot C. L. 1805, *Traité de Fortification Souterraine*, Paris.

Ginatempo M., Giorgi A. 1996, *Le fonti documentarie per la storia degli insediamenti medievali in Toscana*, in VARIOUS AUTHORS (edited by), *Archeologia Medievale*, XXIII, Firenze, pgs. 1-22.

Giorgetti D. 1985, *L'acquedotto romano di Bologna: l'antico cunicolo ed i sistemi di avanzamento in cavo cieco*, in Regione Emilia-Romagna (edited by), *Acquedotto 2000. Bologna, l'acqua del Duemila ha duemila anni*, vol. I, Casalecchio di Reno, (Bologna) pgs. 37-107.

Girotto L. 2002, *1866-1918. Soldati e fortezze tra Asiago ed il Grappa. Storia ed immagini dello "Sbarramento Brenta-Cismon" dal risorgimento alla prima guerra mondiale*, Novale (Vicenza).

Giuntella A. M. 1998, *Note su alcuni aspetti della ritualità funeraria nell'alto medioevo. Consuetudini e innovazioni*, in Brogiolo G. P., Wataghin G. C. (edited by), *Sepolture tra IV e VIII secolo. 7° Seminario sul tardo antico e l'alto medioevo in Italia centro settentrionale (Gardone Riviera 24-26 Ottobre 1996)*, Mantova, pgs. 61-75.

Giussani A. 1903, *Il Forte di Fuentes*, Como.

Gleria E. 1995, *I cliff dwelling anasazi del SO americano*, in Federazione Speleologica Veneta (edited by), *Speleologia Veneta*, Federazione Speleologica Veneta, vol. 3, Vicenza, pgs. 110-113.

Gorfer A. 1985, *I castelli del Trentino*, vol. 1, Trento.

Gorfer A. 1987, *I castelli del Trentino*, vol. 2, Trento.

Hill C., Forti P. 1997, *Cave Minerals of the World. Second Edition*, National Speleological Society, U.S.A.

Hogg I. 1982, *Storia delle fortificazioni*, Novara.

Kadarè I. 1993, *I tamburi della pioggia*, Farigliano (Cuneo).

Kircheri A. 1678, *Mundus subterraneus*, Amsterdam.

Judson S., Kahane A. 1963, *Underground drainageways in southern Etruria and northern Latium*, Pap. British School Rome, pgs. 75-93.

Laureano P. 1993, *Giardini di pietra. I Sassi di Matera e la civiltà mediterranea*, Torino.

Laureano P. 1995, *La piramide rovesciata. Il modello dell'oasi per il pianeta Terra*, Torino.

Laureano P. 2001, *Atlante d'acqua. Conoscenze tradizionali per la lotta alla desertificazione*, Torino.

Le Goff J. 1989, *Costruzione e distruzione della città murata. Un programma di riflessione e ricerca*, in De Seta C., Le Goff J. (edited by), *La città e le mura*, Roma, pgs. 1-10.

Lilliu G. 1980, *L'oltretomba e gli dei*, in Sanna D. (edited by), *Nur*, Cinisello Balsamo (Milano), pgs. 105-139.

Livy T., *Ad Urbe condita*, Moreschini C. (edited by), Milano 1997.

Lo Schiavo F. 1996, *Miniere e metallurgia in Sardegna: la ricerca archeologica dal presente al passato*, in Piola Caselli F. and Piana Agostinetti P. (edited by), *La miniera, l'uomo, l'ambiente*, Firenze, pgs. 187-206.

Luciani R. 1984, *Il criptoportico del Palatino*, in Luciani R. (edited by), *Roma sotterranea*, exhibition catalogue, Roma, pgs. 148-155.

Luca I., Salzotti F., Valenti M. 2001, *L'esperienza dell'insegnamento di archeologia medievale a Siena nel campo dell'informatica applicata*, in De Marchi M., Scudellari M., Zavaglia A. (edited by), *Lo spessore storico in urbanistica*, Mantova, pgs. 31-64.

Lucrezi Berti C. 1973, *Il culto di S. Michele nelle grotte d'Abruzzo e del Molise*, in Museo di Speleologia "V. Rivera", *Quaderni del Museo di Speleologia "V. Rivera" L'Aquila*, No. 2, L'Aquila, pgs. 185-194.

Maggi S. 2003, *Le ferrovie*, Capannori (Lucca).

Machiavelli N., *Il Principe*, Firpo L. (edited by), Torino 1974.

Marselli G. A. 2003, *Da Levi a Friedmann: cultura e modernizzazione* , in Rami Ceci L. (edited by), *Sassi e templi. Il luogo antropologico tra cultura e ambiente*, Roma, pgs. 88-102.

Mascione C., Patera A. (edited by) 2003, *Materiali per Populonia 2*, Quaderni del Dipartimento di Archeologia e Storia delle Arti Sezione Archeologica – Università di Siena, Firenze.

Mauceri L. 1939, *Il castello Eurialo*, Catania.

Mesnil du Buisson C. 1939, *Du siège d'Avaricum a celui de Dura-Europos*, in VARIOUS AUTHORS (edited by), *RA*, XIII, pgs. 60-72.

Mesnil du Buisson C. 1944, *Les ouvrages du siège de Dura-Europos*, in VARIOUS AUTHORS (edited by), *Mémoires de la Société nationale des antiquaires de France*, LXXXI, pgs. 5-60.

Milarepa, *Vita di Milarepa*, Bacot J. (edited by), Torino 1991.

Miller M. E. 1988, *L'arte della Mesoamerica. Olmechi Maya Aztechi*, Milano.

Ministero della Difesa (edited by) 1977, *Sacrari militari della Prima Guerra Mondiale 1915 1918. Pocol, Pian di Salesei, Timau*, Roma.

Minola M., Ronco B. 1998, *Fortificazioni nell'arco alpino. L'evoluzione delle opere difensive tra XVIII e XX secolo*, Ivrea (Torino).

Minto A. 1943, *Populonia*, Firenze.

Moccheggiani Carpano C. 1985, *Le cloache dell'antica Roma*, in Luciani R. (edited by), *Roma sotterranea*, exibition catalogue, Roma, pgs. 164-178.

Moretti M. 1974, *Pittura Etrusca in Tarquinia*, Cinisello Balsamo (Milano).

Motta V. 1981, *L'acquedotto di Milano*, Milano.

Mottana A., Crespi R., Liborio G. 1993, *Minerali e rocce*, Milano.

Moulins J.-J. 2005, *Les défenses de Bizerte 1881-1963*, Langres.

Negroni Cotacchio N. 1981, *Caratteri geografici e strutturali*, in Negroni Cotacchio N. (edited by), *Sorgenti della Nova. Una comunità protostorica e il suo territorio nell'Etruria Meridionale*, exhibition catalogue, Roma, pgs. 158-166.

Nenci C. 2002, *Biografia e cultura enciclopedica dell'estensione del trattato*, in Guerci G, Nenci C., Scazzosi L. (edited by), *Dell'arte de' giardini inglesi*, Città di Castello (Perugia).

Nesti C. 2005, *La legislazione*, in Padovan G. (edited by), *Archeologia del sottosuolo. Lettura e studio delle cavità artificiali*, Notebooks on Medieval Topography (British Archaeological Reports, International Series, S1416), Oxford, pgs. 319-336.

Nicoletti M. 1980, *L'architettura delle caverne*, Bari.

Nicoletti M. 1985, *Architettura e paesaggio nello Yemen del Nord*, Bari.

Nini R. 1997, *I sotterranei della chiesa di San Domenico a Narni*, in Pensabene G. (edited by), *Atti del XVII Congresso Nazionale di Speleologia. Castelnuovo Garfagnana 8-11 Settembre1994*, vol. 1, Firenze, pgs. 343-348.

Ottati D. 1988, *Il ventre di Firenze*, Firenze.

Ottino M. 1996, *Consolidamento e ipotesi di riuso della fortezza di Verrua Savoia*, Politecnico di Torino. Facoltà di Archietettura, Tesi di Laurea anno accademico 1996-1997.

Padovan D., Padovan G., Bordignon L., Ottino M. 1997, *La fortezza di Verrua Savoia*, in Club Alpinistico Triestino – Gruppo Grotte Sezione Ricerche e Studi su Cavità Artificiali (edited by), *Atti del IV Convegno Nazionale sulle Cavità Artificiali, Osoppo 30/31 Maggio - 1 Giugno 1997*, Trieste, pgs. 187-208.

Padovan G. (edited by) 1996, *La fortezza celata. I sotterranei del Castello Sforzesco di Milano*, Vigevano (Pavia).

Padovan G. 1997, *Speleologia in Cavità Artificiali al Forte di Fuentes*, in Pensabene G. (edited by), *Atti del XVII Congresso Nazionale di Speleologia. Castelnuovo Garfagnana 8-11 Settembre1994*, vol. 1, Firenze, pgs. 293-298.

Padovan G. 2000, *La Speleologia in Cavità Artificiali*, in Padovan G., Riera I. (edited by), *Atti del XV Congresso di Speleologia Lombarda*, S. Omobono Imagna Terme 1999, vol. III, Milano, pgs. 11-54.

Padovan G. 2002 a, *Note per la catalogazione e la comprensione delle opere idrauliche sotterranee*, in Giorgetti D. and Riera I (edited by), In binos actus lumina, *Rivista di Studi e Ricerche sull'idraulica storica*, Atti del Convegno Internazionale di Studi su Metodologie per lo studio della scienza idraulica antica (Ravenna, 13-15 Maggio 1999), La Spezia, pgs. 327-352.

Padovan G. 2002 b, *La Civita di Tarquinia: indagini speleologiche. Catalogazione e studio delle cavità artificiali rinvenute presso il Pian di Civita e il Pian della Regina*, Notebooks on Medieval Topography, 3 - BAR S1039, Oxford.

Padovan G. 2002 c, *Milano sotterranea*, in Club Alpinistico Triestino – Gruppo Grotte Sezione Ricerche e Studi su Cavità Artificiali (edited by), *Atti del V Convegno Nazionale sulle Cavità Artificiali. Osoppo 28 aprile -1 maggio 2001*, Trieste, pgs. 407-434.

Padovan G. 2003, *Due noci dure da rompere. I Forti di Demonte e di Tortona alla fine del XVIII secolo: l'organizzazione della difesa, la rete di contromina e l'approvvigionamento idrico*, in Anzanello E., Dal Cin F., Gasparetto P., Gava S. (edited by), *Montello 2002. Atti del 21° Incontro Internazionale di Speleologia (Nervesa della Battaglia, 1-3 Novembre 2002)*, Villorba (Treviso), pgs. 293-365.

Padovan G. (edited by) 2005 a, *Archeologia del sottosuolo. Lettura e studio delle cavità artificiali*, Notebooks on Medieval Topography (British Archaeological Reports, International Series, S1416), Oxford.

Padovan G. 2005 b, *La comprensione del reale*, in Padovan G. (edited by), *Archeologia del sottosuolo. Lettura e studio delle cavità artificiali*, Notebooks on Medieval Topography (British Archaeological Reports, International Series, S1416), Oxford, pgs. 3-8.

Padovan G. 2005 c, *Il territorio e le cavità artificiali: la comparazione e lo studio*, in Padovan G. (edited by), *Archeologia del sottosuolo. Lettura e studio delle cavità artificiali*, Notebooks on Medieval Topography (British Archaeological Reports, International Series, S1416), Oxford, pgs. 73-74.

Padovan G. 2005 d, *Alcuni spunti per le indagini: note di archeologia mineraria*, in Padovan G. (edited by), *Archeologia del sottosuolo. Lettura e studio delle cavità artificiali*, Notebooks on Medieval Topography (British Archaeological Reports, International Series, S1416), Oxford, pgs. 75-100.

Padovan G. 2005 e, *Analisi e sintesi: raccolta e catalogazione dei dati acquisiti*, in Padovan G. (edited by), *Archeologia del sottosuolo. Lettura e studio delle cavità artificiali*, Notebooks on Medieval Topography (British Archaeological Reports, International Series, S1416), Oxford, pgs. 337-354.

Palmisano P., Fanizzi A. 1992, *I Laghi di Conversano. Il fenomeno degli stagni stagionali dei territori carsici pugliesi*, in Federazione Speleologica Pugliese, *Itinerari speleologici*, Serie II, No. 6, Castellana Grotte (Bari), pgs. 35-53.

Panazza G. 1962, *Gli scavi, l'architettura e gli affreschi della chiesa di S. Salvatore in Brescia*, in VARIOUS AUTHORS (edited by), *La chiesa di San Salvatore in Brescia*, Atti dell'ottavo Congresso di studi sull'arte dell'alto Medioevo, Verona-Vicenza-Brescia 5-11 Ottobre 1959, vol. II, Milano, pgs. 5-228.

Panimolle G. 1986, *Gli acquedotti di Roma antica*, Roma.

Parenti R., *Vitozza. Un insediamento rupestre nel territorio di Sorano*, Firenze.

Pavia C. 1999, *Guida di Roma sotterranea*, Roma.

Peirone F. (translation and comments) 1980, *Il Corano*, Milano.

Pellicanò A. 2000, *Del periodo giovanile di Galileo Galilei. Il trattato di fortificazione. Alle radici del pensiero scientifico e dell'urbanistica moderni*, Roma.

Perego L. 2005, *Le ceramiche di Pizzighettone*, in Passi Pitcher L. (edited by), Hic est bonum comedere. *Stoviglie e vettovaglie rinascimentali di una guarnigione di Pizzighettone*, Cremona, pgs. 29-33.

Pergola P. 1997, *Le catacombe romane. Storia e topografia*, Roma.

Pesaro A. 2005, *La ricerca sulle fonti d'archivio nel settore delle cavità artificiali: potenzialità e limiti*, in Padovan G. (edited by), *Archeologia del sottosuolo. Lettura e studio delle cavità artificiali*, Notebooks on Medieval Topography (British Archaeological Reports, International Series, S1416), Oxford, pgs. 219-222.

Pieropan G. 1978, *La Strada delle Gallerie*, Cortina d'Ampezzo (Belluno).

Pipino G. 2003, *Documenti su attività minerarie in Liguria e nel dominio genovese dal medioevo alla fine del Seicento*, in VARIOUS AUTHORS (edited by), *Atti e Memorie*, new series, vol. XXXIX, Società Savonese di Storia Patria, Savona, pgs. 39-101.

Pisani Sartorio G. 1986, *Gli antichi acquedotti di Roma (IV sec. a.C. - VI sec. d.C.): dai pozzi ai condotti*, in VARIOUS AUTHORS (edited by), *Il trionfo dell'acqua. Acque e acquedotti a Roma. IV sec. a.C. - XX sec.*, Exhibition Catalogue for the 16° International Congress and International Aqueduct Exhibition, Roma, pgs. 27-30.

Plato, *Timeo*, Lozza G. (edited by), Milano 1994.

Prayon F. 2000, *L'architettura funeraria*, in Torelli M. (edited by), *Gli Etruschi*, Cinisello Balsamo (Milano), pgs. 334-343.

Ponzio G. 1997, *Brevi lineamenti della fortificazione olandese dal XVI al XVIII secolo*, in Acuto G., Mongiano P. (edited by), *Pietra e acciaio. Quaderni dell'Associazione per gli Studi di Storia e Architettura Militare*, Borgone, (Torino) pgs. 30-36.

Radacich M. 1993, *Le grotte murate*, in Società Alpina delle Giulie (edited by), *Alpi Giulie*, Rassegna di attività della Società Alpina delle Giulie - Sezione di Trieste del C.A.I., No. 87/1, Trieste, pgs. 11-36.

Raveli F., Howarth P. 1988, *I cunicoli etruschi: tunnel per la captazione di acqua pura*, in VARIOUS AUTHORS (edited by), *Irrigazione e drenaggio*, XXXV, 1, Bologna.

Recanatini A. 1997, *Le grotte del Conero*, Jesi (Ancona).

Ricciardi M. 1996, *La civiltà dell'acqua in Ostia antica*, I-II, Roma.

Riera I. 1991, *L'acquedotto romano di Asolo (Treviso)*, in VARIOUS AUTHORS (edited by), *Quaderni di Archeologia del Veneto*, VII, pgs. 181-197.

Riera I. 1993, *Acquedotti e cisterne*, in Ghedini F. and Rosada G. (edited by), *Il sottosuolo nel mondo antico*, Dosson (Treviso), pgs. 15-31.

Riera I. 1994, *Le cisterne*, in Riera I. (edited by), Utilitas necessaria. *Sistemi idraulici nell'Italia Romana*, Milano, pgs. 298-386.

Righetti M. 1987, *Architettura monastica: gli edifici. Linee per una storia architettonica*, in Pugliese Carratelli G. (edited by), *Dall'eremo al cenobio. La civiltà monastica in Italia dalle origini all'età di Dante*, Milano, pgs. 486-575.

Rizzo M. A. 1988, *La riserva archeologico-naturalistica*, in Cristofani C. and Nardi G. (edited by), Caere - *1. Il parco archeologico*, Roma, pgs. 107-110.

Romanelli P 1948, *Tarquinia. Scavi e ricerche nell'area della città*, in VARIOUS AUTHORS, *Notizie degli scavi*, fascicoli 7-12, pgs. 193 271.

Rognoni R. 1983, *Cascina Monastero. Storie di monaci, contadini e operai in S. Maria di Baggio*, Comune di Milano - Consiglio di Zona 8, Milano.

Rossi-Osmida G. 1974, *Le caverne e l'uomo*, Milano.

Rottman G. 2006, *Viet Cong and NVA Tunnels and Fortifications of the Vietnam War*, Botley.

Rottoli M. 2005, *I resti botanici*, in Passi Pitcher L. (edited by), Hic est bonum comedere. *Stoviglie e vettovaglie rinascimentali di una guarnigione di Pizzighettone*, Cremona, pgs. 99-101.

Rovina D. 1997, *Interventi di archeologia postmedievale nella Sardegna centro-settentrionale*, in Milanese M. (edited by), *Archeologia postmedievale: l'esperienza europea e l'Italia*, Archeologia Postmedievale, I, Firenze, pgs. 251-266.

Russo A. 1999, *Il ruolo dell'acqua nei luoghi sacri della Basilicata antica*, in Soprintendenza Archeologica della Basilicata (edited by), *Archeologia dell'acqua in Basilicata*, pgs. 103-126.

Salsa C. 1982, *Trincee. Confidenze di un fante*, Milano.

Samorè T. 1979, *Analisi di incidenti mortali a speleosub e loro prevenzioni*, in C.A.I. Lecco (edited by), *Atti del IX Convegno di Speleologia Lombarda (Lecco, 8-9 December 1979)*, Lecco, pgs. 63-64.

Scazzosi L. 2005, *Per una politica di tutela e conservazione delle architetture vegetali*, in Zerbi M. C., Scazzosi L. (edited by), *Paesaggi straordinari e paesaggi ordinari. Approcci della geografia e dell'architettura*, Milano, pgs. 185-202.

Schliemann H. 1962, *Autobiografia di un archeologo alla ricerca del mondo omerico*, Milano.

Sergeant F. 1914, *Asia. Capitolo VI*, in Johnston J, Burgoyne A., Landon P., Thomson J. (edited by), *Le meraviglie del mondo*, vol. I, Milano, pgs. 221-282.

Setti M. 2004, *La conquista di Eben Emael*, Milano.

Signorelli A., Signorelli B. 2000, *La loro ora più bella: l'assedio di Verrua*, in Amoretti G. and Pettiti P. (edited by), *Atti del Congresso Internazionale di Archeaologia, Storia e Architettura Militare a 40 anni dalla scoperta della Scala di Pietro Micca 1958-1998 (Torino 11-13 Novembre 1998)*, Torino, pgs. 173-198.

Silva E. 2002, *Dell'arte de' giardini inglesi*, Guerci G, Nenci C., Scazzosi L. (edited by), Città di Castello (Perugia).
Simonin L. 1867, *La vie souterraine. Les mines et les mineurs*, reprint, Mâcon 1982.

Società Geologica Italiana 1993, *Guide Geologiche Regionali. Lazio*, Roma.

Stocker U. 1999, *Caverne militari a Nord di San Michele del Carso. Con nota storica edited by Enrico Cernigoi*, in Federazione Speleologica Regionale del Friuli-Venezia Giulia (edited by), *Atti dell'VIII Convegno Regionale di Speleologia del Friuli-Venezia Giulia*, Trieste, pgs. 249-256.

Stopponi S. 1991, *Nuove osservazioni sul pozzo Sorbello e sul suo inserimento nel tessuto urbano di Perugia antica*, in Bergamini M. (edited by), *Gli Etruschi maestri d'idraulica*, Perugia, pgs. 235-246.

Strabo, *Geografia*, Biraschi A.M. (edited by), Milano 1992.

Striffler R. 1994, *Guerra di mine nelle Dolomiti. Lagazuoi-Castelletto 1915-1917*, Trento.

Thucydides, *La guerra del Peloponneso*, Moggi M. (edited by), Milano 1984.

Todaro P. 1988, *Il sottosuolo di Palermo*, Palermo.

Tölle-Kastenbein R. 1990, *Archeologia dell'acqua. La cultura idraulica nel mondo classico*, Milano.

Touring Club Italiano 2002 a, *Atlante stradale d'Italia 1:200.000 Nord*, Milano.

Touring Club Italiano 2002 b, *Atlante stradale d'Italia 1:200.000 Centro*, Milano.

Thum A. 2005, *Il primo ricovero antiaereo del Comune di Milano*, in Terrosu C. (edited by), *Il progetto scuole – Gli adolescenti e la guerra*, Comune di Milano - Settore Musei e Mostre, Milano, pgs. 71-80.

Uggeri G. 1974, *Gli insediamenti rupestri medievali. Problemi di metodo e prospettive di ricerca*, in VARIOUS AUTHORS (edited by), *Archeologia Medievale*, I, Firenze, pgs. 143-175.

Uggeri G. 1976, *Archeologia del bunker*, in VARIOUS AUTHORS (edited by), *Archeologia Medievale*, III, Firenze, pgs. 261-263.

Uggeri G. 2000 a, *Finalità della disciplina*, , in Bonora G., Dall'Aglio P. L., Patitucci S., Uggeri G. (edited by), *La topografia antica*, Bologna, pgs. 15-21.

Uggeri G. 2000 b, *Il contributo della toponomastica alla ricerca topografica*, in Bonora G., Dall'Aglio P. L., Patitucci S., Uggeri G. (edited by), *La topografia antica*, Bologna, pgs. 119-132.

Valbusa U. 1907, *Il Buco di Viso*, estratto dalla *Gazzetta del Popolo della Domenica*, 25 Agosto, Torino.

Vauban Le Prestre S. (marechal de) 1742, *Traité de l'attaque et de la défense des places*, Nouvelle Edition, reprints 2006.

Vegezio F. R., *De re militari*, Angelini A. (edited by), Roma 1984.

Verducci R. 1991, *La chiesa ipogea di Sotterra di Paola*, Reggio Calabria.

Vergani R. 2003, *Gli usi civili della polvere da sparo (sec. XIII-XVIII)*, in Cavaciocchi S. (edited by), *Atti della XXXIV Settimana di Studi dell'Istituto Internazionale di Storia Economica "F. Datini" (Prato, 15-19 aprile 2002)*, Firenze, pgs. 865-878.

Verrini A. 2002, *Le ghiacciaie coperte della Valbormida*, in Società Speleologica Italiana (edited by), *Opera Ipogea*, Società Speleologica Italiana, No. 3, Genova, pgs. 37-51.

Viazzi L. 1976, Le aquile delle Tofane. 1915-1917, Torino.

Viglino Davico M. 1989, *Fortezze sulle Alpi. Difese dei Savoia nella Valle Stura di Demonte*, Cuneo

Viollet-le-Duc E. E. 1996, *Encyclopedie Medievale*, refonte du *Dictionnaire raisonné de l'architecture francaise*, Tours.

Vitruvio P. M., *De Architectura*, Migotto L. (translation), Padova 1997.

von Clausewitz C. 1978, *Della guerra*, translated by Bollati A. and Canevari E., Cles (Trento).

von Eschenbach W. 1981, *Parzival*, translation by Bianchessi G., Torino.

von Kartaschew K. 2002, *La fortezza di Boden. Una corona di forti in Svezia*, in Istituto Italiano Castelli (edited by), Castellum, Roma, pgs. 5-16.

Welin E. 1962, *The tombs at San Giovenale*, in VARIOUS AUTHORS (edited by), *Etruscan culture. Land and people*, Malmö, pgs. 279-288.

Wilke K. P. 2005, *Video e speleo vademecum*, in Padovan G. (edited by), *Archeologia del sottosuolo. Lettura e studio delle cavità artificiali*, Notebooks on Medieval Topography (British Archaeological Reports, International Series, S1416), Oxford, pgs. 233-240.

Xenophon, *Anabasi*, Manfredi V. (edited by), Milano 1984.

Zanon D. 1996, *Le ricerche biospeleologiche nei sotterranei*, in Padovan G. (edited by), *La fortezza celata. I sotterranei del Castello Sforzesco di Milano*, Vigevano (Pavia), pgs. 148-169.

Zanon D. 2003, *Biospeleologia*, in Club Alpino Italiano – Scuola Nazionale di Speleologia (edited by), *Manuale di Speleologia*, Bergamo.

Zanon D. 2005, *Biospeleologia in cavità artificiali*, in Padovan G. (edited by), *Archeologia del sottosuolo. Lettura e studio delle cavità artificiali*, Notebooks on Medieval Topography (British Archaeological Reports, International Series, S1416), Oxford, pgs. 283-318.

Zerbi M. C. 2005, *Il paesaggio nell'approccio della geografia e dell'architettura*, in Zerbi M. C., Scazzosi L. (edited by), *Paesaggi straordinari e paesaggi ordinari. Approcci della geografia e dell'architettura*, Milano, pgs. 15-28.

www.ingramcontent.com/pod-product-compliance
Lightning Source LLC
LaVergne TN
LVHW071203210926
848482LV00015B/719